Cassini at Saturn
Huygen Results

David M. Harland

Cassini at Saturn

Huygens Results

 Springer

Published in association with
Praxis Publishing
Chichester, UK

Dr David M. Harland
Space Historian
Kelvinbridge
Glasgow
UK

SPRINGER–PRAXIS BOOKS IN SPACE EXPLORATION
SUBJECT *ADVISORY EDITOR*: John Mason B.Sc., M.Sc., Ph.D.

ISBN 10 0-387-26129-X Springer Berlin Heidelberg New York
ISBN 13 978-0-387-26129-4 Springer Berlin Heidelberg New York

Springer is a part of Springer Science + Business Media (*springer.com*)

Library of Congress Control Number: 2006940319

Cover design: Jim Wilkie
Copy editing: Alex Whyte
Typesetting: BookEns Ltd, Royston, Herts., UK

Printed in Germany on acid-free paper

To
Eugene Antoniadi, Percival Lowell and Bernard Lyot
who saw so clearly

Other Praxis books by David M. Harland

The Mir Space Station – Precursor to space colonization
The Space Shuttle – Roles, missions and accomplishments
Exploring the Moon – The Apollo Expeditions
Jupiter Odyssey – The Story of NASA's Galileo Mission
The Earth in Context – A Guide to the Solar System
Mission to Saturn – Cassini and the Huygens Probe
The Big Bang – A View from the 21st Century
The Story of the Space Shuttle
How NASA Learned to Fly in Space – The Gemini Missions
The Story of Space Station Mir
Water and the Search for Life on Mars
The First Men on the Moon – The Story of Apollo 11

Creating the International Space Station
with John E. Catchpole

Apollo EECOM – Journey of a Lifetime
with Sy Liebergot

NASA's Voyager Missions – Exploring the Outer Solar System and Beyond
with Ben Evans

Lunar Exploration – Human Pioneers and Robotic Surveyors
with Paolo Ulivi

Space Systems Failures
with Ralph D. Lorenz

Apollo – The Definitive Sourcebook
with Richard W. Orloff

"The day God created Saturn,
He must have been feeling on top of the world, so to speak"

Columnist George Will, upon being shown
Voyager Saturn imagery.

"Of everything revealed by the telescope,
the most meaningful to most people is the rings of Saturn"

Philip Morrison, Massachusetts Institute of Technology.

"We learned more about Saturn in one week
than in the entire span of human history"

Bradford Smith, Voyager Imaging Team Leader, after the
Voyager 1 fly-by.

"Saturn is a lot like a solar system in miniature form"

Wesley T. Huntress, NASA's associate administrator for
space science.

"We cannot see the surface of Titan.
If we are going to learn what this extraordinary and
fascinating body is like, we will need [an] imaging radar"

Hal Masursky, Voyager Imaging Team member.

"On such an ambitious, seven-year-long mission,
we always have to be prepared for the unexpected"

Jean-Pierre Lebreton, Huygens project scientist.

Contents

List of figures

Chapter 3: Saturn revealed

COLOUR SECTION *(between pages 202 and 203)*
A view of Saturn by Voyager 2
Ready for launch: Cassini and Huygens
Cassini sets off
A depiction of Saturn orbit insertion
The 'A' ring
The 'B' and 'C' rings
Miscellaneous ring studies
A portrait of Saturn and its ring system by Cassini
Topographic analyses of the Huygens landing site
Full-horizon views by Huygens during its descent
Quadrant views by Huygens during its descent
Downward-looking views by Huygens during its descent
Iapetus, Saturn's clouds, Enceladus and the Huygens surface view
The 'tiger stripes' of Enceladus
The 'dragon storm' on Saturn and geysers on Enceladus
A view of the ring system with Cassini in Saturn's shadow

Author's preface

To the ancients the only difference between stars and planets was that the stars remained fixed and the planets moved – indeed, the word planet means 'wanderer'. With the invention of the telescope it was found that, in contrast to the point-like stars, the planets show disks. It was soon realised that Saturn is surrounded by a system of rings. Much was discovered during three centuries of observation of the planet and its rings, but its retinue of satellites remained mysterious specks of light. With the dawning of the Space Age, it became possible to dispatch robotic probes to make *in situ* observations of the planets.

When spacecraft flew through the Saturnian system in the years 1979, 1980 and 1981, the results were astonishing. The satellites were finally shown to be miniature worlds in their own right. A generation of planetary scientists have pored over this data and speculated as to the processes that shaped their surfaces. Titan, the largest member of the retinue, has attracted the most attention. Titan is the only satellite of any planet to have a dense atmosphere. Like the Earth's, this atmosphere is mostly nitrogen; but unlike our own it is in cryogenic freeze. Nevertheless, it is laced with organic molecules and may be in a prebiotic state. In a sense, Titan is a 'terrestrial' planet that happens to be in deep freeze in the outer Solar System. Even with such a fascinating moon, it was almost two decades before a new mission was dispatched to study the Saturnian system in depth.

The Cassini–Huygens mission is a joint venture by NASA, the European Space Agency and the Italian Space Agency. It is an endeavour by thousands of people at contractor companies, universities and government facilities across the United States and Europe. Cassini is the most highly instrumented spacecraft ever sent into deep space. Its mission was make a four-year orbital tour, and deliver the Huygens Probe into Titan's atmosphere. In all, 27 scientific investigations and nine interdisciplinary studies drawing data from two or more instruments were to improve understanding of Saturn's interior, atmosphere, magnetosphere, rings, and satellites.

As this book was being written, Cassini had just passed the half-way point in its primary mission tour, with spectacular results. To provide a sense of context, I have summarised what was learned of the Saturnian system prior to the Space Age and by the early spacecraft. When Cassini–Huygens began its epic voyage, I had three items on my wish list of results: hydrocarbon lakes on Titan, geysers on Enceladus and (in reference to Arthur C. Clarke, strictly tongue-in-cheek) a black monolith on Iapetus. The first two have been found ...

David M. Harland
October 2006

Acknowledgements

I would like to acknowledge the help of – in no particular order – Roger D. Launius, formerly Chief Historian at NASA Headquarters in Washington, now at the National Air and Space Museum; Robert W. Carlson, Ellis D. Miner, Marc D. Rayman and David Seidel of NASA's Jet Propulsion Laboratory in Pasadena; Ken Glover; William K. Hartmann of the Planetary Science Institute, Tucson, Arizona; Michael Hanlon; Alex R. Blackwell of the University of Hawaii; Harald Kucharek; Ralph D. Lorenz of the Applied Physics Laboratory at Johns Hopkins University; Athéna Coustenis of the Meudon Observatory in Paris; Paul D. Spudis of the Lunar and Planetary Institute in Houston; Donald and Cynthia Putt of Parnassus On Wheels, in Oregon; Ralf Srama of the Max Planck Institute for Astrophysics in Heidelberg; Seran Gibbard of the Institute of Geophysics and Planetary Physics at the Lawrence Livermore National Laboratory; Linda A. Kelly (née Morabito) of the Planetary Society; Robert H. Brown of the Departments of Planetary Sciences and Astronomy at the University of Arizona; Ione Caley, administrative assistant to Larry W. Esposito of the Laboratory for Atmospheric and Space Physics at the University of Colorado; and Robert T. Mitchell, Cassini Program Manager at JPL. And, of course, Clive Horwood of Praxis.

1

Saturn from afar

NAKED-EYE ASTRONOMERS

From his home on the island of Rhodes in the Aegean, Hipparchus, the greatest of the ancient Greek astronomers, drew up a catalogue of the positions and motions of the objects in the sky. He interpreted the observations as meaning that the Earth was at the centre of everything, and that the planets revolved around the Earth in circles. Claudius Ptolemaeus (more usually called simply Ptolemy), a Greek living in Alexandria in Egypt, observed that the planets did not precisely follow their predicted paths. However, since the circle was regarded as 'perfect' he proposed an 'epicycle' scheme in which each planet pursued a smaller circle about its mean position as it progressed around its orbit.

Having studied mathematics at the University of Cracow, the Polish astronomer Nicolaus Copernicus realised in 1507 that the complexity of Ptolemy's scheme could be banished if it was assumed that the planets revolved around the Sun, with only the Moon going around the Earth. Although Copernicus worked out the consequences of this 'heliocentric' theory and informally circulated it to colleagues, it was not formally published until his death in 1543, as *De Revolutionibus Orbium Coelestium*.

The Danish nobleman Tycho Brahe noted the appearance of a 'new star' in 1572 in the constellation of Cassiopeia. It appeared as bright as the planet Venus for three weeks, but slowly faded and finally disappeared from sight a year later. When Brahe reported his observations in a short book *De Nova Stella*, King Frederick II of Denmark was so impressed that Brahe was assigned the small island of Hven in the channel between Copenhagen and Helsingfors to enable him to establish an 'observatory' to undertake a systematic study of the motions of the planets. As optics had not yet been invented, Brahe developed and refined a wide variety of instruments designed to give accurate measurements of the positions of objects in the sky. In the observatory, which Brahe had named Urania, he had a live-in staff of technicians to assist him with observations. As the work progressed, Brahe received a stream of visiting dignitaries and fellow celestial observers, but when Christian IV assumed the throne support for the project ended. In 1599 Brahe relocated to Prague and continued his work under the patronage of Emperor Rudolph II of Germany.

When Brahe died in 1601, his priceless archive of observations passed to Johannes Kepler, who had, towards the end, served as his chief assistant. At that time, there was no clear distinction between astronomy and astrology, and to supplement his mathematical study of empirical laws of planetary motion Kepler earned his living by casting horoscopes. As an unfortunate sign of the times, Kepler's mother was tried as a witch!

In 1687 Isaac Newton published *Philosophiae Naturalis Principia Mathematica*, promoting his law of Universal Gravitation, from which Kepler's laws followed as a consequence. Brahe had therefore produced a catalogue of exceptionally accurate observations without making any attempt to interpret them, Kepler had provided the empirical analysis without understanding why the planets moved as they did, and Newton had identified the motivating force. Overall, these were remarkable achievements for naked-eye astronomy.

SATURNUS TRIFORMIS

In the same year that Kepler announced the laws that governed the movements of the planets, a paradigm shift revolutionised the study of astronomy.

Galileo Galilei, the son of a musician, was born in Pisa, Italy, in 1564. Although he attended the University of Pisa as a medical student his passion was mechanics, and in pursuing this interest he became the first real experimental physicist since Archimedes of Syracuse, almost two thousand years earlier. He is reputed to have dropped differently sized masses from the top balcony of the Leaning Tower to demonstrate that they would all fall at the same speed – a prediction that was at odds with the conventional view that the heavier ones would fall more quickly. This simple observation was counter to the 'world view' of the Church, which derived from Aristotle, a student of the philosopher Plato. Galileo then accepted the post of professor of mathematics in Padua, which fell under the authority of Venice where the administration was rather more open-minded.

In 1608 Hans Lippershey, a spectacle-maker in Middleburg in the Netherlands, invented a device, utilising two lenses, by which it was possible to make distant objects appear closer. In May of the following year, Galileo heard about this invention and, after building one, set out to improve the design to enable ships on the horizon to be identified a few hours earlier than was previously possible, which gave a significant advantage to Venetian merchants. In November, Galileo indulged his own long-standing interest in astronomy. He began with the Moon, and was the first person to realise that its face was disfigured by vast holes and rugged mountains, which was contrary to the accepted view that all objects in the sky were 'perfect'. His discovery of dark spots on the Sun further undermined this classical viewpoint.

On 7 January 1610, Galileo turned his attention to Jupiter, and, in addition to resolving it as a disk, he noted three small nearby star-like points arranged in a line. Over successive nights, not only did he notice that these three subsidiary objects changed their relative positions, but he also spotted a fourth. As he accumulated more observations, he realised that these objects were circling around the planet, but this was contrary to the accepted view, as taught by Ptolemy, that things moved around the Earth on 'celestial spheres'. Galileo favoured the theory by Copernicus that only the Moon circles the Earth, and that everything else moves around the Sun, and in March 1610 he recorded his observations in a small pamphlet entitled *Sidereus Nuncius*, which he chose to dedicate to Grand Duke Cosimo de Medici, a leading member of one the merchant families that had ruled Florence and Tuscany for some two centuries. As there were four de Medici brothers, Galileo suggested that the Jovian attendants be called the 'Medicean Stars'. While the existence of Jovian satellites did not actually prove Copernicus's theory, it certainly contradicted Aristotle, and Galileo once again found himself in serious strife with the Church. Undeterred, he aimed his telescope at the Milky Way and discovered that it comprises a multitude of faint stars, as indeed do some of the 'nebulosities' listed in the catalogue of stars in Ptolemy's *Almagest*, which extended Hipparchus's catalogue.

In September, having won de Medici's patronage, Galileo was invited to become the mathematician to the Tuscan court in Florence. Soon after arriving, he observed that Venus showed lunar-like phases, which was incontrovertible proof that it was orbiting the Sun, and hammered the final nail into the coffin of the Ptolemaic system.

Meanwhile, in July Galileo had unveiled a real mystery: Saturn appeared to have a substantial attendant on each side of its disk. He wrote to Kepler announcing his discovery of *Saturnus triformis*, saying: "I have observed the most distant planet to have a triple form." However, in the style of the time, he used an anagram cipher, which Kepler misinterpreted as saying that Mars had two moons. Kepler believed that there should be a progression in which the Earth had one moon, Mars had two, and Jupiter – as Galileo had himself discovered – had four, so he was predisposed to this error. "The planet Saturn is not alone," Galileo wrote later to de Medici, "but is composed of three, which almost touch one another and never move nor change with respect to one another. They are arranged in a line parallel to the ecliptic, and the middle one is about three times the size of the lateral ones."

Upon first turning his telescope to Saturn in July 1610, Galileo Galilei was astonished to find that the planet appeared to be three objects in line, as his first drawing shows.

Galileo continued his observation through the rest of the year and into 1611. The appendages or *ansae* (meaning handles) became progressively less noticeable, and in 1612 they disappeared, leaving only the large central object, which, if he had never suspected the presence of the *ansae*, would have appeared thoroughly unremarkable. He was, however, not only baffled but also gravely concerned because his discovery of satellites around Jupiter had attracted criticism from more traditionally minded non-telescopic astronomers and philosophers. Having announced that Saturn had strange *ansae*, he would be ridiculed if he now reported that they had disappeared.

Recalling from Greek mythology that after siring Zeus (Jupiter), Kronos (Saturn) had devoured his subsequent offspring at birth in an attempt to prevent them from supplanting him, Galileo wrote of the disappearance of the *ansae*:

> What is to be said concerning so strange a metamorphosis? Are the two lesser stars consumed in the manner of sunspots? Have they vanished, or suddenly fled? Has Saturn perhaps devoured his own children? Or were the appearances indeed illusions or fraud with which the glasses have so long deceived me, as well as many others to whom I have shown them? Now, perhaps, is the time come to revise the well-nigh-withered hopes of those who, guided by more profound contemplations, have demonstrated the utter impossibility of their existence. I do not know what to say in a case so surprising, so unlooked-for, and so novel. The shortness of the time, the unexpected nature of the event, the weakness of my understanding and fear of being mistaken have greatly confounded me.

Reluctant to report conclusions that would subsequently be shown to be wrong, and suspecting that his early telescopes had in any case been flawed, Galileo made a succession of better instruments and, upon discovering that Saturn looked even more baffling in 1616, he ceased to observe the planet.

Others continued to study Saturn, however. In 1614 C. Schreiner noticed that the *ansae* had reappeared, indicating that Galileo's early observations had been valid. A

In 1616, Galileo found Saturn to be even more puzzling. Although with the benefit of hindsight we readily interpret his drawing as depicting a ring around the planet, he did not recognise the 'appendages' (*ansae*) as such.

progression in the changing appearance of the *ansae* was thence revealed. Firstly, they were thin arms which projected to either side of the seemingly unchanging central disk. A dark gap then appeared within each arm. Over the next seven years this progressively opened up, then the trend reversed and the arms closed again.

When G.B. Riccioli, a Jesuit philosopher in Bologna, examined Saturn in 1640 he

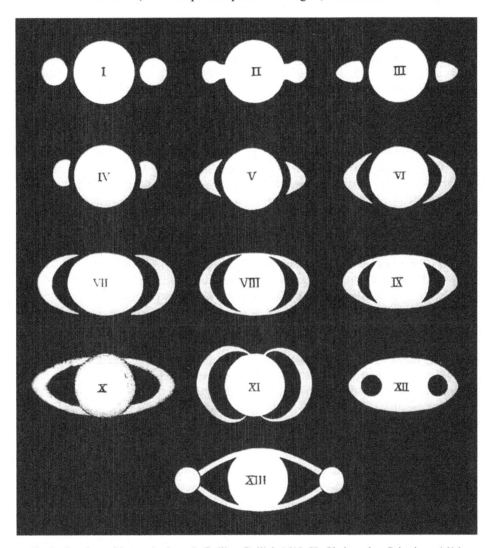

Early sketches of Saturn's rings: I: Galileo Galilei, 1610; II: Christopher Schreiner, 1614; III: G.B. Riccioli, 1641 or 1643; IV to VII: Johannes Hevelius circa 1645 (these are theoretical forms); VIII and IX: G.B. Riccioli 1648 to 1650; X: Eustachio Divini 1646 to 1648; XI: Francesco Fontana, 1636; XII: Giuseppe Bianchini, 1616 and Pierre Gassendi, 1638 to 1639; XIII: Francesco Fontana, 1644 to 1645. (From *Systema Saturnium* by Christiaan Huygens, 1659.)

saw the *ansae* as thin arms. After vanishing in 1641, they reappeared and started a new cycle. From 1647 to 1650 he watched them evolve, then published a book in 1651 summarising the state of knowledge, but this served only to confound the mystery.

GIANT MOON

Christiaan Huygens was born in 1629 into a notable Dutch family with a tradition of diplomatic service to the House of Orange. He studied law and mathematics, and his father provided a stipend to enable him to devote himself completely to his chosen profession: the study of nature. Huygens first worked on mathematical problems, but soon expanded his activities to experiments in mechanics and optics.

In 1655, upon hearing an enthusiastic report by Johannes Hevelius, who had an observatory in Danzig, that a long-focus refractor minimised the bloom of chromatic aberration, Huygens, at that time living in the Hague with his brother Constantin, set out to make a 2-inch-diameter lens having a 12-foot focal length that would provide a fifty-fold magnification. The telescope was soon ready, and on 25 March Huygens aimed it towards Saturn, at that time the most distant and most enigmatic of planets. Although he was frustrated to find that the *ansae* reported by his predecessors were barely visible, he noted that there was a prominent star just 3 minutes of arc away from the planet which, as he continued to observe nightly, was not left behind as the planet pursued its motion against the sky. He inferred from this that Saturn was accompanied by a satellite. As he monitored the companion's motion, he calculated that it orbited the planet in 16 days. In fact, Hevelius had noticed its presence, without realising that it was not a star in the background, as had Christopher Wren, who had been a noted astronomer in England before applying himself to the field of architecture.

Two centuries later, John Herschel eschewed the established practice of referring to the satellites in order of their discovery, and instead assigned them names, naming the senior member of the retinue after Titan, who in mythology was one of Saturn's siblings.

Ironically, Huygens had not felt it necessary to search for more moons. He believed that there was a correspondence between the total number of planets and the total number of satellites, and he equated the six planets – Mercury, Venus, Earth, Mars, Jupiter and Saturn – with the Moon, the four satellites of Jupiter and the newly discovered Titan. He thus congratulated himself for having found the 'final' satellite in the Solar System.

A FLAT RING, NOWHERE CONNECTED

After the exhilaration of discovering what he believed to be the largest satellite in the Solar System on his first night's observations, Huygens built a series of increasingly powerful telescopes. When he had first looked, the *ansae* had been present as very narrow arms projecting to either side of the planetary disk. In early 1656, using a refractor with a 2.5-inch lens, a 23-foot focal length and a magnification of 100, he noted that the *ansae* had disappeared. However, the narrow arms reappeared in October, and using an 'arial' refractor with a focal length of 136 feet, Huygens later noticed that a shadow had been cast on the planet – an observation which showed that the *ansae* were not merely crescent-forms 'alongside' the globe, but formed a structure that continued in front of it.

In 1650 an Italian Jesuit, F.M. Grimaldi, noted that Saturn's globe was not round, but was flattened at the poles, implying that the planet was in a state of

A depiction of Christiaan Huygens observing with his 136-foot-focal-length 'arial' refractor.

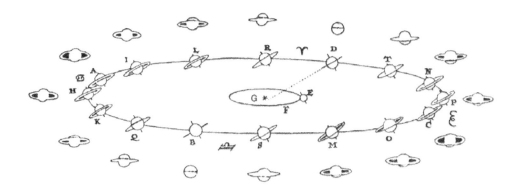

In his *Systema Saturnium* in 1659, Christiaan Huygens showed how an inclined ring
explained how Saturn's appearance varies in a cyclical manner, including how it can
'disappear' as the Earth passes through the ring plane.

rotation. Even though the period of its rotation had yet to be determined, it was
unlikely to be outlandishly long, so Huygens rationalised that if the *ansae* structure
was rotating with the planet the only physical form that could maintain its
appearance was an annular disk. After compiling observations for several more
years, Huygens published *Systema Saturnium* in 1659 in which, after summarising
earlier observations, he announced that Saturn's globe was "surrounded by a flat
ring, nowhere connected to the body of the planet, and is inclined to the ecliptic". In
fact, to assure his claim of precedence in this discovery, on 5 March 1656 he had
appended a note – in cipher, naturally – to a short paper on his discovery of the
satellite the previous year.

With fifty years of observations to consider, Huygens realised that the ring had
been viewed at a variety of angles as Saturn had pursued its 29-year orbit around the
Sun. The plane of the ring is significantly inclined from the ecliptic: for a part of
Saturn's orbit the ring is tilted Earthward and is viewed obliquely from 'above';
during the second part of the orbit it is viewed from 'below'. Huygens realised that
the ring becomes 'invisible' when the Earth passes through its plane – a perspective
that occurs at 14.5-year intervals. The ring had been absent for Galileo in 1612,
absent in 1641, and absent again in early 1656. In 1655, when Huygens had first
inspected the planet, the virtually edge-on ring had projected its shadow onto the
northern hemisphere. After the ring-plane crossing, the shadow had been on the
southern hemisphere. In the first case, the southern face of the ring had been
illuminated and, later, the northern face had been illuminated. William Ball of Exeter
in England had noted the shadow as a thin band on the disk between February and
July 1656, when the *ansae* had been absent. This sequence of observations during this
very dynamic time were crucial to Huygens's insight.

Although Huygens's analysis accounted for all the observations, the suggestion of
a ring was so ridiculous that it was rejected by some of his contemporaries. In 1658
Christopher Wren had suspected that Saturn had an elliptical corona that was in

Christiaan Huygens's insight into the ring-form of Saturn's *ansae* derived from his observation that the shadow cast by the rings appeared in opposite hemispheres in 1655 and 1656, in between which the *ansae* were edge-on and hence invisible.

contact with the globe in two places and rotated with it, but before he could publish his idea he found Huygens's analysis to be more compelling. G.B. Hodierna, a Sicilian priest and mathematician suggested that Saturn was a distended spheroid on which there were two dark patches; however, the dynamics of such an arrangement were untenable. In 1660 Honore Fabri, a Jesuit philosopher, pontificated rather strangely that Saturn had a pair of very large but unreflective satellites close in and a pair of somewhat smaller but reflective ones farther out, but the dynamics of such a system were again untenable. G.P. de Roberval, a mathematician and a founder member of the French Academy of Sciences, argued that Saturn was surrounded by "a torrid zone given off by vapours" which was transparent if present in small quantity, reflected sunlight at the edges if present in medium quantity, and appeared as an ellipse if present in large quantity.

Huygens predicted that after the viewing angle had opened and closed again, the rings would disappear in July 1671 (he was slightly out, because the ring-plane crossing actually occurred in May of that year). By that time, however, there was no doubt of the truth of his interpretation because improved telescopic observations in 1665 had given conclusive proof. Indeed, in 1662 Guiseppe Campani in Italy and Adrien Auzout in France had independently observed the shadow of the planet cast upon the ring, clipping a narrow section out of one of the *ansae*.

Later in life, Huygens founded the Academie Royale des Sciences, and became its first director. His many contributions to mathematics, astronomy, time measurement and the theory of light are regarded as fundamental.

NEW MOONS

Giovanni Domenico Cassini was born in Perinald in Imperia, Italy, on 8 June 1625. After receiving a Jesuit education, he was hired by the Marquis Cornelio Malvasia in Bologna, who calculated ephemerides for astrological purposes. Using the excellent instruments of the Marquis's observatory, Cassini made astronomical observations of exceptional precision and quality. In 1669, with several significant discoveries to his name, and by then holding the chair of astronomy at the University of Bologna, he was invited to assume the first directorship of the Paris Observatory, which was still under construction. Upon relocating to France, he converted his forenames to Jean-Dominique. In light of Huygens's success, Cassini adopted refractors with very long focal lengths which, while unwieldy, facilitated significant discoveries.

On 25 October 1671 Cassini discovered a moon to supplement Huygens's Titan, and found another on 23 December 1672. When John Herschel later introduced a nomenclature based on the mythical Titans, these satellites were named Iapetus and Rhea respectively. Astonishingly, after readily detecting Iapetus, Cassini was unable to recover it until mid-December 1672, after which it vanished once more until early February 1673, at which time he was able to monitor it every night for a fortnight. Eventually, he realised that it was visible only when west of the planet in the sky. It was not until September 1705, using a more powerful telescope, that he managed to detect it east of the planet. At its brightest – at 10th magnitude – it was not much inferior to Titan, but it then faded by two magnitudes on the other side of its orbit. Why should the moon appear five times brighter on one side of its orbit than on the other? Drawing an analogy with the Earth, which has continents and oceans, Cassini speculated that "one part of the surface is not so capable of reflecting to us the light of the Sun". He reasoned that, as in the case of the Earth's Moon which also has light and dark regions, Iapetus's rotation must be synchronised with its orbital period. On the night of 21 March 1684, he spotted another two moons: Tethys and Dione, both of which were extremely faint. In fact, he was able to see them using a long refractor, with its meagre 'light grasp', only because the ring system was almost edge-on. As the rings opened up again, dazzling him, he lost sight of them. Despite improving his telescope, Huygens was never able to see them.

Once Isaac Newton developed the Law of Universal Gravitation in 1687, the fact that Saturn possessed a system of satellites meant that the mass of the planet could be calculated very accurately. Despite having a volume 700 times that of the Earth, it is only 95 times as massive and its density is only 0.69 g/cm^3. If the planet could be placed in a sea of sufficient capacity, it would *float* with fully one-fourth of its bulk above the waterline.

RING DIVISION

G.D. Cassini made the momentous discovery in 1675 of a thin dark line on Saturn's ring. It was presumed that this was a mark *on* the ring which divided the broad bright inner section from the narrower outer section. A year later, he noticed a 'belt' on the planet's disk, just south of the equator. Such latitudinal banding had been discovered on Jupiter in 1630 by Francesco Fontana of Naples, but Saturn's version was more subdued in character.

In 1705 J.J. Cassini (G.D. Cassini's son, and successor as director of the Paris Observatory) ventured that the rings comprised a myriad of meteoroids, but he had no evidence to support this view. The elder Cassini had hired a nephew, G.F. Miraldi, as an assistant. Miraldi's observations of irregularities moving to and fro in the ring plane when it was almost edge-on in 1714 was the first evidence of rotation.

Although G.D. Cassini's discovery of a dark division in Saturn's ring was made in 1675, he produced this drawing the following year. Notice also the depiction of the band on the planet.

After reviewing all the evidence, he concluded that the dark line was a gap separating two concentric rings, each of which was a rigid structure turning synchronously with the planet. Proof that Cassini's Division, as the line became known, was indeed a gap came in October 1852, when William Lassell in England noted that a slice of the planet's disk was visible through it. Considering the obliqueness of the line of sight where the ring system crosses in front of the globe, it was a remarkable observation. In fact, such a line of sight is feasible only when the ring system is open wide *and* when the Sun is at the same angle to the ring plane as the viewing angle, allowing the slice of the globe visible through the gap to be illuminated; at other times, the part of the globe observed through the gap is in the ring system's shadow.

THE ASTRONOMICAL UNIT

In 1767, at the age of 18, Pierre Simon Laplace was appointed as professor of mathematics in Paris, where he devoted the majority of his attention to a comprehensive study of the manner in which the planets perturb one another gravitationally, with a view to determining whether the observed arrangement of the Solar System was stable. He concluded that while the eccentricities of the individual orbits would vary over time, the system would adjust to compensate, and was therefore stable. In addition, using Newton's law of gravitation, he was able to use the magnitude of these perturbations to calculate that the Earth was about 150 million kilometres from the Sun – a value that was designated the Astronomical Unit (AU) because, when combined with Kepler's empirical laws of orbital motion, it gave the basis for a scale by which to measure the Solar System. Upon applying this scale, it became evident that Saturn orbits at the astonishingly large distance of 1.4 billion kilometres. The subtended angles indicated that the planet's diameter was of the order of 100,000 kilometres and the ring system was about 270,000 kilometres in diameter. Laplace published his collected researches in *Celestial Mechanics* in five instalments between 1799 and 1825.

HERSCHEL'S DISCOVERIES

Born in Hanover in 1738, Friedrich Wilhelm Herschel anglicised his forenames to Frederick William upon relocating to England in 1757. As an accomplished musician, he took the post of organist at the Octagon Chapel in Bath in 1767. His passion was astronomy though, and a few years later he was routinely giving eight music lessons during the day and then observing the sky at night. In 1774, after several frustrating years of making and using long and unwieldy refractors, he switched to reflectors and made one with a mirror that was 6 inches in diameter and had a focal length of 7 feet. This proved to be superior to any telescope available to his professional contemporaries.

On 13 March 1781, while methodically charting the sky for double stars whose parallax might enable him to determine the distances to the stars as a step towards his ultimate objective of gauging the structure of the galaxy, he observed a small sea-green disk that was not listed on his chart. Assuming it to be a tail-less comet, he monitored its motion against the stars for several nights and then reported his discovery, which was found to be a new planet. He tried to name it *Georgium Sidus* in honour of King George III, but continental astronomers objected and following the suggestion of J.E. Bode in Berlin it came to be known as Uranus. With an invitation to join the Royal Society, an annual allowance and a grant from the Crown to build a monster telescope, Herschel was able to pursue his astronomical work on a full-time basis, and a knighthood followed in due course as a recognition of his success. Of all the planets, he was most interested in Saturn. Between 1789 and 1805 he presented six papers about it to the Royal Society.

Having observed the Galilean satellites of Jupiter and concluded that the slight variations in their brightnesses were correlated with their orbital periods – indicating that their rotations were synchronised with their orbital motions as in the case of the Earth's Moon – Herschel set out to monitor Saturn's satellites. On 28 August 1789, when testing his 48-inch-diameter 40-foot-focal-length reflector, Herschel inspected Saturn while the rings were virtually edge-on and detected a new moon – Enceladus – circling with a period of 1 day 8 hours 53 minutes. Actually, he had glimpsed it on 19 August 1787 with his 20-foot telescope, but had not been sure and had postponed the matter to devote his efforts to the construction of his new telescope. In fact, he had noted Enceladus again in July 1789, but had assumed it to be Tethys. It was only when he saw all six satellites in a line in August that he was certain he had discovered something new.

Remarkably, Herschel reported the first discovery of a moon of Saturn for over a century as a postscript to the formal submission of a presentation which he had read to the Royal Society two months earlier, entitled *Catalogue of a second thousand new nebulae and clusters of stars*, viz.:

> P.S. The planet Saturn has a *sixth satellite* revolving round it in about 32 hours 48 minutes. Its orbit lies exactly in the plane of the rings, and within that of the first satellite [Titan]. An account of its discovery with the 40-foot reflector ... will be presented to the Royal Society at their next meeting.

As Herschel continued to monitor the ring-plane crossing utilising his 20-foot reflector, he discovered another new satellite on 8 September 1789. After seeing it again on 14 September, he used his new telescope on 17 September to confirm its existence. This moon – Mimas – was very difficult to observe because it was orbiting just 48,000 kilometres beyond the ring system. Indeed it could only be seen when it was beyond the edge-on line of the rings. Nevertheless, with careful monitoring over several nights he was able to determine that it had an orbital period of 22 hours 37 minutes. When he finally delivered his 'postscript' to the Royal Society, he was able to spring a surprise by entitling it *An account of the discovery of a sixth and seventh satellite of the planet Saturn*. They were just detectable as specks of light using the 20-foot telescope, but were clear in the 40-foot telescope. A steady stream of astronomers with smaller instruments paid Herschel a visit in order to see the new moons for themselves.

When viewing the system edge-on Herschel observed the rings as a thin broken line extending to each side of the disk, with small concentrations of light embedded in it. Miraldi had noted such irregularities when the system had been edge-on in 1714. The back and forth motions of the spots of light enabled Herschel to determine the rotational period of the ring system. The 'end points' of the brightest light spots indicated that they were just inside the line of Cassini's Division, and rotated in approximately 10 hours 32 minutes 15 seconds. From the sharp shadow that the rings cast upon the globe Herschel inferred that the ring system was a solid structure. As the planet's oblateness was most pronounced at the time of ring-plane crossing, he further deduced that the axis was perpendicular to his line of sight and that, consequently, the plane of the rings coincided with the equator, which, in retrospect is, of course, the only possible stable orientation. Viewing the planet 'face on', he measured the equatorial diameter as 120,000 kilometres and the polar diameter as 10 per cent less. Herschel also calculated that the planet's equatorial plane is inclined 26.75 degrees to its orbit, which in turn is inclined by some 2.5 degrees to the ecliptic. While the northern face of the rings was presented, Herschel noted that Cassini's Division maintained the same breadth and sharpness of outline, and its 'colour' was indistinguishable from the gap between the ring and the planet. After the ring-plane crossing, he inspected the southern face of the rings and found that the feature was exactly the same, which meant that there was (a) an identical marking on each side of a single ring, (b) a strip on the ring which was transparent, or (c) as was most likely, a gap separating a pair of concentric rings, the inner one being slightly brighter than the outer one, as Miraldi had previously inferred.

Herschel observed Iapetus for two years, monitoring its variations for 10 cycles, thus refining its 79-day 8-hour orbital period, and in 1791 announced that the persistence of its light curve over an interval of a century and a half served to confirm Cassini's suspicion that the moon's rotation was synchronous. Herschel pointed out that its brightness was greatest when it was approaching Earth, and faintest when it was receding from Earth. We therefore view the trailing hemisphere when it appears bright, and the leading hemisphere when it appears dark. One early suggestion was that Iapetus had an irregular shape, but its evident size implied that it was spherical, and prompted Herschel to agree with Cassini that – for some peculiar reason – one hemisphere is much less reflective than the other. Having traced the light curve in such detail, Herschel was able to deduce from the phasing that by far the largest part of the surface was of the darker character. The lighter and darker hemispheres were oriented so that when the satellite displayed 'full' phase it would appear dichotomous if viewed from Saturn. Herschel also inferred from the manner in which the detail in the light curve recycled that there were no transient variations to suggest that the moon had an atmosphere. Herschel also discovered that the plane of Iapetus's orbit was inclined at 14.75 degrees to that in which most of the inner moons orbited the planet.

The attention of most observers was attracted by the rings and the moons; the planet itself was bland. There was a subdued latitudinal banding similar to Jupiter's, but it was not as striking and it was less 'dynamic'. Nevertheless, Herschel resolved irregularities in the shapes of the bands, and by timing their passage across the disk during the winter of 1793/1794 he found that, as the extreme oblateness implied, the planet does indeed rotate rapidly – with a period of 10 hours 16 minutes. In 1796 J.H. Schröter, a magistrate in Lilienthal near Bremen, who had established his own observatory, bought a Herschelian telescope and hired K.L. Harding as an assistant. Schröter observed a number of small spots on Saturn's globe, but did not time their motions to derive a rotational period. Although Saturn's rotation is almost as fast as Jupiter's, Saturn has barely 30 per cent of Jupiter's mass and its envelope is therefore not so strongly attracted. The gravitational attraction at Saturn's visible 'surface' is only marginally greater than that at the Earth's surface, but the 35-kilometre-per-second escape velocity has enabled Saturn to retain its primordial hydrogen. The rapidly rotating equatorial zone is thus particularly distended, giving the globe a form which Herschel graphically described as having "square shoulders".

Despite his contributions to planetary astronomy, this did not form the primary thrust of Herschel's astronomical work, which was nothing less than to measure the distribution of the stars in space in order to infer the construction of the heavens.

ORIGIN OF THE RINGS

The French mathematician P.S. Laplace realised in 1785 that if Saturn's rings were a solid body, then the 'tidal' effects of the planet's gravitational field would have disrupted them.[1] Instead, he argued that the rings were composed of a multiplicity of very narrow ringlets. Interestingly, the Edinburgh maker of reflecting telescopes, James Short, had remarked to J.J. Lalande several decades earlier that he had once seen many divisions on the rings, but no one else had reported such finely detailed observation.

In 1796, Laplace mathematically investigated the rotation of a sphere of gas, and realised that if the bulk of the mass was concentrated towards the centre of rotation, and the periphery was rotating very rapidly, the outer part would become distended and ultimately tend towards a disk. If the angular momentum attained a critical value then the edge of the disk would detach itself and form a ring, in the process reducing the angular momentum of the disk. If the disk continued to contract by gravitational attraction, it would shed a succession of concentric coplanar rings. He argued that as the central mass went on to form a star, the rings would condense to form a series of planet. This became known as the 'nebular hypothesis' for the origin of the Solar System. As the planets formed, the same process would shed rings which condensed to form satellites. Surely, he argued, the concentric rings round Saturn were striking proof of this hypothesis.

ASTEROIDS

Upon contemplating the wide gap between the orbits of Mars and Jupiter, Johannes Kepler speculated that there must be an undiscovered planet in this zone. J.D. Titius of Wittenberg noted in 1766 that there was a consistent progression in the sizes of the orbits of the planets. The relative distances from the Sun of the six planets that he knew matched the values derived by adding 4 to the series 0, 3, 6, 12, 24, 48 and 96, except that the entry for 24 was mysteriously vacant (see Table 1.1). In 1772 J.E. Bode of Berlin, reviving Kepler's thesis, proposed that there must indeed be an as yet undiscovered planet corresponding to this gap in what became known as the Titius–Bode 'law'.

Table 1.1 The Titius–Bode 'law'

	Series	TB law	Actual	AU
Mercury	0	4	3.9	0.39
Venus	3	7	7.2	0.72
Earth	6	10	10	1.00
Mars	12	16	15.2	1.52
(?)	24	28	–	–
Jupiter	48	52	52.0	5.20
Saturn	96	100	95.4	9.54

On 21 September 1800, F.X. von Zach convened a meeting at J.H. Schröter's observatory in Lilienthal of a few friends who shared an interest in searching for this 'missing' planet. It was agreed that the object would probably lie near the ecliptic, so letters would be sent to colleagues urging that over the coming year they methodically search a series of zones of sky in an effort to locate it. The group nicknamed itself the 'Celestial Police'.

However, on 1 January 1801 Giuseppe Piazzi of Palermo in Sicily, who was not actually a 'policeman', and was nine years into the task of drawing up his own star catalogue, saw a star-like object that was moving slowly across the sky from night to night. After following its westward motion for several weeks he watched it come to a halt, then reverse itself, just as the outer planets do as a result of the changing line of sight with the faster-moving Earth. Unfortunately, the object was soon lost in the twilight as the Sun intervened. However, in Germany, the mathematician J.C.F. Gauss managed to process the meagre data and calculate the object's orbit, proving that it was in the gap between Mars and Jupiter. After it had emerged from superior conjunction, it was recovered by Zach on 31 December. In fact, the next day (and precisely a year after it had been noted) it was also recovered by Heinrich Olbers in Bremen. As its mean solar distance of 2.77 AU was in such excellent agreement with the Titius–Bode 'prediction' of 2.8, the matter was considered settled, and the object was named Ceres in honour of the patron goddess of Sicily.

On 28 March 1802, however, Olbers, having reverted to searching for comets, was rather surprised to discover a second object in the same part of the sky and following a similar orbit. William Herschel estimated that Ceres was no more than 260 kilometres across, so it was not much of a planet. Since they appeared at the highest magnification only as 'star-like' points he suggested that they be designated 'asteroids'. Piazzi countered that 'planetoids' was a better term because they were clearly unrelated to stars. In an attempt to explain why there was more than one object, Olbers suggested that the planet that originally orbited at that distance from the Sun had been shattered by a collision with a comet. After K.L. Harding, one of Schröter's assistants, discovered the third one on 2 September 1804, Olbers found the fourth (his second) on 29 March 1807. After a few years without another sighting, the Celestial Police disbanded itself. In 1830, however, K.L. Hencke started his own search, and was rewarded with his first asteroid in 1845 and his second two years later. Others joined in, and by 1850 there were ten members of the 'asteroid belt'. At first, they were eagerly sought, but after the advent of photography the discovery rate accelerated so dramatically that they were derided as the 'vermin' of the sky. In 1853, U.J.J. Leverrier showed by consideration of possible perturbations that the total mass of the belt was no more than the planet Mars. E.E. Barnard estimated the diameters of the brightest members in 1894 and confirmed that they were very small. In fact, we now know that their total mass is less than 10 per cent that of the Moon, which is itself an insignificant fraction of the overall mass of the Solar System.

THE 'MISSING' MOON

In 1815, Harvard College began to consider the erection of an observatory and W.C. Bond was sent to Europe to inspect the workings of similar institutions. Upon his return, he submitted his report and promptly established his own observatory at Dorchester in order to develop instruments and methods. When the Harvard College Observatory was completed in 1844 with a fine 15-inch refractor made in Germany by Merz-Mahler, Bond was appointed as its first director.

Noting that the distribution of Saturn's satellites followed a pattern reminiscent of the Titius–Bode progression, astronomers pondered the intriguing 'gap' between Titan and Iapetus. William Lassell, a wealthy brewer, had built a 24-inch reflector for his private observatory in Liverpool in England, and in 1846, several weeks after the planet Neptune had been discovered, he spotted a large satellite. Fired up with enthusiasm, he began to search for Saturn's 'missing' moon. On 16 September 1848 Bond, working with his son, G. P. Bond, noted a suspicious speck of light in the plane of the rings. Lassell also saw it two nights later. On 19 September they all confirmed that it was moving along with the planet. Once its 21-day 7-hour 28-minute orbit was calculated, it was found to orbit just beyond Titan. It was given the name Hyperion, but was so faint that John Herschel speculated that it might actually be one of several bodies moving around the planet in a sort of asteroid belt, although no such objects were evident.

THE LURE OF THE RINGS

Upon fleeing Napoleon's invasion of his native Germany in 1808, F.G.W. Struve entered the University of Dorpat (now Tartu, Estonia) and, a decade later, accepted the directorship of the observatory there. The 9.5-inch Fraunhofer refractor was the largest instrument of its type in the world. In 1826 Struve introduced a nomenclature for Saturn by which the ring exterior to Cassini's Division was labelled 'A', and the somewhat brighter interior ring was labelled 'B'.

On 25 April 1837, J.F. Encke, director of the Berlin Observatory, noticed a thin dark line near the outer edge of the 'A' ring; this later became known as Encke's Division. Even when the rings were most 'open', it took a fine telescope, excellent atmospheric 'seeing' and an experienced eye to resolve the short crescent in the cusps of the *ansae*. Subsequent observations showed that it undergoes considerable variation in distinctness, and often disappears for some time.

In 1850 Eduard Roche of Montpellier in France calculated that a body that was spiralling in towards a planet would be disrupted by gravitational tides when it strayed within 2.44 times the planet's radius, as measured from the planet's centre. For Saturn, this radius is 146,000 kilometres, and as the outer edge of the 'A' ring is 136,000 kilometres the entire structure is therefore within the critical radius. In fact,

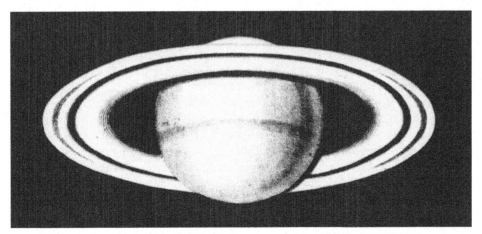

In 1837 J.F. Encke of the Berlin Observatory perceived a division in the 'A' ring.

Roche's study assumed that the two bodies had similar composition and density. Since Saturn's density is a mere 0.69 g/cm^3, the result implied that the rings could not be a coherent fluid structure. Roche's conclusion was that the ring system formed when a satellite spiralled in too close to the planet and was disrupted, which was reminiscent of the myth of Kronos devouring his offspring. Others, however, interpreted Roche's calculations as supporting the analogy with the 'nebular hypothesis', by providing a rationale for why the rings were inhibited from coalescing to form a satellite. Mimas, the closest of the moons, lies just outside the Roche radius, and so has every reason to exist as a coherent body.

A third ring
On 15 November 1850, with the rings wide open, W.C. Bond, perceived with the Harvard College Observatory's 15-inch refractor a dusky inner ring which extended half way in towards the planet as a continuation of the 'B' ring. Two weeks later, W.R. Dawes observed the same feature using a 6-inch refractor at his observatory at Wateringbury in England. When William Lassell paid Dawes a visit on 3 December, Dawes said it was "like a crepe veil covering a part of the sky within the inner ring". Lassell monitored it through the remainder of that night to ensure that it was not an illusion, wrote a report for the newspaper, and settled down to breakfast in a happy spirit, but upon opening Dawes's copy of *The Times* he found a report of Bond's discovery. Inspired by Dawes's remark, Lassell referred to it as the 'Crepe' ring, but in line with the nomenclature it became known as the 'C' ring. Micrometer measurements indicated that the 'A' ring was about 16,000 kilometres wide, Cassini's Division was 5,000 kilometres wide, the 'B' ring was 25,000 kilometres wide, and the 'C' ring extended inwards about 16,000 kilometres, or half of the way down to the planet's cloud tops.

When Czar Nicholas I of Russia established the Pulkova Observatory south of St. Petersburg in 1839, and provided it with a 15-inch Merz refractor, he assigned as director F.G.W. Struve, who brought with him his Dorpat assistant, his son O.W. Struve. A thorough search of the archives by the junior Struve revealed drawings of what had been referred to as a 'dark equatorial belt'. However, in retrospect this was almost certainly the shadow of the dusky ring on the planet. It had been noted by Guiseppe Campani in 1664, Robert Hooke in 1666, Jean Picard on 15 June 1673 and John Hadley in the spring of 1720. William Herschel appears to have glimpsed it in the *ansae*, but not realised its significance. J.G. Galle at the Berlin Observatory noted its presence in the *ansae* on 10 June 1838 as a veil-like extension of the main system across half of the dark space towards the planet, but he had not seen its shadow on the planet, and in any case Galle's report to the Berlin Academy of Sciences produced few comments from his contemporaries. As the rings opened to their maximum extent, the 'C' ring became very obvious. The mystery became how such an experienced and magnificently equipped observer as Herschel could have failed to recognise it. That he had *not* seen the 'C' ring strongly implied that it had been faint at that time. For it to have been simultaneously and independently discovered at this time, it must have brightened over a short timescale.

Using the 6-inch refractor of the East India Company's Madras Observatory in

India, W.S. Jacob noted on 24 August 1852 that the 'C' ring was translucent because it appeared somewhat brighter when transiting the planet's disk, than to either side. In late October of that year, this was noticed independently by William Lassell, who had temporarily set up his 24-inch reflector on the island of Malta. However, G.P. Bond at the Harvard College Observatory, working with his assistant, C.W. Tuttle, saw no evidence of translucency until November 1853, when they noticed the limbs of the planet showing through the 'C' ring.

In the first half of the nineteenth century astronomers had effectively relegated Saturn to the second rank because, while the rings were beautiful, the disk was bland, and they had concentrated their efforts on the more 'dynamic' planets, such as Mars and Jupiter. However, stimulated by the discovery of the 'C' ring, mathematically-minded astronomers eagerly began to explore the minutiae of the relationships between the rings and the retinue of satellites.

Shrinking rings

In 1851 O.W. Struve announced a detailed analysis of the rings. After correlating all the available observations dating back to those of Huygens, he concluded that the inner edge of the 'B' ring was shrinking by about 1 second of arc per century. This meant, he argued, that the material of which the ring was composed was spiralling in towards the planet at a rate of 100 kilometres per year. At this rate, the planet would accrete the ring system in a few hundred years. This startling result meant that astronomers were rather lucky to catch such a transitory phenomenon. To test the hypothesis, in 1851 Struve made a series of micrometer measurements. Upon doing so again in 1882 he found that the dimensions of the ring system had not significantly changed, and was obliged to dismiss his earlier conclusion; evidently the measurements by his predecessors had not been very accurate. In fact, Huygens had introduced the use of a micrometer for making precise measurements, but an optical illusion had made bright objects on a dark background appear slightly larger than they were, and he had thus overestimated the diameter of the ring system.

Fine structure

The mechanical stability of the ring system had been studied mathematically by P.S. Laplace, who had proposed that it is a multiplicity of extremely thin ringlets. During a lengthy study of the system under a range of illuminations, William Herschel found little supporting evidence. In fact, his only suspicion of fine structure was reported five years before Laplace advanced his theory: on four nights in 1780, when the rings were wide open and he could trace them all the way around he had noted in his log that he had perceived a division on the inner ('B') ring.

However, L.A. Quetelet, director of the Brussels Observatory, reported seeing a very fine line divide the 'A' ring when using a 10-inch refractor in Paris in December 1823. In England, Henry Kater, the Vice President of the Royal Society, had a 6.25-inch reflector of excellent quality, and on 17 December 1825 he "fancied that I saw the outer ring separated by numerous dark divisions, extremely close, one stronger than the rest dividing the ring about equally", from which he concluded that the 'A' ring "consists of several rings". The 'B' ring showed "no such appearance". Kater suspected the presence of such lines again in January 1826, but they were not in evidence the following month, and so he concluded that they were "not permanent" – a view that was reinforced when the rings were lacking in detail on an exceptionally clear night in 1828. John Herschel and F.G.W. Struve had independently attempted to verify Kater's sightings in 1826, using larger telescopes, but to no avail.

On 29 May 1838 Francesco de Vico of the Collegio Romano noted the presence of three dark lines, one in the middle of the 'A' ring and the others on the 'B' ring. On 7 September 1843 W.R. Dawes and William Lassell, utilising a 9-inch reflector at Lassell's observatory in Liverpool with very good seeing, saw a fine line dividing the 'A' ring. James Challis, director of the Cambridge Observatory, reported lines on the 'A' ring in 1842 and 1845 using the excellent 11.5-inch Northumberland refractor. On 21 November 1850 Lassell, now using his 24-inch reflector, noted a line near the outer edge of the 'A' ring, which Dawes, using his 6.5-inch refractor, independently saw two nights later, and again several times during the next week. Meanwhile, G.P. Bond at the Harvard College Observatory noted a line near the inner edge of the 'B' ring on 10 October, 11 November and 7 January 1851. Intriguingly, Bond did not see the marking on the 'A' ring reported by Lassell and Dawes, and they did not see his line on the 'B' ring. On 20 October 1851, in excellent seeing, C.W. Tuttle at Harvard noted the 'B' ring to be "minutely subdivided into a great number of narrow rings", starting close to its inner edge and extending over about one-third of its width, adding that "the rings and the spaces between them were of equal breadth". A week later, on 26 October 1851, Dawes saw the 'B' ring "arranged in a series of [four] narrow concentric bands, each of which was somewhat darker than the next exterior one". Warren De la Rue, who had an exceptional 13-inch reflector, observed the rings very clearly in 1856 and saw considerable structure, including in the transition from the 'B' to 'C' rings.

In 1856 Warren de la Rue drew Saturn's rings with several fine subdivisions. Many of his contemporaries regarded such detail as illusory, however. Notice also his depiction of banding on the planet.

Reflecting upon the occasional fine structure in the rings, and the often irregular line of the planet's shadow on the rings, G.P. Bond suggested that the ring material might be a fluid. His colleague Benjamin Peirce agreed, and proposed that they were streams of a fluid whose density was somewhat greater than that of water – that is, exceeding 1.0 g/cm^3.

Although Roche's conclusion applied to a body having the *same* bulk density as its primary, it was evident that companions with greater density could penetrate the critical radius until the gravitational tides overcame their structural integrity. It was therefore not straightforward to decide whether the rings were material that was prevented from coalescing to form a satellite when Saturn condensed out of the solar nebula, or were the debris of a moon that was recently disrupted after straying too close to the planet. The alternatives had a significant consequence: if the rings were a satellite that never formed, they would be composed of 'pristine' relics of the solar nebula, but if they were the debris of a shattered satellite the material would have been thermally 'processed' to some degree.

In 1855 J.C. Maxwell proved mathematically that the rings could be neither fluid nor gaseous bodies, because their rotation would stimulate the formation of waves, which would be disruptive. He also ruled out Laplace's proposal of a multiplicity of exceedingly fine ringlets. In 1857 Maxwell concluded that they could *only* comprise a myriad of small fragments, each of which was orbiting the planet in accordance with Kepler's law that the square of the orbital period is proportional to the cube of the mean distance from the centre of rotation. Therefore, the material at the inner edge of the system travels more rapidly than that farther out. In fact, the possibility that the rings were composed of small bodies had been suggested by J.J. Cassini in 1715

A drawing by E.L. Trouvelot of the Harvard College Observatory in 1872, at a time when the inner 'C' ring was so translucent that it was only partially visible crossing the planet's disk. Notice also the irregular profile of the planet's shadow line on the rings.

and, with considerable conviction, by Thomas Wright in England in 1750, but neither had been able to argue the case mathematically. Nevertheless, the idea had gained sufficient support for Mr Filda to reflect in 1776: "It is thought that the ring around Saturn consists of a great number of Saturnian satellites in close proximity." Maxwell proposed that the impermanent structure in the rings corresponded to 'waves of disturbance' traversing the disk as the gravitating particles influenced one another, and were therefore similar to surface ripples. In the long term, Cassini's Division, the only 'real' gap in the rings, might be impermanent. The irregular and changing line of the planet's shadow on the rings was interpreted as evidence that some particles were in slightly inclined orbits. The 'C' ring must be composed of the same material as the main rings, but be more sparsely populated so that sunlight is reflected less efficiently and the material appears darker. Nevertheless, the particle density was sufficient to cast a hazy shadow on the planet. Indeed, even when the 'C' ring had been too faint to be seen, its shadow had been seen crossing the globe and misinterpreted as a dark equatorial belt. In line with O.W. Struve's hypothesis, it was presumed that the 'C' ring was material in the process of spiralling down towards the planet, gradually eroding the ring system. In 1855 Cambridge University had offered its Adams Prize to the first person to resolve the nature of the rings, and it was promptly issued to Maxwell for his penetrating analysis.

Daniel Kirkwood in America conducted a mathematical study of the asteroid belt in 1866, by which time almost 100 objects had been catalogued. He discovered that there were statistically significant 'zones of avoidance' with mean distances from the Sun matching orbits whose periods were in resonance with Jupiter's. Evidently, the planet's immense gravitational field was perturbing the asteroids in their orbits. In 1867 he turned his attention to Saturn's rings and concluded that Cassini's Division is a 'zone of avoidance' due to perturbations by the moons beyond. He showed that material at this radius would have an orbital period one-half of Mimas's period, one-third of Enceladus's period, one-fourth of Tethys's period and one-sixth of Dione's period. The resonance with Mimas alone would be sufficient to 'sweep' this gap clean. In 1871 Kirkwood found that the orbital period of Encke's Division was three-fifths that of Mimas. Evidently, the simpler the fraction, the stronger the perturbation; thus while Cassini's Division is broad and stable, Encke's is a temporary thinning of the material. Kirkwood noted that material orbiting at the transition between the 'B' and 'C' rings would have a period one-third of that of Mimas. The ephemeral nature of the fine ring structure made it possible to reconcile the view of Bond and Lassell that the 'C' ring was a continuation of the 'B' ring, with occasional subsequent reports of a narrow gap. In January 1888, a few months before Alvan Clark's 36-inch refractor was commissioned at the Lick Observatory in California, J.E. Keeler, visiting, sneaked a preview to test the optics, and was delighted to find a narrow division near the outer edge of the 'A' ring. Although referred to in America as 'Keeler's Gap', this actually marked the return of Encke's Division, which had been absent for some time.

The asteroidal 'zones of avoidance' analogy reinforced Maxwell's assertion that Saturn's rings are composed of a myriad of particles. In 1888 Hugo von Seeliger in Munich made photometric measurements to test whether the rings were composed of

On 7 January 1888 J.E. Keeler tested the Lick Observatory's 36-inch refractor on
Saturn, and resolved a very thin division near the outer edge of the 'A' ring.

discrete particles. He concluded that only such a configuration would display the
observed surface brightness in varying angles of solar illumination. He ventured that
the 'particles' ranged up to a metre or so across. The albedo and phase variations of
the reflected sunlight indicated that they do *not* have smooth surfaces. In fact,
measurements of ring brightness with phase angle implied that the material in the
primary rings – the 'dense' part of the structure – occupies just 6 per cent of the
available volume. In 1933 E. Schönberg in Breslau undertook a similar analysis, and
concluded that while there may be a significant number of large objects, the *mean*
size must be only a few microns. In the 1870s, H.C. Vogel in Germany had noted
that the spectrum of the ring system was very different to that of the planet, being a
simple reflection of the solar spectrum, which indicated that the rings are free of
vapours. The high albedo implied that the material is water ice. It was therefore
concluded that the rings comprise a few large snowballs orbiting in a multitude of
tiny ice crystals.

Albert Marth, who studied under F.W. Bessel and assisted William Lassell when
he took his 48-inch reflector to Malta in the 1860s, was subsequently appointed as
director of the Markree Castle Observatory in Ireland, where he routinely published
the ephemerides of satellites to assist other observers on a worldwide basis. Iapetus is
the only large satellite able to pass through the shadow of the ring system, because its
orbit is both remote and inclined. At Lick Observatory, E.E. Barnard decided to
observe this rare occurrence in order to profile the density of the material across the
rings. Unfortunately, Saturn was inconveniently positioned for the 36-inch refractor,
so he had to employ a 12-inch. The sequence would start shortly before sunrise on 2
November 1889. As Saturn rose, he saw Iapetus emerge from the planet's shadow.
Between the planet and the inner edge of the 'C' ring's shadow, its brightness was
essentially constant, implying that this zone is more or less clear. It then faded
progressively for 75 minutes as it penetrated the 'C' ring's shadow, but it was not

entirely lost until it entered the shadow of the 'B' ring. The rising Sun terminated the observations. There was evidently no gap between the 'C' and 'B' rings at this time, and the ramp in the light curve was striking. Why there should be such a dramatic change in the density at that radius was not evident. Although it was frustrating that the dawn had interrupted the light curve, M.A. Ainslie, who had a 9-inch refractor at Blackheath, and J. Knight, who had a 5-inch refractor at Rye in Sussex, monitored a 7th magnitude star passing behind the 'A' ring on 9 February 1917, thereby providing a measure of the transparency across a chord of that ring. On 28 February 1919, W.F.A. Ellison employed the 10-inch Grubb refractor of the Armagh Observatory in Ireland to monitor Iapetus in eclipse, catching it after it had already cleared the 'B' ring and was illuminated by light through Cassini's Division. Shortly after noting its passage into the 'A' ring's shadow, Ellison was called away from the telescope for 50 minutes, but on his return he was able to monitor the remainder of the moon's passage through the shadow as it never completely disappeared, thereby confirming that the 'A' ring was translucent. William Reid in South Africa observed a star occulted by the 'B' ring on 14th March 1920, and found it to be much less transparent than the 'A' ring. Between them, therefore, these observers had provided a more or less complete profile of the system. The question was: Why were there such sharp transitions in density?

In 1895, in a paper entitled *A spectroscopic proof of the meteoric construction of Saturn's rings*, J.E. Keeler of the Allegheny Observatory, Pittsburg, reported measuring the radial velocities across the ring system employing the Doppler effect.[2] A particle at the inner edge of the 'B' ring *should* travel at 21 kilometres per second in a Keplerian orbit with a 7.5-hour period. A particle at the outer edge of the 'A' ring should travel at 17 kilometres per second in a 13.75-hour orbit. Keeler's measured velocities were 20 and 16 kilometres per second. Slightly different, but supportive velocities were obtained a few months later by W.W. Campbell at Lick Observatory and by H.A. Deslandres at the Meudon Observatory in Paris. Differential rotation *proved* that the rings are composed of a myriad of discrete particles pursuing individual orbits around the planet.

On 2 November 1889, E.E. Barnard made the first observation of Iapetus passing through the shadow of the 'C' ring. Using Tethys and Enceladus as references, he composed this light curve of Iapetus's brightness.

Despite Maxwell's dismissal of fine structure as inconsequential 'density waves' crossing the surface of the ring system, some observers continued to record their appearance. Sidney Coolidge at the Harvard College Observatory noted four or five narrow lines on the 'B' ring on 20 March 1856, and Asaph Hall suspected the presence of lines on the 'B' ring while using the 26-inch Alvan Clark refractor of the US Naval Observatory in Washington on several occasions in 1875 and 1876. On 18 March 1884 Henri Perrotin and Norman Lockyer, jointly using the 15-inch refractor of the Nice Observatory in France, reported fine dark lines. Far more remarkable, however, was the report by E.M. Antoniadi, a Greek astronomer working as an assistant at Camille Flammarion's observatory at Juvisy in France, who, on 18 April 1896, not only saw a trio of widely spaced lines on the 'B' ring, but also half a dozen *radial* markings beyond Cassini's Division on the inner part of the 'A' ring. He promptly wrote to Leo Brenner and Philip Fauth urging that they look. On 26 April they were able to confirm the structure on the 'B' ring, but other than Encke's Division the 'A' ring was featureless. Although Antoniadi noted similar radial markings the following year, because Maxwell had *proved* that radial structure was impossible the reports were dismissed.

In the early-to-mid-twentieth century, T. Cragg, A.P. Lenham, O.C. Ranck, T.R. Cave and Leo Brenner, all of whom were skilled observers with excellent telescopes, reported seeing fine divisions. Percival Lowell, the founder of the observatory in Flagstaff in Arizona which bears his name, reported a number of fine 'divisions' in 1909, 1913 and 1914, and in 1915 he announced that for a time Encke's Division had 'doubled'. Cave saw a similar effect in 1954, but a few years later it had completely closed and was only an abrupt change in the brightness across the 'A' ring.

In 1943 B.F. Lyot recorded numerous ring divisions that he had observed using the 24-inch refractor at the Pic du Midi Observatory. Unfortunately, Lyot died soon afterwards, but his assistant, Audouin Dollfus, published the observations in

On 18 April 1896, in addition to three subdivisions in the 'B' ring, E.M. Antoniadi was astonished to see radial markings on the 'A' ring.

In 1915, when the ring system was as 'open' as it ever gets, Percival Lowell and his assistant E.C. Slipher recorded at least 10 fine divisions in the 'B' ring, which Lowell described as being "conspicuously striped". (Courtesy Lowell Observatory; Memoir No.2, 1915.)

L'Astronomie in 1953. In fact, Lyot saw so much fine structure during moments of exceptional seeing while using a very high magnification and, indeed, a micrometer to measure it, that a generation of less fortunate observers dismissed it as illusory.

Harold Jeffreys pointed out in 1947 that Roche's derivation of the radius within which a satellite could not exist had been based on the assumption that the satellite had the same density as its primary, which in the case of Saturn is abnormally low. Jeffreys extended the theory, and found that a solid object could exist well within 2.44 radii as long as it had sufficient structural integrity to resist the tidal stress. By way of a test, he calculated that if an icy body some 200 kilometres in diameter was disrupted after straying inwards to the mean radius of the ring system, it would have broken into a fairly small number of substantial pieces but would not have produced a myriad of tiny fragments, and he therefore concluded that the rings are *not* the debris of a satellite that strayed too close. Nevertheless, he concluded that the 0.31 g/cm^3 surface density of the ring system suggested that it consists of lumps of ice so, if it is material that was inhibited from forming a satellite, it is likely to be 'cometary' in composition. After a spectroscopic study, G.P. Kuiper of the University of Chicago reported that the particles were either water ice or frost-covered rock.

In 1954 Kuiper made the first serious investigation of Saturn using the recently commissioned 200-inch Hale reflector on Mount Palomar. Upon inspecting the ring system using a very high magnification on a night of ideal seeing, he concluded that Cassini's Division was the only gap; Encke's Division was an abrupt change in the intensity on the 'A' ring; and the 'B' and 'C' rings were juxtaposed. Although he saw hints of three features in the 'B' ring, there was *no* evidence of structure.

In the mid-1960s, R.M. Goldstein of NASA's Jet Propulsion Laboratory used the large spacecraft communications antenna at Goldstone in California as a radar to characterise the surfaces of the Moon, Mars and Venus. In 1972, he was astonished

In 1943 B.F. Lyot made this drawing of Saturn's ring system using the Pic du Midi Observatory's 24-inch refractor during 'perfect' seeing using a high magnification and a micrometer to measure the fine detail. Although his contemporaries dismissed such fine structure as illusory, the Voyager missions spectacularly proved him correct.

to find that Saturn's rings reflected microwave energy with an efficiency of 60 per cent. Mars, in comparison, reflected 8 per cent, the Moon 5 per cent and Venus 1.5 per cent. He inferred from this that, in addition to ice, there must be a substantial number of rocks containing a high proportion of metals in the ring system, many of them several metres across and with irregular surfaces. The presence of such 'evolved' materials implied that the rings could *not* be 'pristine' material left over from the formation of the planet. In a polarisation study published in 1929, B.F. Lyot had noted that the light reflected from the 'B' ring was polarised in a similar manner to that from silicate, whereas the 'A' ring was something for which he had not been able to find a suitable match. The polarisation measurements by Audouin Dollfus at Pic du Midi in 1958 implied that the 'B' ring particles had a mean radius of a few centimetres, were significantly non-spherical and were aligned non-randomly. Bit by bit, the nature of the ring material was being revealed, but it was a tortuous process and in some cases the facts appeared to be contradictory.

By the 1960s it was evident that Saturn has an equatorial diameter of about 120,000 kilometres and an oblateness of 10 per cent; that the 'C' ring starts some 10,000 kilometres above the cloud tops and extends outwards for another 17,600 kilometres, where it blends into the dense 'B' ring, although they are sometimes separated by a gap; the 'B' ring extends outwards for 25,600 kilometres to Cassini's

Division, which is 5,000 kilometres wide; and the 'A' ring is 16,000 kilometres wide. Although the rings are rendered invisible to small telescopes for several weeks during a ring-plane crossings, a large telescope will lose them only for a day or so. At such times, clumps can often be seen in the distribution of material. Estimates of the thickness of the rings had been progressively reduced over the years, decreasing from hundreds of kilometres to a mere 10 kilometres.

SATURN'S ATMOSPHERE

In 1704, in *Opticks*, Isaac Newton pointed out that 'white' light is a combination of a rainbow of colours, and that a glass prism will refract it. However, for some reason he either failed to notice, or chose to ignore, the thin dark lines that populate the solar spectrum. They were first remarked upon by J. Fraunhofer in Munich in 1814, but their origin was not recognised until 1859 when G.R. Kirchhoff and R.W Bunsen of the University of Heidelberg noticed that the lines that appear dark in the solar spectrum corresponded to lines which are bright in the spectra of incandescent gases.

In 1863, P.A. Secchi, the Italian pioneer of astronomical spectroscopy, spotted dark bands towards the red end of the spectra of Jupiter and Saturn, and concluded that their atmospheres were "not yet cleansed" of primordial gases as the Earth's had evidently been in its early history. Several years later, in London, William Huggins, who compared the realisation that a spectrum could reveal the chemical composition of a celestial object to "coming upon a spring of water in a dry and dirty land", mounted a spectroscope on an 8-inch refractor and independently discovered the lines towards the red end of Saturn's spectrum.

When P.J.C. Janssen in France first inspected Saturn's spectrum, he noticed that while it resembled that of Jupiter it had a distinctive line redward that he referred to simply as "the red line" because its origin was a mystery. In 1867 he transported his telescope to the 9,800-foot summit of Mount Etna on Sicily, the tallest volcano in Europe, in order to observe from above most of the tropospheric water vapour, and reported aqueous vapour in Saturn's atmosphere. However, his method involved fitting a prism onto his telescope and making visual comparisons between the planet and the face of the Moon, which was presumed to be arid. Unfortunately, using a micrometer, a visual inspection and documentation of the absorption features along the dispersion could take an hour or more, during which time local atmospheric conditions could easily change, and to make any comparison it was often necessary to wait several hours for the other object to achieve the same elevation, during which time conditions could alter further. As a result, little real progress was made until the introduction of photography enabled the spectrum to be recorded within minutes and compared at leisure; but to record a high-dispersion spectrum required the light grasp of a large telescope, and even then the state of the atmosphere could change while waiting for an object to achieve the requisite elevation. This was overcome by recording spectra in the laboratory for direct comparison. In 1875 Janssen established the Meudon Observatory in Paris and installed an excellent 33-inch refractor supplied by the Henry brothers to study the planets. In 1905 V.M. Slipher at the Lowell Observatory – established in 1894 by Percival Lowell at Flagstaff, Arizona – made a photographic study of the red absorption bands in Saturn's spectrum. There was evidently an as-yet-unidentified chemical in the planet's atmosphere that was leaving its imprint on the reflected solar spectrum. In 1909 he secured spectra using an emulsion that was sensitive into the near-infrared, and discovered more absorption features.

Orbiting twice the distance of Jupiter's orbit from the Sun, Saturn receives only a quarter of the insolation. In 1860 G.P. Bond determined that Jupiter radiates twice as much energy into space as it receives from the Sun. He reasoned that the giant planet must be in the process of contracting, transforming gravitational potential into heat, and concluded that the interior must be a very hot gas. In 1865, in *Saturn and its System*, the prolific populariser of astronomy, R.A. Proctor, wrote: "Jupiter is still a glowing mass, fluid probably throughout, still bubbling and seething with the intensity of the primaeval fires, sending up continuous enormous masses of cloud, to be gathered into bands under the influence of the swift rotation of the giant planet." Saturn, being less massive, was thought to be at an 'earlier' stage of development. In 1882, in the second edition of his book, Proctor wrote: "Regarding the cloud phenomena of the giant planets as generated by internal forces whose real secret lies deep below the visible surface of the cloud belts, we see that these forces must be of tremendous energy, must produce enormous changes in the cloud-laden atmosphere, with effects extending widely both vertically and laterally, and imply enormous heat in the whole frame of each planet." In fact, the giant planets were regarded as 'failed' stars. "Over a region of hundreds of thousands of square miles in extent, the flowing surface of the planet must be torn by sub-planetary forces. Vast masses of intensely hot vapour must be poured forth from beneath, and, rising to enormous heights, must either sweep away the enwrapping mantle of cloud which had concealed the disturbed surface, or must itself form into a mass of cloud." This idea was so popular that in 1885 A.M. Clerke wrote in *History of Astronomy during the Nineteenth Century*: "the chief arguments in favour of the high temperature of Jupiter apply, with increased force, to Saturn, so that it may be concluded, without much risk of error, that a large proportion of the bulky globe … is … heated vapours, kept in active and agitated circulation by the process of cooling."

In 1923 Harold Jeffreys published the first of a series of papers showing that the 'failed' star model was untenable because the outer atmospheres of the giant planets are very *cold*, not hot. In 1924 he suggested that deep within their gaseous envelopes there were solid cores mantled with thick layers of ice. The details were different in each case, but for Saturn, whose great volume was offset by a very low density, he calculated that the envelope comprised about 20 per cent of the radius. He presented his results to a meeting of the British Astronomical Association in 1926, prompting considerable debate. The rarefied character of Saturn's envelope was indicated by an occultation in which the starlight was seen to progressively fade for a considerable distance 'within' the planetary limb. In 1926, W.W. Coblentz, C.O. Lampland and D.H. Menzel undertook radiometry at the Lowell Observatory employing a vacuum thermocouple to directly measure temperatures, and found that the 'visible' surface of Saturn was –150°C, some 15 degrees cooler than Jupiter. Although diehards put forward *ad hoc* models to explain why only the outermost layers were cold, and the interior was 'heated vapours', this in reality marked the end of the 'failed' star model of the giant planets.

The mystery of the redward absorption bands remained until 1931, when Rupert Wildt analysed Slipher's spectra. From the fact that the near-infrared absorption was stronger than that in the visible range, he inferred that the longer-wavelength absorption was induced by transitions in the vibrational states of molecular ammonia and methane. This was confirmed in the laboratory in 1933 by Theodore Dunham at the Mount Wilson Observatory. With this insight, and the recent measurements of the temperatures of the reflecting surfaces of Jupiter and Saturn, it was realised that Saturn's chilly outer envelope should show relatively less molecular ammonia and more methane, because the ammonia should have frozen out to form clouds of icy crystals. In 1929 B.F. Lyot had published polarisation measurements indicating the presence of clouds. It was the over-concentrated methane in the upper atmosphere that was producing the strong absorption features.

Saturn's banded structure is subdued in comparison to Jupiter's. In fact, Saturn at its most 'active' bears a striking resemblance to Jupiter at its most quiescent. The banded structure is also variable. At the turn of the century, E.E. Barnard observed a dark 'polar cap' and no less than five dark belts, four of which were in the northern hemisphere (this asymmetry was due to the ring system masking his view of the southern hemisphere). Several years later, when the rings were edge-on, he saw only two dark belts on each side of the equator. R.W. Wood of Johns Hopkins University noticed that in pictures taken in 1915 by the 60-inch reflector of the Mount Wilson Observatory the 'polar cap' and dark belts were conspicuous in violet light but virtually indistinguishable in infrared light – and limb darkening was also less pronounced. Evidently, the two emulsions were showing features at different depths in the planet's atmosphere.

Asaph Hall used the US Naval Observatory's 26-inch Alvan Clark refractor to observe Saturn intensively from 1875 to 1889. After finding the disk to exhibit little variability, on 7 December 1876, while checking Iapetus, he was surprised to find "a white spot on the ball of the planet". He was able to track the spot's motion around the equatorial zone until 2 January 1877, at which time it disappeared. A number of early observers had reported seeing small spots, but this was the first time that such a prominent spot had been seen. Having tracked it for 60 revolutions, Hall calculated a rotational period for the spot of 10 hours 14 minutes 24 seconds, which was in remarkable agreement with William Herschel's estimate of 10 hours 16 minutes, determined on the basis of irregular patterns in the latitudinal bands.

In 1891 A.S. Williams in England using a 6.5-inch reflector tracked another spot, and computed a similar period. In fact, over the next several years Williams tracked a number of small bright and dark spots in the equatorial zone and noted a continuously declining rotational period which he interpreted as evidence of long-term variability in the winds which were sweeping along the discrete features. Specifically, Williams concluded that "the great equatorial atmospheric current of Saturn was blowing 66 miles an hour more quickly in 1894 than it was in 1891". However, E.E. Barnard had often observed Saturn during this period using the 36-

A photograph of Saturn taken by the Mount Wilson Observatory's 60-inch reflector showing Cassini's Division separating the 'A' ring from the 'B' ring, the fact that the 'B' ring is the brightest part of the system, the 'C' ring silhouetted against Saturn's disk, and the shadow cast by the planet on the far side of the ring.

inch Lick refractor, and had recorded none of these spots. On the morning of 16 June 1903, when Barnard aimed the 40-inch refractor of Yerkes Observatory at Saturn he was pleasantly surprised to find a remarkably bright spot at 36 degrees north latitude. He had been sceptical of reports of both white and dark spots by earlier astronomers with much smaller telescopes. Unfortunately, cloud prevented further observations until 24 June, but after following the spot he derived a period of 10 hours 39 minutes 21 seconds – somewhat longer than the period measured by Hall for his equatorial spot. Meanwhile, W.F. Denning, using his 10-inch refractor in Bristol, had independently discovered another white spot on 9 July 1903 with a period of 10 hours 37 minutes 56 seconds. This spot had also been seen by Catalan astronomer José Comas-Solá through his 6-inch refractor, who derived a period of 10 hours 38 minutes 24 seconds. T.E.R. Phillips in England saw a bright spot at latitude 36 degrees south with a similar period in 1910, and in America G.W. Hill derived a period for Saturn's equator of 10 hours 14 minutes 24 seconds. The variance in the derived rotational periods implied that, like Jupiter, Saturn's atmosphere rotates differentially, with the rate being fastest at the equator. In 1937, when the rings were edge-on and the axis was perpendicular to the line of sight, J.H. Moore at Lick used the Doppler effect and directly measured the rotational period as 10 hours 2 minutes – a figure that was significantly shorter than any determined by tracking spots in the equatorial zone. The period increased continuously with latitude, being approximately half an hour longer at 36 degrees, and almost an hour longer at 57 degrees.

On 3 August 1933, while observing Saturn through his 6-inch Cooke refractor in London, Will T. Hay, a comedian/actor with an enthusiasm for astronomy, noted a white spot in the equatorial zone. Its existence was promptly confirmed by W.H. Steavenson. It circled the planet in 10 hours 13 minutes. Over the next few weeks it became progressively more stretched out in longitude, the leading section became

A 'white' spot was discovered near Saturn's bright equatorial belt by W.T. Hay in 1933, who recorded its appearance in this drawing.

diffuse and the tail darkened. By mid-September it had merged into the equatorial zone. W.H. Wright believed it to be "a mass of matter thrown up from an eruption below the visible surface, encountering a current travelling with greater speed than the erupted matter, which was carried forward by the current while still being fed from the following end".

SO MANY MOONS

Asaph Hall of the US Naval Observatory in Washington published the results of an analysis in 1884 which showed that the orbits of Saturn's inner satellites Mimas, Enceladus, Tethys, Dione and Rhea were near-circular and coplanar with the ring system, but the outer moons were not. As in the Jovian system, the inclinations of the orbits of the satellites increase with distance from the planets. Furthermore, the orbits of the outer moons, Hyperion in particular, have significant eccentricities and hence must perturb one another. Titan clearly dominates the outer system. It is the only satellite to show a disk. During the ring-plane crossing of 1892, L.W. Underwood in America was struck by the fact that its apparent diameter "so far exceeded the apparent thickness of the ring that it gave the appearance of a beautiful golden bead moving very slowly along a fine golden thread". Hyperion is much too small to influence its neighbours significantly. Although remote, Iapetus is larger, and its orbital plane is inclined by 14.75 degrees, so it prompts the marginally inclined orbits of Titan and Hyperion to rotate. The accurate derivation of Titan's orbit in 1834 by F.W. Bessel, a pioneer of astrometry, had enabled this precession to be measured.

In 1888 W.H. Pickering used a photographic telescope at Harvard to conduct a systematic search for Saturnian satellites, but found nothing. Upon being hired as an assistant professor of astronomy by Harvard University in December 1890 he set up the Arequipa Outstation high in the Peruvian Andes in 1891, mainly to observe the imminent favourable opposition by the planet Mars but also to pursue the hunt for Saturnian satellites at the ring-plane crossing of 1892, but the 13-inch astrograph was inadequate for the task.

Meanwhile, on 9 September 1892, while visually inspecting Jupiter with the 36-inch refractor of the Lick Observatory during that planet's exceptionally favourable opposition, E.E. Barnard discovered Amalthea orbiting very close to the surface. It was the first Jovian satellite to be discovered since Galileo turned his primitive telescope towards the planet and saw the four large moons. For this discovery, Barnard was awarded the Lalande Gold Medal of the French Academy of Sciences in 1892, the Arago Gold Medal in 1893 and the Janssen Gold Medal in 1900. As events transpired, this was the last satellite of any planet to be discovered visually.

The discovery of Phoebe
When the 24-inch astrograph funded by Catherine Bruce (and therefore referred to as the Bruce Telescope) was installed at Arequipa in 1897, Pickering renewed his search again. In April 1899, he found indications of a new satellite on plates taken on 16, 17 and 18 August of the previous year. It was at the surprisingly large distance of 12.8 million kilometres from Saturn. He named it Phoebe, and estimated its orbital period as 490 days. When confirmation was not forthcoming, he "began to wonder if the images on the plates of 1898 might not after all have been defects, or faint stars recurring by a curious coincidence in exactly the proper places to represent the motion of a satellite". In 1900 he spotted it again, but it was even farther from the planet. By September 1902, Phoebe was found to be present on 42 of 60 plates which had been taken by the Bruce telescope of Saturn's vicinity, so it was possible to refine its orbit. However, on a series of plates exposed at Arequipa in April/May 1904, Phoebe was not at the predicted positions.

After re-analysing the data, Pickering announced in late 1904 that Phoebe orbits in a *retrograde* manner, doing so in an unusually eccentric orbit that varies between 9.8 and 15.6 million kilometres with a period of 546.5 days. P.J. Melotte took a series of photographs of Phoebe using the 30-inch reflector of the Royal Observatory in Greenwich over a four-month period during the ring-plane crossings in 1907 and tracked the moon as it moved clear of Saturn, out to apoapsis, and a little way back in again.

Significantly, not only is Phoebe's orbit retrograde, it is inclined. (By definition, an inclination exceeding 90 degrees is retrograde.) The inclination of Phoebe's orbit to the system's equatorial plane is 150 degrees. In effect, therefore, the divergence from the system's equatorial plane is 30 degrees. Even by this measure, Phoebe's orbit is the most inclined of all the satellites. However, Saturn's equatorial plane is inclined 26.75 degrees to its orbit, which is itself inclined some 2.5 degrees from the ecliptic, so Phoebe's orbit is actually within a few degrees of the ecliptic, and this strongly suggests that it is an interplanetary interloper that was captured when it strayed too close to the planet. This could also explain the retrograde motion.

Although Phoebe could be an asteroid, in light of Saturn's position in the Solar System it is more likely to be a cometary nucleus, and if this is the case it is a remarkably large one.

A.C.D. Crommelin said of Phoebe: "There is no question that the discovery of Phoebe reflects the greatest credit on Professor Pickering. It was no mere accident, but the result of a deliberate search for additional satellites which he had been carrying on for many years. Even after the existence of the satellite is known, it is a tedious matter to identify it on a photograph, but to have discovered it in this way – one little grey dot among myriads of others – is indeed astonishing." In fact, the determination of Phoebe's orbit represented the transition from visual to photographic astronomy. Even when he knew precisely where to look, E.E. Barnard was *only just* able to see it visually in 1904 at 17th magnitude using the 40-inch refractor at Yerkes Observatory.

Themis
On 28 April 1905 Pickering reported the discovery of another satellite, which he named Themis. On the basis of several sightings he inferred that it orbited between Titan and Iapetus, in the region occupied by Hyperion, with about the same orbital period. Interestingly, he concluded that the plane of its orbit was inclined at the same angle as Hyperion's, but in the opposite sense. However, when confirmation was not forthcoming Themis was dismissed as a 'false alarm'. In fact, the misidentification is readily understood, because Saturn was passing through the constellations of Scorpio and Sagittarius at the time, and was therefore viewed against a particularly dense field of stars. The object that Pickering noted on 13 plates could easily have been an asteroid which, while travelling in the same direction of the planet, soon overtook it.

Interrelationships
With its system of rings and more moons than any other planet, Saturn continued to fascinate mathematically-minded astronomers. In the 1880s M.W. Meyer amplified Kirkwood's analysis of the ring system in terms of resonances with the satellites. In fact, the relationships between the moons themselves showed that the entire system is *tightly* integrated.

Considered in terms of their distance from the planet, several 'groups' of moons were perceived:

- Mimas, Enceladus, Tethys, Dione, Rhea
- Titan, Hyperion
- Iapetus
- Phoebe

Saturn is much larger than the Earth, but to gain a sense of perspective, note that Dione is about as far from the centre of Saturn as the Moon is from the centre of the Earth. Looked at another way, the ring system and the first five large moons orbit within the volume of space that our Moon has all to itself. Furthermore, while the Moon takes a month to orbit the Earth, Mimas circles its primary in just 22.6 hours.

In fact, even Titan, orbiting beyond this inner group, requires only 16 days to complete one orbit. H.L. d'Arrest in Berlin identified resonances in the periods of the orbits of this inner group, with Mimas paired with Tethys, and Enceladus with Dione. That is, Mimas takes twice as long as Tethys to complete an orbit and Enceladus takes twice as long as Dione. The outer satellites are similarly interrelated. The period of Hyperion's orbit is five times that of Rhea's. Hyperion is also paired with Titan: for every three of Hyperion's orbits, Titan completes four. Titan is paired with Iapetus: in the time that Iapetus takes to complete one orbit, Titan completes five. By Kepler's laws of orbital motion, the period of an orbit is related to the radius of its orbit, so it was apparent that gravitational interactions between the various satellites over the years had refined their orbits until these stable resonances were achieved. O.W. Struve's son, and assistant at the Pulkova Observatory, K.H. Struve, made a thorough study from 1889 to 1894 using the observatory's 30-inch refractor, and found that pairs of moons repeat positional relationships, although perturbations from the other moons make the patterns evolve over long periods. For example, in addition to Enceladus making two circuits for every one by Dione, their conjunctions occur when Enceladus is at or near periapsis. Similarly, Mimas makes two orbits for every orbit by Tethys with the conjunction occurring slightly above the equatorial plane and more or less midway between their ascending nodes, with the actual point drifting to and fro with a period of 70 years. Furthermore, the major axis of Rhea's orbit was revealed to be controlled by Titan which, although far beyond, is the most massive member, with the alignment drifting to and fro about its mean position over a period of 38 years. Using such perturbations, K.H. Struve calculated the masses of the individual moons with a high degree of accuracy.

Titan is the only member of Saturn's retinue large enough to present a disk, but efforts to measure its diameter showed considerable variation. Nevertheless, it was soon evident that Saturn's primary satellite was comparable to Jupiter's Ganymede. In 1930, in the absence of evidence to the contrary, K.H. Struve's son, Georg Struve, at the Berlin Observatory, firstly assumed that the albedos of the other moons were the same as Titan's, then used photometric observations made by P. Guthnick and E.C. Pickering to infer the size and bulk density of each moon in order to facilitate a comparative analysis of the system of satellites. His analysis implied that the inner group were predominantly water ice. Furthermore, the bulk densities of Mimas, Enceladus and Tethys were less than unity – the density of water – implying that they were loosely consolidated. The greater densities for Titan and Hyperion meant that they were mixtures of rock and ice, and hence similar to the Galilean satellites of Jupiter. Orbiting outside Titan, and with an intermediate density, Iapetus appeared to be in a class of its own. One striking observation was that while the densities decrease with distance from Jupiter, the trend seemed to be reversed in the Saturnian system.[3]

After inspecting the small inner satellites spectroscopically, G.P. Kuiper suggested that Mimas and Enceladus, and perhaps also Tethys, were "condensations" of water ice and other frozen volatiles such as ammonia, rather than rocky bodies mixed with and coated with water ice. V.I. Tscherednitschenko in Russia came to a similar conclusion in 1951. Thus, contrary to Georg Struve's assumption that all the moons have the same albedo, the surfaces of the inner moons are more reflective and increasing their albedos reduces their diameters and increases their densities.

HINTS OF OUTER AND INNER RINGS

Although the ring-plane crossing of 17 April 1907 was unobservable as Saturn was too close to the Sun in the sky, the crossings of 4 October 1907 and 7 January 1908 were more favourable, and W.W. Campbell using the 36-inch refractor at the Lick Observatory, Percival Lowell using his 26-inch refractor, and E.E. Barnard using the 40-inch refractor at the Yerkes Observatory made independent studies of the bright clumps in the edge-on system, noting that one was in the 'B' ring close to Cassini's Division and that, rather surprisingly, the other was deep in the 'C' ring.

On 5 and 7 September 1907 Georges Fournier inspected Saturn using the 11-inch refractor at Jarry Desloges' observatory on Mont Revard in Savoy, France, and saw "a luminous zone" *beyond* the 'A' ring. There was no sign of it when he looked again on 11 September. On 5 October 1908 M.E. Schaer of the Geneva Observatory, using a 16-inch reflector, noticed a dusky ring crossing the disk of the planet exterior to the 'A' ring. He observed it for several nights, then made his announcement. Whereas Fournier had seen a luminous zone, Schaer saw a dusky feature, so it was not clear that they had seen the same thing. The Royal Observatory at Greenwich in England promptly turned its 28-inch refractor towards Saturn and on 10 October, despite the fact that the Moon was bright, W. Boyer noted that the 'A' ring had a dusky outer edge. For the next two nights, T. Lewis reported something similar. On 15 October, Arthur Eddington wrote that with good seeing "there appeared to be a narrow dusky ring surrounding the bright ring, visible on the northern edge; I could not detect it on the southern edge, nor where it crossed the disk of the planet". However, E.E. Barnard, using the 40-inch refractor, had not noted anything suspicious in 1908. When Schaer reported that the feature was prominent in January 1909, Barnard inspected the ring system on 12 and 19 January, and his negative results prompted others to conclude that if the tremendous light grasp of the 40-inch could not reveal it then surely there could be nothing to see. Nevertheless, R. Jonckheere in Lille, France, reported seeing it in November 1911. A prolonged period during which there were no further reports prompted speculation that it was of highly variable intensity. In April 1952 it was reported as no more than "very vague" by R.M. Baum. Over the next few years it was seen by several observers, including R.R. de Freitas Mourao in 1958 using the 18-inch Cooke refractor of the Rio de Janeiro Observatory, but Patrick Moore, who was observing at that time using the 33-inch refractor of the Meudon Observatory in Paris, saw nothing. Notwithstanding the scepticism, W.H. Haas later designated this the 'D' ring, even though its outside position did not really match the spirit of the nomenclature.

In 1969 Pierre Guerin in France secured photographs hinting at material between the 'C' ring and the planet.[4] A thin gap between this material and the 'C' ring was dubbed the 'French Division'. Although the designation 'D' had been earmarked by

Haas for the mysterious dusky feature outside the 'A' ring, it was officially assigned to this inner zone.

Significantly, the Goldstone radar study detected faint 'echoes' from *beyond* the visible ring system, indicating that there is rocky debris between the outer edge of the 'A' ring and the innermost moon, Mimas, as visual observers had from time to time reported.

SATURN'S INTERIOR

By the 1930s, thanks largely to Harold Jeffreys, the idea that the giant planets were 'failed' stars had finally died off. In 1938 Rupert Wildt posited a detailed model for their internal structure. Arguing by analogy with the manner in which the Earth was presumed to have gravitationally separated, he proposed that these planets should have undergone a similar process, and have formed rocky cores. As their masses are sufficient to retain even the most lightweight of elements, they must have envelopes of hydrogen and helium. All this was consistent with Jeffreys' argument, but Wildt went further and said that in each case the rocky core must be englobed by a shell of frozen volatiles such as water and carbon dioxide. In the case of Saturn, by assuming 'reasonable' densities for these layers (i.e. 6, 1.5 and 0.3 g/cm^3, respectively), he was able to use the known bulk density and the moment of inertia to scale the internal structure: the 45,000-kilometre-diameter core was proportionally smaller than Jupiter's, the 12,000-kilometre-thick icy shell was thinner than Jupiter's, but the 25,000-kilometre-deep envelope was deeper than Jupiter's due to the fact that Saturn's weaker gravity cannot compress its envelope as tightly. He inferred that the low bulk density reflected the fact that although the hydrogen–helium envelope contributed most of the volume, it contained little of the mass. Evidently, Saturn rotates very rapidly because its mass is concentrated in its core.

Initially, Wildt's model was welcomed, but R. Kronig calculated in 1946 that the interior pressure of Jupiter must be sufficient to force hydrogen to adopt a metallic form. Furthermore, geophysical studies had revealed that there had not been sufficient time for the Earth's interior to have gravitationally separated as much as Wildt had been led to believe, and the process would be even slower in a larger planet. Working on the basis that the interiors of the giant planets are chemically homogeneous, and that any structure is due to pressure-induced changes in state, W.H. Ramsey performed calculations in 1951 which showed that the hydrogen in the centre of these planets must be metallic, and hence considerably denser than Wildt had presumed. The fine details were amplified in 1958 by W.C. DeMarcus. In Saturn's case, Ramsey calculated that the metallic-hydrogen core is about 36,000 kilometres in diameter. Within it might be a nugget of rock 27,000 kilometres in diameter at a temperature of 15,000°C. Later, as a refinement, it was suggested that the deepest part of the hydrogen envelope is in a liquid state, and the outer part in a molecular state.

THE DISCOVERY OF JANUS

Saturn was at opposition on 19 September 1966. The first of a series of ring-plane crossing occurred on 2 April 1966, but Saturn was close to the Sun in the sky then, and by the time astronomers were able to make observations the rings had begun to open up again. However, between the crossings on 29 October and 17 December astronomers were able to view the southern side of the system which, because it was not illuminated, meant that the rings were virtually invisible. When Saturn's rings were edge-on in December, Audouin Dollfus at Pic du Midi took a series of plates and spotted a new moon. It orbited so close to the outer edge of the 'A' ring that it could be seen only when the rings were 'absent', so he did not expect confirmation until 1980 (the next ring-plane crossing). In fact, Dollfus was unsure whether there were two satellites in the same orbit, but deciding that he could reasonably vouch for only one he named it Janus. Upon checking his records, Patrick Moore realised that he had noted it several times between July and November while observing with the 10-inch refractor at the Armagh Observatory in Ireland, but had thought it to be Mimas or Enceladus.

A TIMELY PARADIGM SHIFT

"We long to see the actual texture of the rings," wrote the astronomer and popular author R.S. Ball in 1886 in his best-selling *The Story of the Heavens*. But how could this ever be achieved? When G.D. Cassini was appointed to the directorship of the Paris Observatory, one enthusiastic optician had proposed developing a 1,000-foot-long arial refractor with which – it was calculated – it should be possible to see the inhabitants of the Moon. Of course, this telescope was never built. It was even more difficult to conceive of a telescope capable of resolving the fine structure of Saturn's rings. However, with the invention of the rocket, some astronomers started to dream of shooting into space in order to inspect the planets for themselves.

"Utter bilge", scoffed Richard van der Riet Woolley in 1956. As the Astronomer Royal, his voice carried considerable authority. "Space travel is inevitable," insisted Kenneth W. Gatland, a member of the council of the British Interplanetary Society. The United States announced that it would place a satellite into orbit of the Earth as part of the International Geophysical Year, which would actually run for 18 months from mid-1957 through 1958. On 4 October 1957, however, the Soviet Union placed Sputnik into orbit. The Space Age had begun. Within a decade, probes had landed on the Moon, and men were eager to follow them. It was evident from the manner in which the results from the first probes to make fly-bys of Venus and Mars revolutionised our thinking about these planets that the 'robotic explorer' represented a paradigm shift as profound as the invention of the telescope.

Telescopes have come a long way since Galileo's time. Commissioned in 1896, the 40-inch refractor of the Yerkes Observatory is still the most powerful such instrument in the world.

In the Space Age, robotic exploring machines (in this case Mariner 4) carry cameras and a variety of scientific instruments to make *in situ* observations of the planets.

2

First close look

SPACE PARTICLES AND FIELDS

On 31 January 1958 America's first satellite, Explorer 1, was sent into an elliptical orbit ranging out to an altitude of 2,000 kilometres. The Geiger–Müller counter that it carried revealed that electrically charged particles circulate in the Earth's magnetic field. The instrument's principal investigator was J.A. Van Allen of the University of Iowa, and this radiation became known as the Van Allen belt. In 1962, as Mariner 2 departed the Earth's vicinity to make a fly-by of Venus, it found that interplanetary space is pervaded by particles that flow from the Sun as a 'solar wind'. It was then realised that the radiation within the Earth's 'magnetosphere' originated from this wind, and that auroral displays occurred when dense streams of particles forced their way into the open 'cusps' above the magnetic poles. In the 1960s NASA's Ames Research Center, which is located south of San Francisco, sent a series of extremely successful Pioneers into solar orbit, some slightly inside the Earth's orbit and some just outside it, carrying suites of 'particles and fields' instruments to report the state of the solar wind. One discovery was that the strength of the solar wind draws the Earth's magnetosphere downstream to a considerable distance in a 'magnetotail'.[1]

When Ames set out to develop a spacecraft to venture farther out into the Solar System, it chose Jupiter as the objective. The space agency's budget was in decline, and rather than request funding for a new programme Ames proposed it as an 'advanced' mission on the reasonable basis that it was wise to use the successful track record as leverage – after all, the focus would be on particles and fields. 'Pioneer Jupiter' was authorised in February 1969, with Charles F. Hall as project manager.

The earlier probes had been drum-shaped and had rotated several times per minute, both for stability and to sweep their sensors in three-dimensions in order to measure the trajectories of the charged particles caught up in the magnetic field in the immediate vicinity. Although the new spacecraft would undertake similar measurements, the mission requirements obliged a different configuration. In the earlier design, there was an array of photovoltaic cells around the drum to make electricity from sunlight, but as insolation decreases in proportion to the square of

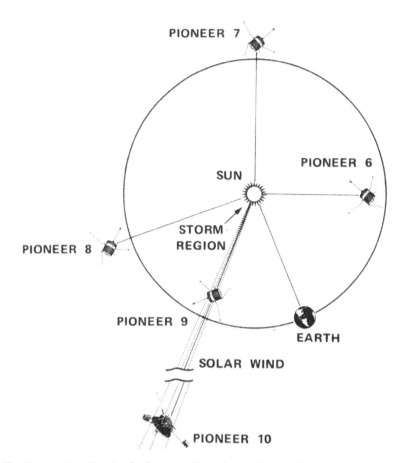

The Pioneers heading for Jupiter coordinated with their predecessors in the inner Solar System to study the solar wind.

the distance from the Sun, this would not be practicable in the outer Solar System. The only option was to use *Radioisotope Thermo-electric Generator* (RTG) power cells that produce power by the natural radioactive decay of non-weapons-grade plutonium dioxide, transforming it into electricity using solid-state thermoelectric converters. It had two, mounted on short booms, splayed at 120 degrees, providing a total of 140 watts. Because the transmitter would radiate only 8 watts of power, it would need a large narrow-beam antenna to communicate from so far away, even at a 2,048-bit-per-second data rate. The overall design therefore assumed the form of a cluster of instruments mounted at the rear of a large dish antenna. As the Pioneer family were not fitted with computers capable of extended autonomous operations, they had to be commanded from Earth, so as the new spacecraft slowly rotated to monitor the particles and fields in its environment its antenna would be constantly aimed at the Earth to enable its commands to be received and its data to be reported. As far as possible, off-the-shelf systems were used to minimise development costs.

IMAGING PHOTOPOLARIMETER

GEIGER TUBE TELESCOPE

METEOROID DETECTOR
SENSOR PANEL

ASTEROID – METEOROID
DETECTOR SENSOR

HELIUM VECTOR
MAGNETOMETER

PLASMA ANALYZER

TRAPPED RADIATION
DETECTOR

COSMIC RAY
TELESCOPE

FLUX GATE
MAGNETOMETER

INFRARED RADIOMETER

CHARGED PARTICLE
INSTRUMENT

The configuration of the Pioneer spacecraft dispatched to the outer Solar System.

Instrumentation

The 260-kilogram spacecraft had a 28-kilogram payload of scientific instruments. The *Magnetometer* had an optical sensor which monitored a cell filled with helium in which magnetic fields passing through the instrument induced electrical discharges. A lightweight boom would project it 6.6 metres from the spacecraft's axis, away from the body of the spacecraft and the 'noisy' RTGs. It was to measure the strength and direction of the magnetic field washing over the vehicle, both in interplanetary space and within the Jovian magnetosphere, and note the transition from one environment to the other.[2] E.J. Smith of the Jet Propulsion Laboratory – operated for NASA by the California Institute of Technology – was the principal investigator.

The *Plasma Analyser* sampled through an aperture in the high-gain antenna. Solar wind particles were to be passed through a pair of plates, across which a sequence of voltages would be applied to derive the energy spectrum. The principal investigator was J.H. Wolfe of Ames.

The *Charged Particle Detector* comprised two pairs of 'particle telescopes', one pair to study the solar wind, the other pair for the harsher Jovian environment. During the interplanetary cruise, one was to measure the composition of cosmic rays in the energy range 1 to 500 MeV and the other was to measure protons and 'alpha particles' (helium nuclei) with energies from 400 keV to 10 MeV to distinguish ions of light elements in the solar wind from those in cosmic rays. In Jovian space, fission induced in a foil of thorium struck by protons exceeding 30 MeV would measure the flux of protons, and a solid-state electron current detector would measure the flux of electrons exceeding 3 MeV that were believed to be responsible for the decimetric radio emissions. J.A. Simpson of the University of Chicago was the principal investigator.

The *Cosmic Ray Telescope* employed three solid-state detectors: one to measure the fluxes of electrons in the range 50 keV to 1 MeV and protons in the range 50 keV to 20 MeV; a second to measure the flux of protons in the range 56 to 800 MeV; and a third to measure the fluxes of nuclei of the ten lightest elements (up to neon). The principal investigator was F.B. McDonald of the Goddard Space Flight Center in Greenbelt, Maryland.

The *Geiger Tube Telescope* was to use a set of Geiger–Müller tubes to measure the intensity, energy spectra and angular distribution of the electrons and protons as the spacecraft flew through the Jovian radiation belts. It was a much improved form of the instrument on Explorer 1. The Geiger Tube Telescope was supplemented by a *Trapped Radiation Detector* which had five detectors: two scintillation counters to detect the ionisation trails of electrons less than 5 keV and protons less than 50 keV passing through them; an electron-scatter detector to count electrons in the range 100 to 400 keV; a solid-state diode to detect protons in the range 50 to 350 keV; and a Cerenkov counter to detect the flashes as electrons in the range 500 keV to 12 MeV passed through it. C.E. McIlwain of the University of California at San Diego was the principal investigator.

The distribution of dust in space was to be investigated by two instruments. The *Meteoroid Detector* employed 13 sensor panels arranged in a circle on the rear of the high-gain antenna dish, each of which contained 18 sealed cells pressurised with gas. The rate at which a cell depressurised upon being punctured would be proportional to the size of the hole. It would be able to detect strikes by particles with masses as small as one-billionth of a gram. The principal investigator was W.H. Kinard of the Langley Research Center in Hampton, Virginia. The other instrument employed a remote-sensing technique. The *Meteoroid–Asteroid Detector* had a cluster of four 20-centimetre-diameter reflecting telescopes illuminating photomultiplier tubes. The 8-degree fields of view of the telescopes overlapped slightly, so if any three reported a near-simultaneous increase in brightness this would be interpreted as a reflection, and the range, path and velocity of the source would be computed. It was hoped that as it passed through the asteroid belt this instrument might measure the distribution of larger bodies as well as dust. The principal investigator was R.K. Soberman of the General Electric Company in Philadelphia.

The *Ultraviolet Photometer* had photocathodes to detect neutral hydrogen and helium by their 1,216-Ångström and 584-Ångström emissions, respectively. The instrument was to measure the distribution of these gases in interplanetary space, and measure how Jupiter's upper atmosphere scattered solar ultraviolet to compute the amounts of atomic hydrogen and helium in order to determine the 'mixing rate' of the atmosphere. The observations would test fundamental theories, because the

'Big Bang' theory of the origin of the Universe made a specific prediction about the overall ratio of hydrogen to helium, and theories of how Jupiter formed from the solar nebula and evolved had been based on assumptions that could not easily be tested by terrestrial observations. D.L. Judge of the University of California at Los Angeles was the principal investigator.

The *Infrared Radiometer* utilised a 76-millimetre-diameter Cassegrain telescope to illuminate an array of thin-film bimetallic thermocouples. It was to determine the temperature across Jupiter's cloud tops in the 14- to 56-micron wavelength range. It had long been known that Jupiter radiates more energy than it receives from the Sun. Although its infrared emissions are strongest between 20 and 40 microns, the Earth's atmosphere is opaque. Studies at shorter wavelengths had indicated the distinctive latitudinal structure of the atmosphere to be rather more pronounced than at visual wavelengths, so close-up observations during a fly-by would not only measure the overall 'energy budget' but also yield clues as to the thermal structure and chemical composition of the upper atmosphere. Guido Münch of the California Institute of Technology was the principal investigator.

As with their predecessors, the new spacecraft were 'spinners' which rotated for stability and to optimise the spatial resolution of their sensors. This would enable the

A diagram showing how the narrow-angle photopolarimeter served as a 'spin–scan' imager as the Pioneer spacecraft rotated. An image was built up a line at a time by tilting the instrument's field of view and (on Earth) joining adjacent scan strips.

Ultraviolet Photometer and Infrared Radiometer to scan Jupiter. However, imaging is difficult from a rotating spacecraft. In interplanetary space this would not matter as there would be nothing specific to look at, but it was unthinkable to send a vehicle to Jupiter and *not* take pictures of it. In the absence of a stabilised scan platform, Ames added a narrow-angle scanning *Photopolarimeter* and developed a method of joining a succession of scan strips to assemble an image. This 'spin–scan' technique would serve as a rudimentary imaging system. The principal investigator was Tom Gehrels of the University of Arizona.

As the spacecraft passed behind the limb of an object, the manner in which its signal was attenuated would provide information about the environment close to the body. The principal investigator for the Occultation Experiment was A.J. Kliore of NASA's Jet Propulsion Laboratory. In addition, detailed tracking of the spacecraft during its encounters would enable the masses of the objects – planets and satellites – to be measured, and so yield clues as to their internal structures. No spacecraft instrumentation was required. The tracking would be facilitated by measuring the Doppler on the spacecraft's radio signal. J. D. Anderson of JPL was the principal investigator for this Celestial Mechanics Experiment.

PATHFINDING

It was standard procedure at that time to build two identical spacecraft in order to provide a degree of redundancy in case one was lost at launch, or failed thereafter. If only one survived dispatch, it would report on the solar wind beyond the orbit of Mars, pass through the asteroid belt, make a fly-by of Jupiter as a reconnaissance of its magnetosphere, then fly on out of the Solar System. If the first achieved its primary objective, there would be an option of expanding the backup's mission. The launches were to be phased to enable the results of the first Jovian encounter to be fully assessed and fed into the final in-flight planning for its successor.

Even as Ames was preparing its Pioneers for launch, planning was underway at JPL in Pasadena, several hundred kilometres to the south, for a 'Grand Tour' of the outer Solar System by a spacecraft that would utilise a close fly-by of Jupiter as a 'gravitational slingshot' to reach Saturn, and then, in turn, the planets beyond. This had emerged from a study by G.A. Flandro in 1965, in which he had explored the use of slingshots.[3] In fact, as Mariner 4 had only just made the first fly-by of Mars, it took visionary planning to envisage a 12-year mission. The Grand Tour name had actually been coined by G.A. Crocco at the International Astronomical Union's Congress in Rome in 1956 – a year before the launch of Sputnik started the Space Age – when he had presented a study of multiple-planet trajectories.

In a sense, Jupiter's tremendous gravitational field can serve as the 'doorway' to the outer Solar System. The majority of the energy involved in placing an object into space is devoted to climbing from the Earth's 'gravitational well', so low-Earth orbit has been referred to as being 'halfway to anywhere in the Solar System'. As long as the positions of the planets are conducive to using slingshots, once a spacecraft has been set on course for Jupiter, it requires *no more fuel* to reach the planets orbiting farther out. Furthermore, although flying from one planet to the next, in sequence, greatly increases the *distance* that has to be travelled to reach one in the outer Solar System compared to the 'direct' route employing half of an elliptical transfer orbit, the accelerations resulting from the encounters *en route* shorten the transit *time*. Of course, a planet loses precisely the same amount of energy as a spacecraft gains but, being so much more massive, a planet's orbit is negligibly affected. In providing us with ready access to the outer Solar System, Jupiter is serving up the elusive 'free lunch'.

Simply reaching Jupiter was a major issue. There were doubts as to whether a spacecraft could safely pass through the asteroid belt. The known large objects were not a serious concern, because they were few in number and spaced far apart. The real worry emanated from the small rocks and specks of micrometeoroid debris that seemed likely to litter that entire zone. Striking a piece of rock as small as the size of a pea at a relative speed of several dozen kilometres per second would likely be catastrophic. Even if the spacecraft passed through the asteroid belt unscathed, there were dangers in the Jovian environment. By sheer serendipity, in 1955 Jupiter had been found to be a strong and variable radio source, which implied that it is surrounded by an intense belt of charged particles. A major 'unknown' in planning the Grand Tour fly-by, therefore, was how deeply could a spacecraft penetrate the Jovian magnetosphere without being disabled by radiation. At JPL, the Grand Tour planners hoped that Ames's Pioneers would provide the information to plot a safe course for their own mission. In fact, JPL suggested closing to a planetocentric range of 1.3 radii, with the Pioneer barely skimming the giant planet's cloud tops. In the absence of firm data – and fearing the worst – Ames decided that its first spacecraft would approach no closer than 3 radii to provide some 'hard' data on the Jovian environment, and if the radiation proved to be tolerable then the second spacecraft would be sent in much closer to explore the possibility of a slingshot for Saturn.

JOVIAN FLY-BYS

On 2 March 1972 an Atlas rocket lifted off with Pioneer 10. After the Centaur stage had boosted it away from the Earth, a small solid-rocket accelerated the spacecraft further to 50,000 kilometres per hour, faster than any previous probe, in order to fly a billion-kilometre 'fast track' route to Jupiter. It soon crossed Mars's orbit, and entered uncharted territory. In July 1972 it officially entered the asteroid belt, but the boundary was rather arbitrarily defined. This was a very nervous time for the people who expected the spacecraft to fall silent at any moment, but it emerged unscathed from the other side in February 1973. During the interplanetary cruise, the particles and fields instruments mapped the magnetic field and reported how the strength of the solar wind varied in response to solar storms and with increasing distance from the Sun; monitored the interactions between the plasma and the electric and magnetic fields; measured the size, mass and velocity of dust, particularly while traversing the asteroid belt; and monitored the flux of cosmic rays which were passing through the Solar System from their mysterious sources far beyond.

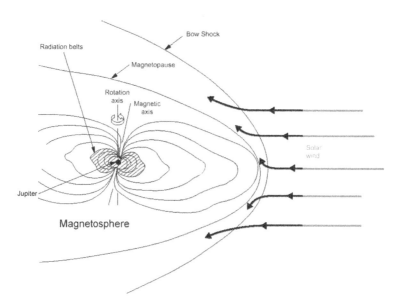

As the solar wind is compressed against the Jovian magnetosphere it forms a 'bow shock' of turbulent plasma, and draws the magnetosphere into a tail extending far downstream.

On 26 November 1973 the instruments noted an abrupt change in the state of the electromagnetic environment. This was the 'bow shock' where the solar wind struck the sunward side of the Jovian magnetosphere. Although this had been predicted, the fact that it was 8.5 million kilometres from the planet was a considerable surprise. A day later, Pioneer 10 crossed the magnetopause and entered the magnetosphere. It became evident that the magnetosphere spans at least 100 planetary radii, making it the largest discrete structure in the Solar System. If it could be seen, it would be the largest object in the Earth's sky. No longer exposed to the solar wind, the spacecraft proceeded to map the intensity, direction and structure of the planet's own magnetic field. Over the next few days, the fluxes and energies of the trapped charged particles increased. After a week, by which time the spacecraft was very close to the planet, the radiation was so intense that several of its instruments were reporting 'off-scale-high', having been saturated. On 3 December, Pioneer 10 flew 130,000 kilometres above the Jovian cloud tops, which, in terms of planetocentric distance, was slightly less than 3 radii.

The Photopolarimeter scanned narrow strips of Jupiter's disk as the spacecraft rotated. The resulting imagery not only far exceeded the best telescopic pictures, the 500-kilometre surface resolution revealed fine structure in the atmosphere which not even dedicated observers had glimpsed during rare moments of perfect seeing. The Infrared Radiometer, whose field of view became tighter than the planet's disk only for an hour or so in the run up to closest approach, measured the temperature field at the cloud tops, thereby confirming that the planet emits more energy than it receives from the Sun.[4] The Ultraviolet Photometer recorded 'glows' around Jupiter

A NASA artist's depiction of Pioneer 10's historic Jovian fly-by.

from neutral hydrogen, and from auroral activity. As the spacecraft slipped behind Jupiter's limb, the way in which its radio signal was refracted and attenuated profiled the physical and chemical properties of the upper atmosphere. Tracking of the fly-by charted the gravitational field, which measured the planet's moment of inertia, in turn providing insight into its interior structure.

In addition, Pioneer 10 secured the first close imagery of the four large satellites. The trajectory had been carefully set up so that the spacecraft would pass behind Io, the innermost of the four, and this revealed that the moon has a tenuous ionosphere. Furthermore, the Ultraviolet Photometer revealed that a large region of space near Io glows. In early telescopic reports, Io had been thought to be too small to retain an atmosphere, so the moon became one of the mysteries of the Jovian system that the Voyagers would investigate.

As a result of its encounter with Jupiter, Pioneer 10 was accelerated away from the Sun, to report on the outer heliosphere.

FIRST TO SATURN

Pioneer 11 was successfully placed on course for Jupiter on 5 April 1973. After its predecessor had "tweaked the dragon's tail" at the end of the year, Ames decided to

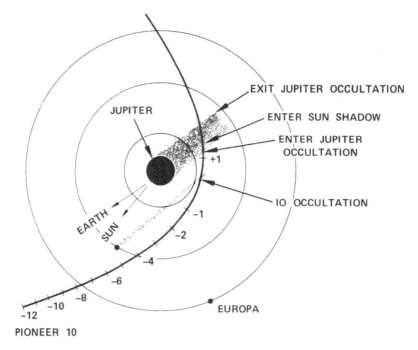

EXIT JUPITER OCCULTATION

ENTER SUN SHADOW

ENTER JUPITER OCCULTATION

IO OCCULTATION

JUPITER

EARTH

SUN

EUROPA

PIONEER 10

A polar projection of Pioneer 10's route through the Jovian system, contrived so that it would be occulted by Io.

try a gravitational slingshot to send Pioneer 11 on to Saturn. The slingshot technique was demonstrated on 5 February 1974 by Mariner 10, which used a close fly-by of Venus to become the first – and so far only – mission to Mercury, the innermost of the planets.

Although Saturn orbits the Sun farther out than Jupiter, it would be on the far side of the Solar System when Pioneer 11 reached Jupiter in 1974. To reach Saturn, the vehicle would have to fly sufficiently close for the tremendous gravitational field to virtually double the trajectory back on itself to redirect the spacecraft right across the Solar System. Such a deflection would involve making a much closer fly-by than before, but it was evident from the intensity of the radiation close to Jupiter that the spacecraft would not survive such a deep penetration. However, the charged particle radiation was concentrated in the equatorial zone, and Pioneer 10 had suffered such a dosage because its trajectory was near the Jovian system's equatorial plane. Ames therefore decided to have Pioneer 11 pass 'beneath' the most intense radiation on the near side of the planet, skim over the south pole and be deflected around the far side so that it would race through the radiation in the equatorial belt on the far side on a steeply inclined trajectory. If all went well, Pioneer 11 would be subjected to a *lower* dosage than its predecessor. As a bonus, this fly-by would also provide a different perspective on the inner magnetosphere. The spacecraft would emerge in an elliptical solar orbit that was steeply inclined to the ecliptic and would meet Saturn five years later. Without Pioneer 10's reconnaissance, the navigational requirements for such

The 'spin–scan' photopolarimeters of Pioneer 10 (left) and Pioneer 11 (right) built up images of Jupiter's disk, revealing its atmospheric structures in unprecedented detail.

an ambitious two-planet mission would not have been practicable, so this was an excellent use of the notional 'backup' spacecraft.

On 2 December 1974, as Pioneer 11 skimmed a mere 43,000 kilometres over the Jovian cloud tops, its imagery not only documented the high-latitude zones, but the closer fly-by gave an even better view of the intricate structure of the atmosphere.

The navigation on the way in had been excellent, and the spacecraft, now dubbed 'Pioneer Saturn', emerged on course for Saturn. Its particles and fields instruments reported on the state of the solar wind during the cruise through uncharted territory far above the ecliptic. In early 1976, some 150 million kilometres 'above' the orbit of Mars, the trajectory was at its greatest departure from the ecliptic. This information was a significant secondary objective as all the previous spacecraft in the series had sampled the wind in the plane of the ecliptic. In fact, this part of the mission was to stimulate a proposal to fly a pair of particles and fields spacecraft out to Jupiter on trajectories that would send them over opposite poles, to emerge in emerge in elliptical solar orbits almost perpendicular to the ecliptic, one above and the other below, and report on the solar wind emanating from the Sun's polar regions. However, in the event, only one spacecraft for this International Solar Polar Mission was built; it was launched in 1990 as Ulysses.

Despite having employed a trajectory designed to minimise the radiation dosage, Pioneer 11 did not escape Jupiter unscathed. It began to suffer spurious commands, and after months of analysis it was realised that the Meteoroid Detector had also been damaged, so it was switched off. As the instruments had, in turn, been temporarily deactivated during testing to isolate the fault, the Plasma Analyser suffered from chill and initially failed to restart, but after many attempts, it was reactivated in late 1977.

Planning for Saturn
The dilemma for the Saturn-encounter planners lay in chosing the most appropriate

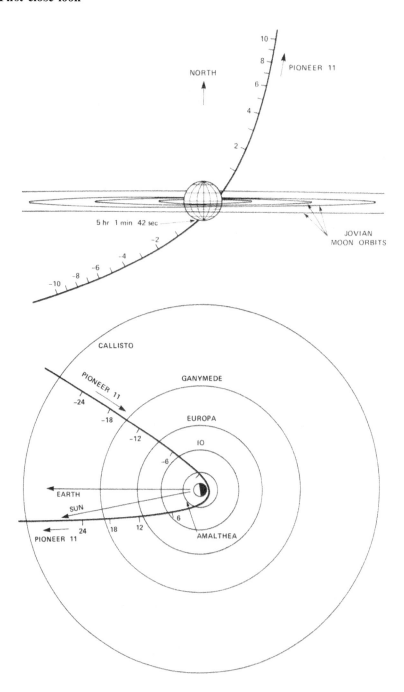

By passing under Jupiter's south polar region, Pioneer 11 was able to 'double back' on a trajectory steeply inclined to the ecliptic.

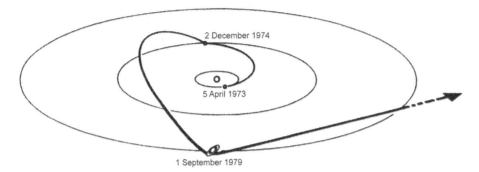

2 December 1974

5 April 1973

1 September 1979

After Jupiter, Pioneer 11 flew high above the ecliptic on a five-year cruise to Saturn, on the far side of the Solar System.

trajectory for the spacecraft to fly through the system. Pioneer 11 had accomplished its primary mission at Jupiter, so Saturn, as C.F. Hall explained, was "gravy", and because it was not constrained by the need to undertake a slingshot for another target the scientists were free to accept a degree of risk in pursuing specific scientific objectives.[5]

As before, JPL wished to explore the region through which it hoped to send one of its own spacecraft on the Grand Tour, and this slingshot called for crossing the ring plane at a planetocentric distance of about 2.9 radii, which was just beyond the 'A' ring, and then closing to about 1.4 radii of the planet. This time, however, Ames was willing to risk damage to the spacecraft by crossing the ring plane at 1.15 radii, then passing within 1.06 radii of Saturn, which would skim a mere 9,000 kilometres above the cloud tops and would provide a very accurate map of the gravitational field in order to investigate its internal structure. If, as seemed likely, Saturn possessed a magnetic field, then charting the inner magnetosphere in this way would yield information which the later JPL missions – which would be required to keep their distance from the planet – would not be able to supply. An option of flying through the apparently safe gap of Cassini's Division satisfied no one. The debate therefore centred upon the relative dangers of the inner 'D' ring versus the outer 'E' ring.[6] Of course, matching the uncertainties experienced by Pioneer 10 when penetrating the Jovian magnetosphere, the trailblazing nature of this flight meant that key decisions had to be made in the absence of 'hard' data because no one really knew what the ring-plane environment was like – despite more than three centuries of telescopic study. The nub of the issue was the mean sizes of the 'particles' in the various parts of the ring system. A flake of ice less than 1 millimetre across would be unlikely to do serious damage to this particular spacecraft, even striking at 112,000 kilometres per hour. A piece of ice larger than 10 millimetres, however, would almost certainly be crippling. By any reasonable theory, the particle population should be inversely proportional to size – that is, most of the material should be at the small end of the size range. A large object would be more likely to disable the spacecraft, but there should be fewer such particles and they should be widely spaced, so there ought to be a reasonable prospect of the spacecraft passing between them. The real danger came from particles ranging in size from 1 to 10 millimetres, in essence from something the size of a grain of sand to the size of a pea, because these would be fairly numerous and fairly densely packed.

Another aspect of the decision involved Titan, Saturn's largest satellite, which would be on the far side of its orbit during Pioneer 11's flight through the Saturnian system. The scientific objectives would be:

- to measure the temperature of Titan's atmosphere;
- to accurately determine the satellite's mass;
- to obtain spin–scan images of its visible surface; and
- to improve information on its orbit and ephemeris.

The 'inner' option would produce a very close fly-by of Titan on the way out, but if Pioneer 11 was disabled in crossing the ring plane this opportunity would be pre-empted. The fly-by of the 'outer' option would not be as close, but at least the spacecraft stood a more reasonable chance of surviving to report its observations of this fascinating moon. On 8 November 1977 C.F. Hall formally recommended aiming for the 'French Division' between the 'B' and 'C' rings, but several weeks later Thomas Young, the Director of Planetary Programs in Washington, decided in favour of the 'outer' option because it was "essential" to determine whether a spacecraft *could* survive passage through the ring plane on the trajectory required for the Grand Tour. Irrespective of whatever happened to Pioneer 11, the 'outer' option would yield invaluable insight: if it was lost, the Grand Tour would have to be reassessed, and perhaps even abandoned; if it survived its passage through the ring plane, the Grand Tour would be able to proceed with a reasonable expectation of making a successful slingshot at Saturn.

In July 1978, Pioneer 11 fired its thrusters to refine its trajectory to pursue the 'outer' option. Planning for this historic second encounter was complicated by the fact that Saturn would be at superior conjunction on 11 September 1979. With the spacecraft on the far side of the Sun, the solar corona would degrade the radio signal. Indeed, if a solar storm blasted plasma across the line of sight, the signal might become unreadable. As the angular separation would diminish by about 1 degree per day, it was deemed best to use this manoeuvre to advance the encounter by several days, to 1 September. Hence, Pioneer 11 would be able report particles and fields data during the approach and throughout the fly-by, but only for about a week on the way out before its signal was lost. Furthermore, to provide a degree of redundancy in the receipt of the crucial data at the ring-plane crossing, this event was scheduled to occur when two of the Deep Space Network antennas had a line of sight.

To terrestrial observers in September 1979, Saturn's rings were inclined at a mere 2 degrees, displaying the illuminated southern face, and they were 'closing' towards edge-on in 1980. Pioneer 11, however, was approaching the planet from north of the ecliptic, and from its perspective the ring system was inclined 6 degrees and showing the non-illuminated northern face. Thus, as the spacecraft approached, it would view the system in silhouette, with the most opaque sections of the rings appearing dark. It would see the sunlit face only after crossing the ring plane. As the fly-by loomed, there was every expectation that the data would revolutionise our understanding of the ringed planet. As a particles and fields platform, Pioneer 11's scientific objectives at Saturn were similar to those undertaken at Jupiter, namely:

- to map the planet's magnetic field (presuming that it had one) and determine its intensity, direction and structure;
- to determine the energy spectra of the electrons and protons along the spacecraft's trajectory through the system;
- to map the interaction of the Saturnian magnetosphere with the solar wind;
- to determine the structure of the planet's upper atmosphere where molecules were expected to be electrically charged to form an ionosphere;
- to measure the temperature of Saturn's atmosphere;
- to map the thermal structure of the atmosphere by infrared observations coupled with radio occultation data;
- to obtain spin–scan images of the planet and the ring system;
- to make polarimetry measurements of the planet, the ring system and some of the satellites;
- to determine how the spacecraft's trajectory was perturbed as a result of flying close to the planet and its major satellites, in order to refine estimates of their masses;
- to obtain information to enable the orbits and ephemerides of the planet and its major satellites to be refined; and
- to investigate the ring-plane environment to determine whether it would be safe to fly a Voyager spacecraft through it.

As Saturn is twice as far away from Earth as Jupiter, the spacecraft was able to transmit at only one-quarter the data rate that had been possible at Jupiter – that is, a mere 512 bits per second. With solar interference to contend with, only the largest Deep Space Network antennas were capable of receiving the signal.

Approaching Saturn
Weak, long-wavelength (kilometric) radio busts had been detected from Saturn, so it was presumed to have a magnetic field. However, there was debate over whether this would be sufficiently strong to ward off the solar wind and form a magnetosphere of trapped radiation. Those researchers who thought it could, expected Pioneer 11 to find the bow shock somewhere between 50 and 20 radii from the planet. On 30 August, as the spacecraft crossed the 50-radii line, it was still in the solar wind. The first indication of a change was early on 31 August, at 24 radii. The wind was gusty, however, and an hour and a half later the magnetosphere was compressed and the spacecraft found itself once more in the solar environment. Another 12 hours passed before it finally crossed the magnetopause, by which time it had closed in to a mere 17 radii. The Plasma Wave Spectrometer sensed the electrical waves washing back and forth. These waves occur at frequencies that can be heard by the human ear, so

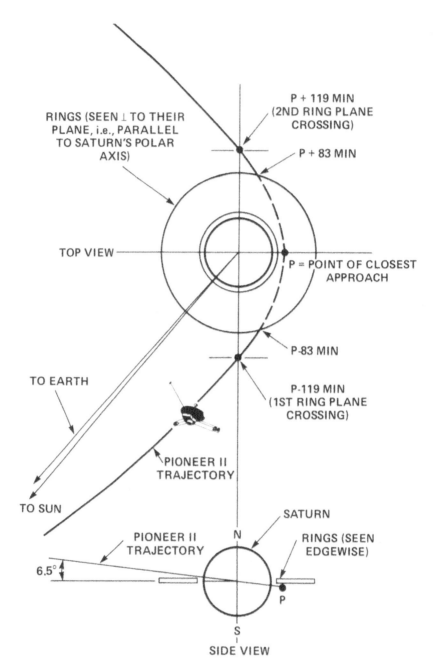

RINGS (SEEN ⊥ TO THEIR
PLANE, i.e., PARALLEL
TO SATURN'S POLAR
AXIS)

P + 119 MIN
(2ND RING PLANE
CROSSING)

P + 83 MIN

TOP VIEW

P = POINT OF CLOSEST
APPROACH

P-83 MIN

TO EARTH

P-119 MIN
(1ST RING PLANE
CROSSING)

PIONEER II
TRAJECTORY

TO SUN

SATURN

PIONEER II   N
TRAJECTORY

RINGS (SEEN
EDGEWISE)

6.5°

P

S

SIDE VIEW

Pioneer 11's route through the Saturnian system, contrived so that it would test the route through the ring plane that would later permit a Voyager spacecraft to attempt the Grand Tour of the outer Solar System.

this data is readily transformed to audio to enable the researchers to listen to what the spacecraft's 'electronic ears' heard as it flew through the highly dynamic electromagnetic environment.[7,8]

On 31 August, as Pioneer 11 closed to within 1 million kilometres of Saturn, it built up a spin–scan image of the ring system off to one side of the globe, which caused an immediate sensation. Not only did the silhouetted view show considerable structure in the ring system, and the presence of a very thin ring (later designated the 'F' ring) 3,000 kilometres beyond the 'A' ring, but also what appeared to be a new moonlet 35,000 kilometres farther out. The 'F' ring was so narrow that the spin–scan imaging system could not resolve it all the way around; it was evident only as a short section of arc at the cusp of the *ansa* where the line of sight passed through a longer column of the material. The 'B' ring, which is the brightest part of the system on the sunlit face, was so dense that from the spacecraft's perspective it appeared dark. The 'A' ring was not so opaque, and the fact that some parts of it passed more light indicated that there was some structure within it.[9] The 'C' ring was bright because the fine material it contained forward-scattered sunlight efficiently. The fact that Cassini's Division was bright from this perspective indicated that it contained a significant amount of fine material – contrary to the belief that it was swept completely clear. The imagery revealed considerably more structure than had been conjectured. Instead of the expected smooth sheets of icy particles, each of the 'classical' rings was found to comprise narrow ringlets. In 1943 B.F. Lyot, using the 24-inch refractor at the Pic du Midi Observatory with excellent seeing high in the French Pyrenees, had observed 11 divisions and hints of finer structure as variations in brightness on the 'B' ring, and while his drawings had been received sceptically by his contemporaries, he had been proved spectacularly correct.

The major gaps in the ring system could be accounted for by the presence of the satellites beyond, in particular Mimas, but they could not explain the fine structure. After J.C. Maxwell had proved that the rings were composed of a myriad of discrete particles, each of which pursued its own orbit around Saturn, it had been presumed that the transient fine structure that was reported from time to time was a 'density wave' effect inducing a fluctuation that propagates through the system. However, while this could potentially explain *everything*, in fact it explained *nothing* specific, so the fine structure in the ring system was a delightful mystery for a new generation of theoreticians.

As Pioneer 11 closed in, its Ultraviolet Photometer detected a bright hydrogen glow from the planet's illuminated hemisphere, a fainter glow from a large region of space tracing out Titan's orbit, and a dim glow from the 'B' ring, suggesting that the ring material is slowly degassing.

Through the ring plane
The crossing of the ring plane at 2.87 planetary radii, just outside the visible edge of the 'A' ring', was predicted for 09:02 on 1 September. In fact, because the system is so incredibly thin, the spacecraft darted through in a fraction of a second. With the planet on the far side of the Solar System, the spacecraft's signal took 86 minutes to reach the Earth, so the nail-biting moment for Ames was 1028. As the final minute

A computer-enhanced image of the non-illuminated side of the ring system by Pioneer 11 on 31 August, at a range of just under 1 million kilometres. This 'discovery' image revealed not only detail in the 'A' ring but also the presence of the 'F' ring beyond, and what appeared to be a new satellite. The 'B' ring is opaque. The 'C' ring and Cassini's Division appear bright because there is fine material within them that is efficiently forward-scattering sunlight.

ticked away, the engineers, scientists and managers in the control room fell silent, all listening anxiously to the signal, wondering whether it would be abruptly terminated. The scene was broadcast live by the TV networks. The signal continued through the predicted moment. Then, as David Morrison, Tom Young's deputy, later reflected, there were "scattered cheers and many sighs of relief". There was jubilation at JPL, because the Grand Tour trajectory had been shown to be viable. Some of the Ames researchers later voiced "a few rueful comments" lamenting the lost opportunity of the 'inner' option.

A few minutes after the ring-plane crossing, at 2.53 radii, the particles and fields instruments noted Pioneer 11's passage through a significant magnetospheric wake. As Saturn's inner magnetosphere rotates with the planet, it sweeps across the more slowly travelling satellites and so the magnetic anomalies *lead*, rather than *follow* the satellites. In fact, the data suggested that the spacecraft had passed a few thousand kilometres directly in front of an object 200 kilometres in diameter. Could this be a second new moonlet? The orbital period of an object at this distance from Saturn would be 17 hours. It was soon established that the moonlet that had been imaged earlier was not only orbiting at the same distance, but would have been exactly where the second object was encountered at the time Pioneer 11 was present; it was

A NASA artist's depiction of Pioneer 11, the first spacecraft to encounter the ringed planet. Note the absence of structure in the ring system.

therefore the same moonlet, and the spacecraft had almost collided with it! If it had struck the moonlet and fallen silent, the JPL planners would have been obliged to reconsider their plan to use this trajectory for the gravitational slingshot to Uranus. In a sense, therefore, Pioneer 11's most significant achievement in the Saturnian system was to *survive*.[10] Prior estimates of its chance of passing safely through the rings had varied from a mere 1 per cent (the pessimists) to 99 per cent (the optimists). The radar observations[11] of rocky material beyond the 'A' ring had been particularly worrisome.

It had been predicted that charged particles flowing back and forth along Saturn's magnetic field lines would be absorbed where the lines were intersected by the main ring system. The peak intensity for low-energy magnetospheric electrons had been encountered at about 7 radii, mid-way between the orbits of Rhea and Dione.[12] The predictions were accurate: as Pioneer 11 continued to close on the planet after passing through the ring plane, the counts of charged particles of all energies fell to zero at 2.292 radii, as it passed into the 'magnetic shadow' of the 'A' ring, and for a while it documented the most *benign* radiation environment in the Solar System.[13]

Half an hour after crossing the ring plane, Pioneer 11 had closed to 1.35 radii, its closest point of approach. Although as it flew 20,880 kilometres above the cloud tops the imagery confirmed the atmosphere to be subdued in comparison to Jupiter, it showed considerably more structure than had ever been glimpsed telescopically. The Infrared Radiometer mapped the temperature field. The mean temperature of 93K at the cloud tops was warmer than could be accounted for by insolation, so Saturn, like Jupiter, has an 'energy budget' sufficient for it to radiate slightly more energy than it receives from the Sun.

One surprise was that the axis of Saturn's magnetic field is virtually coincident with the planet's rotational axis; the offset being only 0.7 degrees.[14] The fields of the Earth and Jupiter are tilted by 11.7 and 9.6 degrees respectively, and so it had been presumed that this must be an inevitable consequence of the generation mechanism. This and the fact that the centre of activity is displaced 2,400 kilometres northward of Saturn's centre sent the theorists scurrying back to their 'drawing board'. Unlike the Jovian moons, therefore, Saturn's satellites are not subjected to an environment in which the polarity of the field flips as an inclined rapidly rotating magnetosphere sweeps past them. Because Saturn's magnetic field is not as strong as Jupiter's, its magnetosphere is neither as large nor as able to resist the gusts in the solar wind, and it dramatically inflates and deflates in response. In fact, its boundary washes back and forth over Titan's orbit, at 20 radii, so this satellite is often exposed to the solar wind when passing sunward of the planet; Hyperion, Iapetus and distant Phoebe are always in the solar environment except when passing through the magnetotail. The aurorae result from charged particles flooding in through the magnetosphere's polar cusps to excite the ionosphere, and so are similar to the terrestrial phenomenon.

Two hours after the fly-by, Pioneer 11 slipped back through the ring plane, this time at 2.78 radii, where its particles and fields instruments recorded evidence of yet another tenuous ring (later designated the 'G' ring). It had been a remarkable initial reconnaissance. Having survived the rings unscathed, the spacecraft was able to head for its final objective.

A GLIMPSE OF TITAN

Titan is the only member of Saturn's retinue to show a disk. Early efforts to directly measure its apparent diameter were subjective, and produced a wide range of values. E.E. Barnard and Percival Lowell independently measured it using micrometers, and calculated its diameter as 4,150 kilometres. As recently as the dawning of the Space Age, its diameter was given as 5,680 kilometres. In this range, it was clearly a close rival to Ganymede, the largest of Jupiter's Galileans.

Given an estimate of Titan's mass from studies of the orbital motions, its escape velocity would be only a little greater than the 2.4 kilometres per second of our own satellite. However, whereas the Moon is airless, the fact that Titan is so much colder would make any molecules of gas so sluggish that they might be unable to escape to space. In 1908, Catalan observer José Comas-Solá reported seeing 'limb darkening', implying the presence of an appreciable atmosphere. However, the fact that he made similar reports about each of the Galileans prompted considerable scepticism and, in all likelihood, his observations were flawed. Nevertheless, his reports stirred James Jeans in the UK to undertake a mathematical study of 'escape processes' on these satellites, concluding in 1925 in the case of Titan that, in spite of its relatively small size, as long as its temperature was in the range 60–100K it could have retained a primordial atmosphere acquired from the solar nebula. Measured on a fundamental scale, the weight of molecular hydrogen (H_2) is 2 units, and that of molecular oxygen (O_2) is 32. Jeans calculated that a gas exceeding 16 units would not have been able to exceed the escape velocity. Given reasonable assumptions for the composition of the solar nebula at Saturn's distance from the Sun, Titan would have acquired significant amounts of ammonia (NH_3), molecular nitrogen (N_2), methane (CH_4) and the noble elements argon and neon. Ammonia would freeze at such temperatures, so Titan's internal composition probably includes ammonia ice. Although all the other candidates would remain in the gaseous phase, the fact that molecular nitrogen, argon and neon are inert meant that they were not readily detected spectroscopically. Only methane seemed likely to be detectable. After the University of Chicago teamed up with the University of Texas to operate the 82-inch reflector of the McDonald Observatory on Mount Locke, Texas, G.P. Kuiper turned the newly commissioned instrument to Titan in 1942 and observed distinctive limb-darkening on the yellowish-orange disk. In the winter of 1943–1944, infrared observations using an emulsion that was sensitive out to 0.7 micron revealed a spectrum similar to that of Saturn, with distinctive absorption bands for methane and a hint of bands for ammonia, so the moon did indeed possess an atmosphere of the primordial type inferred by Jeans.[15]

Despite the tiny size of Titan's disk, skilled observers using large telescopes at times of excellent seeing often reported albedo variations. On nights of exceptional seeing throughout the 1940s, B.F. Lyot and Pic du Midi colleagues made a series of drawings of the moon using the 24-inch refractor at very high magnification, and recorded a variety of splotches and light and dark bands. As Titan was believed to rotate synchronously, the fact that these markings were clearly not correlated with the 16-day orbital period meant that they could not be surface markings. Lyot

Between 1943 and 1950, B. F. Lyot and colleagues at the Pic du Midi Observatory viewed Titan using the 24-inch refractor at high magnification showing a variety of 'splotches' and 'bands'. Because the detail did not integrate into a consistent map, they concluded that it represented atmospheric structure (and in 1944 G.P. Kuiper established spectrographically that the moon does possess an atmosphere).

proposed that they were transient atmospheric features and, if so, this implied not only that the atmosphere was sufficiently 'optically thick' to hide albedo variations on the surface, but also thick enough to support an active weather system.

In 1946, Kuiper developed a sophisticated electronic detector that was sensitive farther into the infrared than contemporary emulsions for faint objects (2.5 microns as opposed to 0.7 micron for film) and mounted it in a spectrograph. After testing it at the Yerkes Observatory, he installed it on the McDonald Observatory's 82-inch reflector. His rich results highlighted the requirement for further laboratory testing to chart the chemicals producing absorption in this newly opened spectral window. In 1948, after considering all the evidence, he published his conclusion that Titan has an atmosphere of methane with a density comparable to that of Mars's atmosphere, which at that time was believed to have a surface pressure of 0.1 bar. In 1952 Kuiper established that Titan is unique in Saturn's retinue in having methane absorption in its spectrum. In the 1950s, most professional astronomers were obsessed with the large-scale structure of the Universe and paid scant attention to the Solar System, so Kuiper and Audouin Dollfus, Lyot's protégé, had the planetary field more or less to themselves.

In 1972 L.M. Trafton of the University of Texas renewed infrared studies in the 1- to 2-micron range and found surprisingly strong absorption indicating either that the methane abundance was at least an order of magnitude greater than that observed by Kuiper, or that a large amount of an as-yet-unidentified gas was present.[16,17] If the methane molecules were static they would all absorb light at precisely the same wavelength. However, they actually move around and collide with one another, so the Doppler effect from the distribution of speeds 'broadens' a line and the amount of absorption as a function of wavelength is dependent upon the ambient pressure. The profile of the spectral band therefore provides a pressure measure. If Kuiper had been right about the methane pressure, then the 'collisional broadening' of the 1.1-micron absorption feature implied that the methane molecules were being jostled by another gas which was present in considerably greater concentration. This was consistent with Lyot's observations of atmospheric structures. A polarisation study in 1973 confirmed the presence of cloud particles, which lent further support to the 'dense atmosphere' thesis.[18,19] Also in 1973, J.J. Caldwell of Princeton began to develop a model in which methane was the primary constituent. He predicted that conditions at the surface were 86K and 20 millibars.[20] Measurements of the surface temperature had proved to be complicated, and radio and infrared techniques in the mid-1960s had produced a variety of figures in the range 165–200K, which were not supportive of this model. F.J. Low's discovery in 1965 of an infrared excess[21,22] prompted Caldwell's colleague, G.E. Danielson, to propose a temperature inversion in the upper atmosphere.[23] However, D.M. Hunten of the University of Arizona countered that Trafton's data probably meant that considerably more gas was present than could be in the form of methane, and that methane is actually a *minority* constituent in a dense atmosphere. The majority constituent could clearly not be ammonia, as this would long since either have frozen on the surface or have been dissociated by solar ultraviolet, liberating the nitrogen and hydrogen.[24] Molecular nitrogen is difficult to detect spectroscopically from Earth, but it *could* be the majority constituent. He predicted that the conditions on the surface could be an astonishing 20 bars at 200K.[25,26]

A compromise was achieved in 1980 when W.J. Jaffe and T.C. Owen, using the National Radio Astronomy Observatory's Very Large Array of radio telescopes in Socorro, New Mexico, measured the 'emission temperature' of Titan's surface at a wavelength of 6 centimetres and found it to be 87K.[27] They inferred that nitrogen contributed no more than 2 bars to the atmospheric pressure, and further speculated that although much of the hydrogen that had been liberated by photodissociation of ammonia must have been lost, the hydrogen that remained in the lower atmosphere would create a 'greenhouse effect' in which sunlight absorbed at visible wavelengths was re-emitted in a cascade of longer-wavelength photons, which would act to warm the surface.[28,29]

On 2 September 1979, as Pioneer 11 withdrew from Saturn, its trajectory took it within 363,000 kilometres of Titan. A spin–scan image of the northern hemisphere showed a mottled orangey disk so optically thick that it hid the surface completely.[30] It was an enticing glimpse, but the spacecraft's instruments were not designed for remote-sensing planetary satellites. The particles and fields instruments noted Titan's wake in the planet's magnetosphere, but there was no indication of an intrinsic magnetic field for the moon.[31]

The departing spacecraft encountered the magnetopause early on 3 September, but the solar wind was still gusting and the bow shock washed back and forth several times before it finally exited on 8 September, by which time the planet was so close to the Sun in the sky that the radio signal was very difficult to receive.

Although Pioneer 11 reconnoitred the route through the ring system that the JPL mission planners hoped to employ, the alignment of the outer planets was not yet conducive to the Grand Tour, so the exit trajectory sent it out of the Solar System in the opposite direction to Pioneer 10.

Ames's hardy Pioneers were therefore the first to pass through the asteroid belt, the first to investigate the Jovian magnetosphere, and the first to reach Saturn. Their

As Pioneer 11 withdrew from Saturn, it passed within 363,000 kilometres of Titan. This 'spin–scan' image of Titan's northern hemisphere displays a mottled orangey disk, but no sign of large-scale atmospheric structure or surface detail.

achievements, however, were soon overshadowed by the more sophisticated robotic explorers that followed.[32]

FOLLOW-UP

For some time, the identity of the moonlet with which Pioneer 11 had almost collided remained a mystery. The object that Audouin Dollfus had noted during the ring-plane crossing of December 1966, and named Janus, had been estimated to orbit at 2.65 radii, but its ephemeris was uncertain. The situation was complicated by the fact that Dollfus had speculated that he might have been observing *two* objects at different locations at the same orbital radius.[33] Certainly, Pioneer 11 saw no evidence of Janus at its predicted location. The situation would not be able to be resolved until the system was viewed edge-on once again and the object(s) could be recovered. As a result of the Pioneer 11 fly-by, interest in Saturn was at a fever pitch when the ring system was edge-on in 1980, and there was an unprecedented search for new moonlets using large telescopes equipped with the latest advance in imaging technology: the CCD camera.

On 6 February 1980 B.A. Smith at the University of Arizona detected an object that was labelled S/1980S1. It was seen at various times by several observatories and its orbital period was measured as 16.67 hours. Its motion was found to be consistent with the object imaged by Pioneer 11. On 26 February, D.P. Cruikshank of the University of Hawaii spotted another object, which was referred to as S3. It was found to have virtually the same orbit, but was on the far side of the planet. Evidently, Dollfus had been correct in his suspicion that he had been seeing two objects. However, with the uncertainty in Janus's ephemeris it was impracticable to determine which of the pair had been named by the International Astronomical Union. Pending an official ruling, they were referred to as S10 and S11 because Janus (whichever it was) was the tenth confirmed discovery and the other (whichever it was) must be the eleventh.[34,35] Once the decision had been made, and one named Janus and the other Epimetheus, it was concluded that Pioneer 11's close encounter had been with Epimetheus.

The discoveries continued apace. On 1 March 1980, P. Laques and J. Lecacheux of the Pic du Midi Observatory found one orbiting in Dione's orbit, but leading it by 60 degrees.[36] In 1772 fellow Frenchman J.L. Lagrange had developed the mathematics of such systems, identifying several stable geometries. However, such arrangements are stable only if the co-orbiting object is small. This new moonlet occupied Dione's leading Lagrangian point. After initially being dubbed 'Dione-B', it was labelled S/1980S6 and subsequently named Helene. No such arrangements were known within the Jovian system, but there are two groups of asteroids – collectively known as the Trojans – that orbit the Sun in Jupiter's leading and trailing Lagrangian points.

All these new moonlets were added to the target list for the next spacecraft, the first of which was already approaching Saturn.

3

Saturn revealed

THINKING BIG

In 1970, following up the trajectory work by G.A. Flandro, NASA's Jet Propulsion Laboratory began to design a sophisticated 'Mark 2' version of its highly successful Mariner series of planetary probes. The plan was to dispatch two pairs of vehicles to investigate the outer Solar System. The first pair would employ Jovian slingshots to reach Saturn, whereupon they would be deflected on to distant Pluto. The second pair would exploit Jupiter to visit first Uranus and then Neptune. Although funding for the development of this new vehicle was not forthcoming, JPL was permitted to modify its existing design to follow up Pioneer 11 with visits to Jupiter and Saturn. Working within this restricted budget, the engineers made every effort to ensure that if the new spacecraft was still healthy at Saturn, and if additional funding could be secured at that point, it would be capable of an 'extended' mission. In 1977, with the launches imminent, this 'Mariner Jupiter–Saturn' mission was renamed 'Voyager'. It was fortunate, of course, that this window was conducive to eventually pursuing the 'Grand Tour'.

VOYAGER

In contrast to the Pioneers, which were commanded from Earth, the Voyagers would have sophisticated computers that could be uploaded periodically with sequences of activities that they would execute autonomously, and they would have a high degree of fault tolerance to enable them to look after themselves. While they were to study particles and fields, they were also to be capable of remotely sensing a planet or a moon, and so could stabilise themselves. On the interplanetary cruise, and while studying a planetary magnetosphere, they would slowly rotate in order to maximise their perception of the surrounding space, but they would stabilise themselves and aim a scan platform carrying a battery of co-aligned optical instruments in order to investigate a specific target, slowly slewing the platform to compensate for relative motion. The spin–scan imagery had been enlightening, but everyone was eager to see what a 'real' camera would reveal. Voyager had two Attitude and Articulation Control Subsystem (AACS) computers, one of which would be in control at any given time while the other stood by to intervene. This would orient the spacecraft to undertake scientific activities, aim the scan platform and maintain the antenna facing the Earth. A pair of Flight Data Subsystem (FDS) computers, again one serving and the other standing by, would run the science instruments and format their data for transmission to Earth.

A diagram of the Voyager spacecraft.

Instrumentation

The Voyager instrument suite comprehensively addressed two primary themes. The aimed imaging systems mounted on the scan platform were boresighted so that they could observe a given object at a range of wavelengths simultaneously, and the other instruments continued the Pioneer investigation of the particles and fields in space.

Aimed instruments

The *Imaging Science System* (ISS) incorporated a 200-millimetre-focal-length f/3 wide-angle lens boresighted with a 1,500-millimetre-focal-length f/8.5 narrow-angle telescope. The field of view of the 'wide' subsystem was actually only 3 degrees across, comparable to using a 500-millimetre 'telephoto' on a 35-millimetre film camera, and the narrow-angle subsystem's field of view was one-tenth the width, so these were both really telescopic subsystems. Each subsystem had its own camera. Unfortunately, CCD solid-state technology was developed just too late to be exploited by Voyager, and a refined version of the vidicon system developed for the later Mariner spacecraft was used. It had an 11-millimetre-square selenium–sulphur television tube which was 'read out' by a slow-scan process which took 48 seconds to produce an image comprising 800 lines, each with 800 picture elements (pixels), encoding the brightness on a 256-point scale.[1] The images were monochrome, but each camera had an 8-option filter wheel so that colour images could be made (on Earth) from a succession of frames exposed through three filters. The system was considerably more capable than the spin–scan technique of the Pioneers. In addition to general imaging and specialised observations using narrow-pass filters in the visual and ultraviolet range, it would provide 'context' to assist in interpreting the results from the other optical instruments. The principal investigator of the imaging science team was B.A. Smith of the University of Arizona.

The *Photopolarimeter System* (PPS) utilised a 200-millimetre-focal-length f/1.4 Cassegrain telescope and a photoelectric photometer. Three wheels intersected the optical path, one providing four aperture settings ranging from 2.1 to 61 milliradians, one with filters for 8 narrow wavelength bands in the range 0.235 to 0.750 micron, and the other with 8 polarisation options, so that the size of the field, passband and polariser orientation could be set independently for any particular observation. It was to investigate particles in interplanetary space, cloud particles and gases in planetary atmospheres, and the chemical composition and texture of the surfaces of satellites.[2] C.F. Lillie of the University of Colorado's Laboratory for

A diagram of the optical configuration of the wide-angle camera for Voyager.

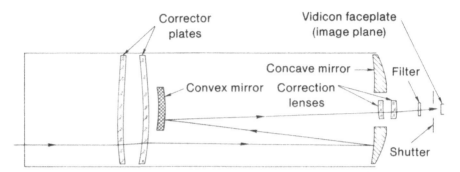

A diagram of the optical configuration of the narrow-angle camera for Voyager.

Atmospheric and Space Physics was the principal investigator until A.L. Lane of JPL took over in 1978.

The *Ultraviolet Spectrometer* (UVS) did not have a lens, it had a series of linear apertures set in line which served as a collimator to produce a field of view 2 by 15 milliradians, then a diffraction grating illuminated a linear array of 128 detectors, each of which measured the brightness on a 1,024-point scale to measure the range 50 to 170 nanometres with a spectral resolution of 1 nanometre (10 Ångströms). It was to investigate ultraviolet 'glows' in interplanetary space and in ionospheres, and use 'limb-sounding' measurements of the extent to which insolation was absorbed during solar occultations to profile the chemical composition of the upper regions of planetary atmospheres.[3] A.L. Broadfoot of the University of Arizona was the principal investigator.

The *Infrared Radiometer, Interferometer and Spectrometer* (IRIS) incorporated a wide-field 'sighting scope' and a 500-millimetre-diameter Cassegrain telescope with a 4.4-milliradian (quarter-degree) field of view. A Michelson interferometer served two channels, one in the range 0.33 to 2.0 microns to measure reflected insolation and the other sensitive out to 50 microns to measure thermal emission. The IRIS was primarily to determine the chemical composition of clouds in planetary atmospheres and investigate the temperatures and pressures as functions of depth.[4,5] R.A. Hanel of the Goddard Space Flight Center in Greenbank, Maryland, was the principal investigator.

Particles and fields sensors
Some of the particles and fields instruments addressed the same themes as those investigated by the Pioneers, but others were new.

The *Magnetic Fields* (MAG) experiment incorporated four magnetometers, two high-field instruments on the body of the spacecraft and two low-field instruments mounted mid-way, and at the far end of a 13-metre boom in order to be clear of the spacecraft's own magnetic field. Each magnetometer sensed the ambient field in three orthogonal axes so that the strength and direction of the field could be measured. The instrument was sufficiently sensitive to measure a magnetic field one-millionth of the strength of that at the Earth's surface. It was to study the magnetic fields present in interplanetary space and planetary magnetospheres. The principal investigator was N.F. Ness of the Goddard Space Flight Center.

In the presence of a magnetic field, the charged particles (electrons and ionised atomic nuclei) of a plasma are subjected to a force, and phenomena arise which have no counterpart in a purely gaseous state. In particular, oscillations in the density of the plasma as the charged particles are herded by the magnetic field produce waves, with resultant fluctuations in the strength of the electric field. The *Plasma Wave System* (PWS) used two 10-metre-long antennas projecting from the body of the spacecraft in a 'V' shape. To determine the direction of the flows resulting from the passage of plasma waves, a wideband waveform receiver and a spectrum analyser measured the variations in the electric field component over the 10 Hz to 56 kHz range.[6] The principal investigator was F.L. Scarf of Thompson Ramo & Wooldridge (TRW) in Redondo Beach, California.

In 1955, Jupiter was serendipitously discovered to be a strong and highly variable radio source in the decimetric and decametric bands, prompting the inference that the planet had a strong magnetic field, and this had been confirmed by Pioneer 10. The *Planetary Radio Astronomy* (PRA) experiment used the same antennas as the PWS to detect radio emissions in two bands, one from 1.2 kHz to 1.2 MHz at intervals of 19.2 kHz and a bandwidth of 1 kHz, and the other from 1.2 to about 40 MHz at intervals of 307 kHz and a bandwidth of 200 kHz. It operated by cycling through the 198 discrete frequencies in six seconds and reporting their intensities on a 256-point logarithmic scale. Measurements of the polarisation and spectral characteristics of emissions from planets would enable the plasma densities in the generation regions to be inferred, so in contrast to the other instruments which sensed the spacecraft's immediate environment, the PRA was the remote-sensing member of the particles and fields suite.[7] The principal investigator was J.W. Warwick of Radiophysics Incorporated in Colorado.

Whereas PWS studied the flows of charged particles indirectly by their electric fields, the *Plasma Science* (PLS) experiment collected them. It had a pair of Faraday cup plasma detectors: one was aimed in the general direction of the Earth and the other was mounted orthogonally. The Earth-facing detector had three apertures which, between them, were open to that entire hemisphere. The side-looking detector had a narrower field of view. They were sensitive to particles with energies ranging from 10 eV to 5.95 keV – the latter moving at about 1 per cent of the speed of light. It was to determine the composition and energy of the particles, as well as their direction of flow in interplanetary space and in planetary magnetospheres.[8] The principal investigator was H.S. Bridge of the Massachusetts Institute of Technology.

The *Low-Energy Charged Particle* (LECP) experiment had two instruments on a rotating mount. The low-energy magnetospheric particle analyser incorporated eight solid-state detectors that could discriminate electrons from ions and, between them, were sensitive to charged particles with energies from 10 eV to 15 keV. It was to investigate the composition of the plasmas in interplanetary space and in planetary magnetospheres.[9] Its name notwithstanding, the Low-Energy Particle Telescope was to investigate the solar wind by extending the energy range to several millions of electron volts (the top end of the PLS range was several thousands of electron volts).

S.M. Krimigis of the Applied Physics Laboratory of the Johns Hopkins University in Maryland was the principal investigator.

The *Cosmic Ray Science* (CRS) instrument utilised two particle telescopes, one to measure the energy spectra of heavy nuclei (those with atomic masses ranging up to that of iron, and for some elements discriminating between isotopes) across a broad energy range, and one to measure the energy spectrum of electrons in the 5 to 110 MeV range, in order to determine the cosmic ray component of the plasmas found in interplanetary space and in planetary magnetospheres.[10] The principal investigator was R.E. Vogt of the California Institute of Technology in Pasadena.

The PLS, LECP and CRS instruments characterised the charged particles striking their detectors and, between them, characterised the fluxes of electrons and nuclei in the space through which the spacecraft flew.

In addition, the spacecraft's X-Band and S-Band radio transmitters provided the basis for a number of *Radio Science Subsystem* (RSS) experiments.[11] The principal investigator was G.L. Tyler of Stanford University's Center for Radar Astronomy. During an occultation, the manner in which the signal was refracted would serve to characterise the occulting object's atmosphere, if it possessed one, and the Doppler effect enabled the object's gravitational field to be charted, its moment of inertia to be calculated, and the state of its interior to be inferred by the manner in which the spacecraft's trajectory was deflected. In a sense, this was 'free' science as it required no specific instrumentation.

Irrespective of whether the Voyagers were rotating or 3-axis stabilised, the 3.66-metre-diameter dish of the high-gain antenna was always aimed at the Earth. The communications system was rated at 115 kilobits per second from Jupiter to handle the output of imagery – a rate almost 100 times that of the Pioneers. At Saturn, the data rate would be 45 kilobits. For redundancy, there were two 23-watt transmitters. A 500-megabit reel-to-reel tape recorder was included to act as a buffer during times when the spacecraft was occulted, or when it was securing data faster than it could transmit it. A trio of RTGs mounted in line on a single boom provided 400 watts of power for the spacecraft's systems and instruments. The instruments accounted for 118 kilograms of the launch mass of 815 kilograms – a rather better ratio than on the outer Solar System Pioneers.

JOVIAN SURPRISES

Voyager 1 was launched on 5 September 1977 upon a Titan III launch vehicle with a Centaur escape stage, the most powerful combination available and, in fact, the final vehicle of this type to run off the production line as NASA phased out 'expendable' rockets in favour of the then-favoured Shuttle-only policy.

As Voyager 1 started its interplanetary cruise, it deployed its various booms and activated its scan platform, whose action was impaired by tiny fragments of debris trapped in the mechanism during assembly, but this was progressively ground down and expelled by a series of slewing exercises. Two weeks out, and almost 12 million kilometres away, the spacecraft returned a historic 'departure shot' capturing for the

On 20 August 1977, a Titan III/Centaur launches the first of the Voyagers.

A polar projection of Voyager 1's route through the Jovian system.

first time the Earth and its Moon in the same frame, both of which were showing the same crescent phase.

In fact, Voyager 1 had been launched a fortnight behind its mate, but because it pursued a slightly faster trajectory it took the lead on 15 December. Although it is rather arbitrarily defined, Voyager 1 emerged from the far side of the asteroid belt on 8 September 1978. A month later, Voyager 2 did likewise. NASA's record was now four-for-four in this respect. Evidently, earlier fears concerning the risk of debris had been overestimated, and the highway to the outer Solar System was open to regular traffic.

As Voyager 1 closed within 100 million kilometres of Jupiter, the resolution of its narrow-angle imagery was better than that from telescopes. On 4 January 1979, it initiated a month-long campaign to investigate the dynamics of the atmosphere. By mid-February, it was able to expand its activities to remote sensing of the Galilean satellites. The magnetosphere had extended sunward about 100 planetary radii when the Pioneers had visited, but the Sun had been fairly quiescent then, and it was now near the peak of its sunspot cycle and the increased pressure of the solar wind had compressed the magnetosphere. Voyager 1 did not sense the bow shock until 28 February, and it was not until 3 March, after having the shock wave wash back and forth, that it finally crossed the magnetopause and entered the inner domain, by which time the craft was at a distance of only 47 radii, and within hours of crossing the orbit of the outermost of the four large satellites. The moon, however, was not in the vicinity; in fact, the spacecraft had a clear run to Jupiter, during which it was able to chart the magnetosphere and document the atmospheric structures on the sunlit hemisphere of the planet. The imagery was later sequenced as a movie showing a 10-hour period as the planet rotated once upon its axis. The high-resolution imagery revealed an astonishing variety of detail and an astoundingly dynamic atmospheric system.[12]

In 1664 Robert Hooke in England noted a spot spanning one-tenth of Jupiter's diameter, just south of the equator. It was discovered independently in 1665 by J.D. Cassini in Italy, who, after moving to Paris, recorded it intermittently until his death in 1712. 'Hooke's Spot', as it was known, was drawn by H.S. Schwabe in Germany in 1831, by W.R. Dawes in England in 1857, by A.M. Mayer in America in 1870, and repeatedly by the fourth Earl of Rosse in Ireland in the early 1870s. It took on a striking red hue in 1878, prompting the new name of the 'Great Red Spot'. After slowly fading in 1882, it reappeared in 1891. Although it briefly fades on occasion, and is often difficult to detect, it has become a permanent feature of the South Tropical Zone as a 26,000-kilometre-wide oval spanning 20 degrees of longitude at latitude 22 degrees south.[13] The Pioneer imagery had hinted at internal structure.[14] Voyager 1 revealed this in intricate detail, documenting its 7-day anti-clockwise rotation. It is an anticyclonic vortex with a central upwelling column that yields to subsidence around its periphery. In obstructing the prevailing latitudinal ('zonal') jet stream, the Great Red Spot leaves an extensive system of eddies in its 'wake' – an intricate structure that had never been suspected by the most fortunate of telescopic observers. The latitudinal banding was discovered to extend to higher latitudes than expected. The 'whistlers' detected at radio-frequencies indicated the presence of electrical discharges in the atmosphere. Although the cameras did not capture any lightning flashes on the dark hemisphere, it was concluded that the discharges likely occur just below the cloud deck forming the visible surface, which implied in turn that there is vigorous vertical circulation at work.

The closest point of approach early on 5 March was at a planetocentric range of 4.9 radii, with Voyager 1 passing 270,000 kilometres above the cloud tops, slightly beyond the evening terminator. The slingshot increased Voyager 1's speed *by* 48,000 kilometres per hour. Six hours earlier, shortly prior to crossing the orbit of Io, the innermost of the Galileans, some long-range images were taken of Amalthea, the moonlet which E.E. Barnard had spotted orbiting close to the planet, showing it to be an irregular body with its long axis aimed towards Jupiter. As the spacecraft flew through the equatorial plane, it secured a long-exposure picture 'over its shoulder' to search for moonlets even closer to the planet, and in doing so it revealed that Jupiter possesses rings. Unlike Saturn's rings, Jupiter's are dark because they comprise small particles of dust which reflect sunlight poorly. They were striking from down-Sun because small particles are efficient at forward-scattering sunlight. This suggested that the Jovian particles are micron-sized specks of dust, rather than ice, the main constituent of Saturn's system of rings.[15]

All the major moons were best seen from the exit trajectory, and Voyager 1 met them in sequence. The point of closest approach to Jupiter had been well inside Io's orbit, and as the craft caught up with this moon it observed its trailing hemisphere

Jupiter's Great Red Spot is a 26,000-kilometre-wide oval that spans 20 degrees of longitude in the South Tropical Zone. In obstructing the prevailing zonal jet streams, it leaves an extensive system of eddies in its 'wake'. The intricate structure in this contrast-enhanced Voyager 1 image had not been suspected by even the most fortunate of telescopic observers. The image was taken on 25 February 1979 from a range of 10 million kilometers, and resolves details as small as 160 kilometers across.

before passing within 21,000 kilometres of its south pole, in the process documenting much of the hemisphere that permanently faces the planet. Telescopic studies had discovered a cloud of neutral atoms and ionised nuclei in a toroidal form centred on Io's orbit.[16] This is so wide that if the Earth were located at the geometric centre, then the Moon's orbit would comfortably fit inside the hole of the torus. The plasma was studied from afar by Voyager 1's Ultraviolet Spectrometer. By passing within Io's orbit, the spacecraft spent several hours inside the torus and the particles and fields instruments made *in situ* measurements. When skimming Io's south pole, the instruments detected a strong flow of charged particles. Electrons stream to and fro along Jupiter's magnetic field lines, forming a pair of 'flux tubes' that connect the moon to the planet. This 1-million-ampère current is by far the most powerful direct current in the Solar System.

Voyager 1 detected strong ultraviolet emission over Jupiter's sunlit hemisphere. Dubbed 'dayglow', this emission meant that the temperature of the thermosphere is 1,000K, but a haze which absorbs ultraviolet light overlies the polar regions of the atmosphere. On the night side, there is visible and ultraviolet auroral emission. In the case of the Earth, the auroral excitation is due to solar wind particles streaming into the polar cusps of the magnetosphere and interacting with the ionosphere, but Jupiter's magnetic field is strong enough to fend off the worst of the solar wind. The arc-like auroral glows are caused by particles originating in Io's plasma torus. Upon being launched in 1990, the Hubble Space Telescope imaged this excitation extending across Jupiter's dayside, and upon being upgraded with an imaging spectrograph it was able to see the spot-like glows where the ends of the flux tubes connect with the atmosphere. Jupiter rotates more rapidly than Io's travel around its orbit, and these 'footprints' stay at the satellite's longitude, but as the excitation slowly diminishes the spot is stretched out into a 'comma'. In fact, Io, which has been described as the 'beating heart' of the Jovian system, is the powerhouse of the magnetosphere.

However, little was known of Io itself. It orbits only 350,000 kilometres above Jupiter's cloud tops, which is approximately the distance between the Earth and the Moon, but while Io is comparable in size to the Moon, the giant planet is 318 times more massive than the Earth. The tremendous gravitational field will draw in material from interplanetary space and accelerate it, so the expectation was that Io would be heavily cratered. When the early low-resolution Voyager 1 imagery showed vaguely circular albedo features these were taken to be craters, but as the spacecraft closed in, it became evident that these dark features were not impact scars at all. Astonishingly, a thorough survey revealed that Io has *no* impact craters. As no object can completely escape impacts, the absence of craters indicated that some process was continuously resurfacing the moon, 'removing' its craters. This process was evidently volcanism. Nevertheless, no one seriously expected to see a volcano in the process of erupting. On 8 March, as Voyager 1 departed the Jovian system, it took an extended exposure to show the position of the crescent moon against the stars for navigation purposes to verify that the slingshot had deflected the trajectory for Saturn. When navigation team member Linda A. Morabito 'enhanced' this image by computer, she spotted a faint mushroom-shaped plume projecting 280 kilometres beyond the limb.[17] It was an active volcano! An anomalous glow on the terminator was another volcano! Its vent site was in darkness, but its plume of dusty gas was so tall that it caught the rays of the Sun. In fact, further analysis identified nine active vents with plumes rising up to 300 kilometres. The IRIS noted infrared emissions from a large number of isolated 'hot spots' which were not producing plumes. Volcanism was rife.

Just before Voyager 1 ventured into the Jovian system, a paper published in the journal *Science* had reported the results of modelling the tidal stresses acting on Io.[18] Europa's orbital period is twice that of Io, and Ganymede's is twice that of Europa. These resonances make Io's orbit slightly elliptical. The eccentricity is just 0.0041, but it has a significant effect. Io maintains one hemisphere facing Jupiter because it rotates synchronously. A 'tidal bulge' rises and relaxes in response to the cyclically

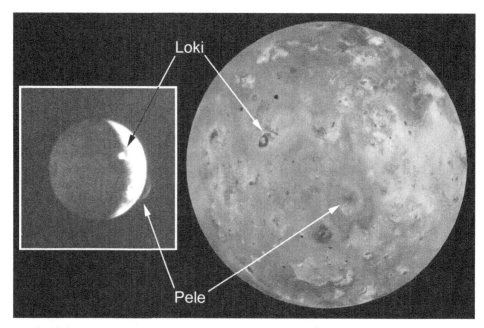

On 8 March 1979, as it withdrew from Jupiter, Voyager 1 snapped an over-exposed image of Io (left) so that JPL's interplanetary navigational engineers could verify the spacecraft's aim for Saturn by measuring the moon's position with reference to the stars. When Linda A. Morabito drew attention to the umbrella-shaped 'plume' on the limb, it was realised that Io is volcanically active. In fact, there was a second plume on the terminator. The limb site (Plume 1) was later named Pele, and the terminator site was named Loki.

varying gravitational force. The induced mechanical stress is converted into heat. The analysis found that Io must derive two to three *orders of magnitude* more heat from this stress than from the decay of its radioactive elements. In addition to implying that the moon's interior should be significantly thermally differentiated, the authors of the paper tentatively predicted that it might be volcanically active. However, they were just as surprised as anyone to find that Io is the *most* volcanically active object in the Solar System. Terrestrial infrared observations made in 1981 determined that Io's heat flow is 30 times greater than that of the Earth.[19]

At 3,100 kilometres in diameter, Europa is the smallest of the Galileans. It was not well positioned for viewing, and Voyager 1 approached no closer than 732,000 kilometres. Europa's intriguingly high albedo of 64 per cent made it one of the most reflective bodies in the Solar System. It had been speculated on the basis of maps drawn by telescopic observers that icy caps may have left only a narrow strip of dark rocky terrain along the equator. In fact, even long-range imagery revealed the moon to be *completely* enshrouded in ice. Ganymede and Callisto were observed too, but the closest approach was in darkness over their anti-Jovian hemispheres, and the highest-resolution imagery was of their dusk terminators on the way in and their dawn terminators on the way out.

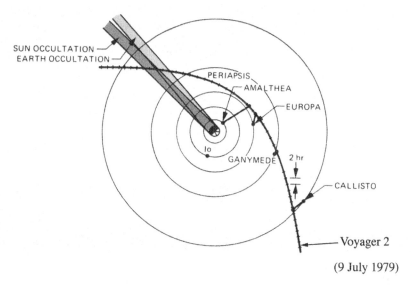

SUN OCCULTATION
EARTH OCCULTATION
PERIAPSIS
AMALTHEA
EUROPA
Io
GANYMEDE
2 hr
CALLISTO
Voyager 2

(9 July 1979)

A polar projection of Voyager 2's route through the Jovian system.

Three months later, Voyager 2 made its approach. When it sensed the bow shock on 2 July 1979, it became evident that the solar wind had abated slightly, and the magnetosphere had re-inflated.[20] The spacecraft's computer had been programmed to follow-up its predecessor's discoveries. This time, most of the satellites were encountered on the way in.[21] First was Callisto on 8 July at 215,000 kilometres, then Ganymede on 9 July at 62,000 kilometres. Fortunately, the hemispheres which had previously been in darkness were now illuminated, so these two satellites were well documented. Callisto was found to be virtually saturated with craters, and there was no indication of any endogenic resurfacing activity. Callisto's ancient surface still bears the scars of its accretion.

Ganymede is not only the largest of the Jovian moons, it is the largest satellite in the Solar System. In fact, being somewhat larger than the planet Mercury, it is really a small planet that is orbiting around Jupiter. In 1849, W.R. Dawes in England made a study of Ganymede and concluded that its most prominent surface feature was a bright 'polar spot'. Further observations were made by E.E. Barnard, Percival Lowell and E.M. Antoniadi. In 1951 E.J. Reese combined their observations as a single map. B.F Lyot produced maps of all four of satellites, but these were not published until 1953, after his death. In terms of the distribution of albedo features, Lyot's Ganymede map was in fair agreement with Reese's. The maps published by Audouin Dollfus in 1961 were generally considered to be the best. Between them the Voyagers documented about 80 per cent of the surface, revealing two main types of terrain. The dark cratered terrain comprised several large circular features and a large number of small polygonal blocks. The largest feature, appropriately named Galileo Regio, was an oval that spanned one-third of the anti-Jovian hemisphere. While this had been seen by telescopic observers, in general the distribution of the

Callisto's dark heavily cratered ancient surface shows no indication of resurfacing.

The sharp terminator in this Voyager 2 view of Europa as a crescent implied that its surface is exceedingly flat. This vantage point documented the 'fractured terrain' on the anti-Jovian hemisphere.

dark features was not well represented on any of the maps. The rest of the surface was a brighter 'sulci' terrain which, in places, seemed to have fractured the darker features in a way that implied that the moon had undergone significant internal thermal processing at some time in its history, resulting in the extrusion of icy fluids: *cryovolcanism*.

Next was a view of Europa's anti-Jovian hemisphere from 204,000 kilometres. The generally reflective surface was revealed to be darkly 'mottled' and criss-crossed by linear features which, although only a few kilometres wide, nevertheless extended for thousands of kilometres. The exceptionally 'sharp' terminator line meant that the range of elevation from pole to pole was only a few hundred metres. The discovery that Europa was as smooth as a cue-ball suggested that it had once been covered by an ocean that had frozen. The paucity of impact craters implied that this icy shell had formed 'recently'. Could there be a deep ocean of liquid water beneath the shell – one with more water than all the Earth's oceans combined? It was a remarkable prospect. For Voyager 2, Io was inconveniently located on the far side of its orbit, but the long-range imagery showed that most of its volcanoes were still active, and a new eruption was now underway.

Ganymede's dark cratered terrain is fragmented into polygons by the lighter-toned ridged and grooved 'sulci' terrain. The large dark oval is Galileo Regio.

Voyager 2's closest point of approach to Jupiter on 9 July was just outside the orbit of Europa. Upon crossing the equatorial plane the next day, it looked back to investigate the ring system and in so doing detected three new moonlets nearby. The brightest part of the ring system lay between 51,000 and 57,000 kilometres above the cloud tops. There was a tenuous 'halo' closer in that extended above and below the equatorial plane. Beyond was another tenuous zone which formed a thick disk, dubbed the 'gossamer' ring. In fact, the rings and moonlets are related. The moonlets Metis and Adrastea, whose orbits are only several thousand kilometres apart, define the outer boundary of the main part of the system. Amalthea orbits about half way out through the gossamer ring and Thebe defines its periphery, at a planetocentric range of 3.11 radii. The suggestion that the rings might be composed of dust blasted off the moonlets by the energetic impacts of micrometeoroids that had been drawn in and accelerated by Jupiter's mighty gravitational field was later confirmed by the Galileo spacecraft.

The particles and fields instruments found that each of the Galilean moons had a 'wake' extending as much as 200,000 kilometres *ahead* of it as the rapidly rotating inner magnetosphere swept past them. As a result of this 'magnetospheric wind', their trailing hemispheres are strongly irradiated with charged particles, an effect that might chemically alter the surface materials. Another dynamic derives from the fact that the axis of the planet's magnetic field is inclined at 9.5 degrees to its spin axis, so the moons in the equatorial plane endure cyclic polarity reversals as the magnetosphere rotates. At about 25 radii on the down-Sun side of the magneto-sphere, the field lines switch from 'closed' to 'open', and the magnetosphere is drawn downstream by the solar wind. It was speculated that this magnetotail might be so lengthy as to extend beyond the orbit of Saturn.

In retrospect, the most significant insight gleaned from the Jovian fly-bys by the Voyagers was that despite being in a frozen realm far from the Sun, the objects that Galileo, the discoverer of the major satellites, had described as "marvellous things", were indeed marvellous in a way that we had not predicted, and each was a miniature world with its own highly distinctive geological history.[22]

ENCOUNTER WITH SATURN

While JPL was determined to preserve the 'option' (as yet unfunded) of flying the Grand Tour, by sending Voyager 1 deep into the Jovian magnetosphere the resulting slingshot had accelerated it so much that it would reach Saturn too early for onward routeing. The two vehicles had been dispatched a few weeks apart – Voyager 2 first – and had arrived at Jupiter a few months apart – Voyager 1 leading – but by not approaching the giant planet so closely Voyager 2 had set off on a slower trajectory to Saturn which was timed to arrive when conditions were just right for the Grand Tour. However, at this stage in the overall mission Voyager 2 was officially backing up its mate. Only if the first spacecraft achieved all of its primary objectives at Saturn would the second be released for this 'extended' mission.

The overall scientific objectives at Saturn were:

- to investigate the dynamics of the planet's outer atmosphere;
- to profile the chemical composition of the outer atmosphere;
- to chart the temperature field at the cloud tops to infer the deeper thermal structure;
- to map the magnetic field;
- to chart the flows of charged particles in the magnetosphere;
- to determine Titan's key physical parameters, to investigate the composition of its atmosphere, and to discover how this interacts with the planet's magnetosphere;
- to determine the geological histories of the icy satellites;
- to determine the detailed structure of the ring system; and
- to search for new rings and moonlets.

Pioneer 11 had established that Saturn has an appreciable magnetosphere with charged particles circulating within it, but lacking the 'beating heart' of Io, it does not pump out decimetric and decametric radio energy. Nevertheless, it had been found to produce very weak long-wave emissions.

In February 1980, Voyager 1's Planetary Radio Astronomy instrument started to detect bursts with wavelengths of several kilometres.[23,24,25,26] In closer to Saturn, a periodicity of 10 hours 39 minutes 24 ($\pm$7) seconds became evident in these very-low-frequency emissions.[27] If the magnetic field was rotating with the core of the planet (as seemed to be the case for Jupiter) then this was a measure of the rotational period uncomplicated by the differential rotation of the atmosphere. The fact that this was slower than the measured rates for spots in the equatorial zone indicated that what A.S. Williams in 1891 had called the "the great equatorial atmospheric current" is 'super-rotating'.

By mid-1980, Voyager 1 had initiated long-range imaging of Saturn. At that time it was viewing the ring system virtually edge-on, but as it closed in the angle opened up. By the end of September, the narrow-angle camera's resolution exceeded that of the best terrestrial telescopes. Of course, terrestrial astronomers were unable to see the rings at that time as they were nearly edge-on. By early October, with the range down to 50 million kilometres, the resolution had improved to 1,000 kilometres per pixel. An astonishing amount of detail could be resolved, revealing some surprises.[28] The 20,000-kilometre-wide 'C' ring showed several distinct gaps, including one 300 kilometres wide that had a narrow, dense, bright ringlet within it which was slightly eccentric, being 90 kilometres wide at its farthest point from the planet and only 35 kilometres wide at its closest. It had been expected that perturbations by the moons orbiting beyond the ring system would have swept the 5,000-kilometre-wide Cassini Division clean, but there was a thin strand of material present *within* it. How could there be material in the 'zone of avoidance'?

On 6 October, dark *radial* markings were spotted on the 'B' ring, rotating "like spokes on a wheel", as explained by R.J. Terrile, the member of the imaging team who discovered them. The increase in rotational period across the width of the 'B' ring exceeds 3 hours, hence a radial feature should become significantly distorted within a matter of minutes. However, the spokes emerged from where the planet's

The sequence (which runs from left to right and top to bottom) shows the rotation of the dark 'spokes' on Saturn's 'B' ring.

shadow fell on the rings and persisted for several hours as they rotated, becoming fainter and less well defined with time, finally losing their integrity only just before re-entering the shadow. A new command sequence was sent to the spacecraft to investigate this phenomenon, and so on 25 October, with the range now down to 24 million kilometres, a 500-kilometre-per-pixel frame was shot every five minutes over a 10-hour period – almost a complete planetary rotation. Over the next week, JPL's Image Processing Laboratory sequenced these frames to create a time-lapse movie showing the spokes rotating. The fact that the motion matched the planet rather than the ring system meant that the spokes were an artefact of the planet's magnetic field which in turn implied that the spokes were charged particles being swept along over the 'B' ring.[29] The force that created the spokes was evidently at work where the planet's shadow fell onto the rings. Analysis suggested that the spokes were particles a few microns wide which were 'elevated' away from the plane by either magnetic or electrostatic forces that were effective in darkness. The spokes emerged well defined, and slowly diffused as they progressively migrated back down to the ring plane in sunlight, only to be elevated again upon re-entering the shadow.[30] As Voyager 1 closed to within a few million kilometres of the planet, it detected lightning-like radio bursts. Some researchers suggested that these emanated from the ring system, the idea being that the material in the spokes was being 'charged up' in the shadow, and the bursts were electrostatic discharges between the clouds of dust as it settled towards the rings upon emerging into sunlight.[31,32,33,34,35,36,37,38,39] Others suggested that the radio bursts were more likely due to atmospheric lightning in the super-rotating equatorial wind stream.[40,41]

In 1896, E.M. Antoniadi had seen radial features on the 'A' ring, but even the accomplished chronicler of Saturnian studies, A.F.O'D. Alexander, had dismissed them as being "probably illusory".[42] Nevertheless, in 1977, Stephen O'Meara, an experienced visual planetary observer, reported radial structure on both main rings. Interestingly, the features seen by the Voyagers were confined to the 'B' ring.

Imagery on 14 October enabled the orbits of the moonlets Janus and Epimetheus to be refined. Their orbits are 30 kilometres either side of a planetocentric distance of 151,450 kilometres. In accordance with Kepler's laws of orbital motion, the lower one travels slightly faster and catches up with the higher one every 4 years or so, at which time they *swap orbits*. This occurs because the trailing moonlet, in the lower orbit, is accelerated and rises; at the same time, the leader is retarded and falls. Since the leader, which was in the higher orbit, accelerates away as it drops into the lower orbit, they do not actually pass. In fact, they probably never come closer than a few kilometres of one another.[43]

On 25 October, while checking the incoming imagery of the rings intended for the spokes movie, S.A. Collins discovered a moonlet just beyond the 'F' ring. The next day he found another one, just inside the ring. Several years previously, upon the discovery of the system of widely-spaced thin rings of Uranus,[44,45] it had been suggested that this delicate structure was maintained by a number of moonlets which 'shepherded' the loose material, but these bodies were hypothetical.[46] Now, it was apparent that the narrowness of the 'F' ring derived from the presence of this pair of moonlets. As the 13th and 14th satellites confirmed in Saturn's retinue, they were

On 16 February 1977, the noted planetary observer Stephen O'Meara drew radial structure on both the 'A' and 'B' rings.

later named Prometheus and Pandora. Prometheus, being 1,250 kilometres inside the ring, travels slightly faster and overtakes Pandora, which lies 1,100 kilometres beyond the ring, every 25 days.

As Voyager 1 closed to within 25 million kilometres, the other optical instruments began to make observations. The IRIS took spectra several times per day in order to determine the composition of Saturn's atmosphere and, in addition to hydrogen and helium, it detected traces of phosphene, methane, acetylene and ethane. As October drew to an end, with the range down to 17 million kilometres, the disk finally began to show dark belts and bright latitudinal zones, and there was evidence of plumes of material billowing up from below. The disk was of such low contrast, however, that the structure was barely discernible unless appropriate filters were used to penetrate the haze.[47] By this time, the Ultraviolet Spectrometer had confirmed the presence of aurorae, so the solar wind was evidently able to force its way into the polar cusps of the magnetosphere.

The amazing complexity of the ring system was becoming more apparent each day. Even with Pioneer 11's 'taster', the scientists were incredulous as the ring imagery streamed in. When the resolution improved to 150 kilometres per pixel, an enhanced image revealed 95 individual ringlets. Furthermore, *five* strands of material were visible inside Cassini's Division, separated by a few hundred kilometres. A few days later, one of these mysterious ringlets was found to be elliptical. From farther out, Cassini's Division had appeared to be sharply defined, but it became difficult to identify its edges as more structure became evident. By the time the resolution had improved to 10 kilometres per pixel, *several dozen* narrow ringlets could be seen in Cassini's Division. Daniel Kirkwood's hypothesis that the gap was swept clean by Mimas's presence clearly had to be reconsidered. One idea suggested that a density wave effect was at work, such that in perturbing a ring particle at the resonant radius Mimas makes the particle's orbit elliptical. This then induces 'bunching' which, in turn, sets up a spiral density wave within which particles collide, lose energy and spiral in, forming a gap just outside the resonant radius.[48] This was a good start, but – as always – the proof would be in the fine detail, and the evidence would be the final arbiter. For example, the outer edge of the 'B' ring was not only found to be elliptical, meaning that the orbits of the particles there are elliptical, but the elliptical form 'rotates' to maintain its short axis aimed at Mimas as the moon travels around its orbit.

As Pioneer 11 had passed Titan at a range of 363,000 kilometres, it had returned a few spin–scan images showing not only that the atmosphere was optically thick but that the moon was completely enshrouded by an obscuring haze. Titan was one of Voyager 1's primary targets. The spacecraft's inbound trajectory to the Saturnian system had been calculated specifically to provide a fly-by of the moon. It started to take pictures of it in early November, in the hope of catching a glimpse of the surface through a gap in the overcast, but the prospects did not look promising. With the resolution now 200 kilometres per pixel, it was realised that the 'F' ring was slightly eccentric, extending 400 kilometres farther from the planet on its semi-major axis. On the planet itself, a profusion of zonal structure was now apparent. Indeed, 24 narrow belts and zones were counted in the southern hemisphere, with abundant evidence of isolated storms. Very little of this atmospheric structure was apparent to telescopic observers, however. On 6 November, the spacecraft performed a small manoeuvre to nudge its closest point of approach 650 kilometres nearer to Titan, onto a trajectory that would later result in an occultation by the ring system after passing Saturn. Imaging showed hints of a banded structure and a darker polar zone on Titan. By now, all the remote-sensing instruments were observing the moon on a regular basis. The Ultraviolet Spectrometer's detection of emission from both neutral atomic and molecular nitrogen from the moon was the first proof that there was nitrogen in the moon's atmosphere. The 'small atmosphere' model developed by J.J. Caldwell's group predicted a methane-dominated atmosphere at a low (20 millibars) pressure, a cold (86K) surface and an upper atmospheric inversion. The 'massive atmosphere' of D.M. Hunten's group predicted a nitrogen-dominated atmosphere with methane as a minority constituent at a very high (20 bars) pressure, sufficient for a methane-, hydrogen- and nitrogen-induced 'greenhouse' in the lower atmosphere which would produce a warm (200K) surface. Observations by the Very Large Array in 1980 had indicated a surface temperature of 87K, in which case if molecular nitrogen was present it contributed no more than 2 bars to the atmospheric pressure.[49] Voyager 1's positive detection of nitrogen therefore ruled against the thin methane-based atmosphere. The fact that

Although Voyager 1's imaging system was far superior to Pioneer 11's 'spin–scan', there were no gaps in Titan's optically-thick atmosphere to reveal the nature of the surface.

the haze layer was evidently both global and permanent lent credibility to a denser atmosphere.

With the improving resolution, the 'F' ring that Pioneer had discovered but had been barely able to resolve, was found to be no more than 100 kilometres wide, with some sections being 'thicker' than others. Some of the concentrations stretched for thousands of kilometres around arcs. "We have no explanation of what causes them, what holds them together, or how long they last," admitted B.A. Smith, leader of the imaging team. As R.J. Terrile was examining one of the new images of the 'F' ring, he discovered another new moonlet about 100 kilometres in diameter, which was later named Atlas. The fact that this moonlet was located 800 kilometres beyond the 'A' ring prompted the realisation that it was this moonlet, rather than a resonance with one of the larger moons beyond, that was responsible for the ring's sharp outer edge.

As the Sun was behind the spacecraft as it penetrated the Saturnian system, the moons, irrespective of where they were in their orbits, were almost fully illuminated and only albedo variations were apparent. Topographic relief would not become evident until the spacecraft was so deep within the system that it could view a moon 'from the side', when relief would be apparent along its terminator. Surface coverage was therefore controlled by a combination of the spacecraft's trajectory through the system and the positions of the moons in their orbits. To maximise the coverage, the spacecraft was programmed to take photographs throughout the two days that it would spend within the system, capturing the moons under various illuminations and resolutions as they pursued their orbits. An early long-range image of Iapetus showed a dichotomy that confirmed the existence of a well-defined boundary to its strangely dark hemisphere.

Voyager 2 was serving as an interplanetary monitoring platform, reporting upon the state of the solar wind *en route* to Saturn. The wind had eased considerably since Pioneer 11's visit the previous year, and it had been predicted that Voyager 1 would meet the bow shock at 40 planetary radii. However, when Voyager 2 reported a gust on 8 November, this range was reduced to between 30 and 35 radii. By 9 November, Voyager 1 was 5 million kilometres from Saturn. The large-scale albedo variations on Titan were now unambiguous. The northern hemisphere was slightly darker than the southern hemisphere, and there was a faint dark equatorial band and a dark 'hood' over the north pole, but the disk was otherwise remarkably bland. "No fine structure of any kind has been seen on Titan," B.A. Smith told the journalists who were eager to see the moon's surface. "We'll just have to wait, and watch daily to see if anything shows up." The Saturnian system was inclining its north pole slightly sunward, so perhaps the gross asymmetry was a seasonal variation in either the production of the haze or in the dynamical processes that keep it aloft. A significant discovery had already been made, however. The Ultraviolet Spectrometer had noted strong hydrogen emission from a diffuse toroidal disk about 4 planetary radii thick, ranging between 8 and 25 radii, corresponding to a distance midway between the orbits of Rhea and Dione out to slightly beyond Titan's orbit. This was significant because it meant that if the hydrogen was from Titan, then some process had to be replenishing the supply, otherwise it would have long since diffused throughout the system.

If the geologists were frustrated that Titan's surface remained hidden, they were fascinated by the longshots of some of the inner moons. Studies of their orbits had established that they had low densities, implying that they were mostly water, and spectroscopy had indicated icy surfaces, but almost nothing was known of their physical state. In gross terms, moving outwards from Saturn, the icy moons could be classified in pairs by the size of their diameters: Mimas and Enceladus (400 to 500 kilometres), Tethys and Dione (about 1,000 kilometres) and Rhea and Iapetus (about 1,500 kilometres). An early view of Rhea's trailing hemisphere at a resolution of 100 kilometres per pixel suggested the presence of bright fuzzy spots and streaks with a filamentary structure. Intriguingly, there were similar patterns on Dione. Because the illumination was face-on these were surface albedo variations. Might they be similar to the sulci on Ganymede? Even though all of these moons were much smaller than Ganymede, such patterns raised the prospect of the moons having once been geologically active, and if so, this was big news. With the solar wind taking a turn for the worse, the prediction for the bow shock was refined to between 29 and 22 radii – the latter value being what it had been the previous year. On 10 November, an early image of the trailing hemisphere of Tethys showed a circular feature 200

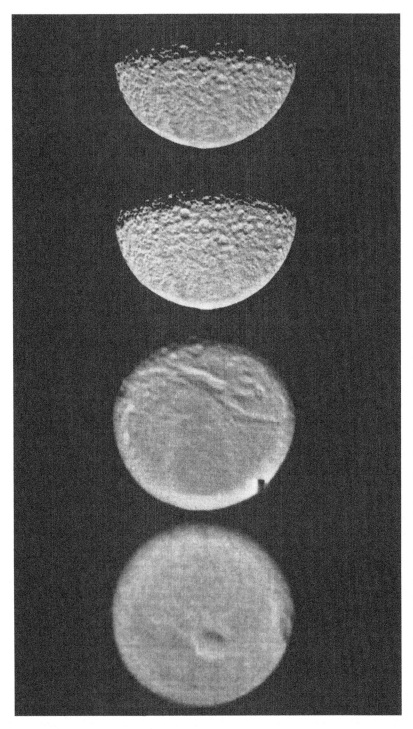

Voyager 1's view of the fully-illuminated trailing hemisphere of Tethys from a range of 2 million kilometres (left) showed a strange circular albedo feature. A view of the Saturn-facing hemisphere from 1.2 million kilometres revealed a tremendous canyon system, and later views of the terminator from half that range revealed the cratered terrain immediately to the west of the canyon (whose southern section is visible on the terminator). North is towards the top in all cases.

kilometres wide near the centre of the disk. Although this *looked* just as the rim of a crater ought to when obliquely lit, the illumination was actually face-on. This had to be an unusual albedo feature. Unfortunately, given the rate at which the moon travelled around its orbit, rotating synchronously as it did so, this feature was in darkness when the geometry was best for imaging, at which time the terminator from pole to pole was shown to be heavily cratered terrain.

Voyager 1 closed within 2 million kilometres of Saturn as 11 November dawned, and its remote-sensing instruments were regularly inspecting Titan. The moon's disk now exceeded the narrow-angle camera's field of view, so 2-by-2 (and soon 3-by-3) mosaics were being assembled. "We cannot see the surface of Titan," reported Hal Masursky as the resolution passed 35 kilometres per pixel without showing any sign of a gap in the haze. "If we are going to learn what this extraordinary and fascinating body is like, we will need a Titan-orbiting imaging radar." At 15:00, as the spacecraft closed to within 500,000 kilometres of Titan, it switched from its 'far encounter' to its 'near encounter' sequence, which would start with the Titan fly-by. The bow shock was finally encountered at 16:50.[50] It washed to and fro several times until 20:13, when the spacecraft crossed the magnetopause at 22.9 radii, at which time Titan was barely three hours distant. A high-resolution mosaic had shown the presence of a thin layer of haze beyond the limb, about 100 kilometres above the optically-thick cover. This extended all the way around the disk, and near the north pole it merged with an opaque dark hood. The absence of such a feature in the south was another aspect of the dichotomy between the two hemispheres. As the range diminished, the IRIS was able to secure infrared spectra of increasingly higher resolution. At 21:41, Voyager 1 was at its closest point to the moon, passing 4,000 kilometres above the main haze at a relative speed of 17 kilometres per second. At 22:25, the scan platform slewed onto the limb to enable the IRIS to measure the composition of the 'detached' layer of haze.[51]

With Titan inside the magnetosphere, Voyager 1's magnetometer had a better chance of determining whether the moon has its own magnetic field. At 22:45, the spacecraft prepared for being occulted by the moon by boosting the strength of its radio signal and ceasing to modulate this with data. At 23:00 the particles and fields instruments noted the spacecraft's flight through Titan's magnetospheric wake, but there was no indication of an intrinsic magnetic field; if the moon possesses one, its strength is less than one-thousandth of the Earth's field. At 23:11 Voyager 1 flew into the moon's shadow. A minute later it passed beyond the moon's limb as seen from the Earth, and the manner in which its radio signal was refracted profiled the physical properties of the atmosphere down to the surface. At 23:22 it crossed through the plane of Titan's orbit, travelling from north to south at an angle of 8.7 degrees. During its approach, the spacecraft had observed the illuminated face of the ring system, but in passing through the plane of the moon's orbit it also crossed the projection of the ring plane. At 23:24 it re-emerged into sunlight and at 23:27 it re-emerged from the far limb. The fact that the occultation ended 43 seconds earlier than expected meant that the spacecraft's trajectory was 'off' by 200 kilometres. This prompted a reprogramming of the scan platform to aim it directly at the smaller targets that were scheduled for inspection closer to the

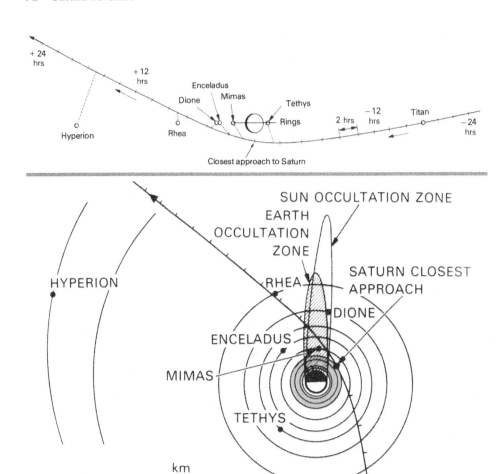

Voyager 1's route through the Saturnian system.

planet. Looking back, the Ultraviolet Spectrometer noted emissions from several detached layers of haze, the highest of which was some 400 kilometres above the top of the optically-thick haze. It was now evident why telescopic observers had had so much difficulty in attempting to measure the diameter of the moon's disk – the limb-darkening and the high altitude haze had made the limb indistinct.

One immediate result from the radio occultation was an accurate measurement of Titan's diameter: the solid body inside the obscuring haze is 5,150 kilometres in width. Given an estimate of its mass from how it deflected Voyager 1's trajectory, it was finally possible to calculate an accurate bulk density of 1.88 g/cm^3, which implied 45 per cent ices. To a first approximation, in terms of diameter and density, this placed Titan between Callisto (4,806 kilometres and 1.86 g/cm^3) and Ganymede (5,286 kilometres and 1.94 g/cm^3). However, each of these three bodies is highly distinctive. As the Galileo spacecraft established during its extended tour of the Jovian system in the late 1990s, Ganymede strongly resembles the Earth in having a differentiated interior and a magnetic field, whereas Callisto is essentially undifferentiated. Having condensed out of the chillier nebula farther from the Sun, Titan ought to have acquired more methane (carbon) and ammonia (nitrogen) ices,[52] and the dissipation of accretional heat would have driven these exotic ices to the surface, whereupon they would have evaporated to create an atmosphere.[53] Dissociation of ammonia would have released nitrogen, some of which would have combined with methane to create a variety of organic molecules. However, telescopic spectroscopy had positively identified only methane. The radio occultation data measured how the temperature, pressure and composition of the atmosphere varied with altitude.[54] The surface pressure clearly exceeded the 20 millibars predicted for a methane-dominated atmosphere. Given the measured diameter, the atmosphere is evidently dense and deep. The 1.5-bar surface pressure is all the more striking because a much deeper column of gas is required to create such a pressure on a body whose surface gravity is just 14 per cent of that of the Earth. In fact, Titan's atmosphere contains 10 times as much gas as the Earth's considerably shallower envelope. Actually, methane is present in more or less the expected amount – the atmosphere has been 'pumped up' by molecular nitrogen. The Ultraviolet Spectro-meter had made the first detection of nitrogen by strong emission from the upper atmosphere where the molecules are excited by the flux of magnetospheric electrons. This shed light on the nature of the photochemical reactions that occur in the upper atmosphere. The IRIS confirmed molecular nitrogen (N_2) as the main constituent, and then methane (CH_4) with some acetylene (C_2H_2), ethylene (C_2H_4), ethane (C_2H_6) and propane (C_3H_8), cyanogen (C_2N_2) and cyanoacetylene (HC_3N), hydrogen cyanide (HCN), molecular hydrogen (H_2) and a whiff of helium.[55,56] The complex hydrocarbons derive from photodissociation of methane, and the nitriles are the result of the C_2H radical's action on unsaturated hydrocarbons and the dissociation of molecular nitrogen in the upper atmosphere by magnetospheric electrons.[57]

Temperature inversions are common in planetary atmospheres. The temperature of the Earth's atmosphere initially declines with increasing altitude, then increases again. In the troposphere, the tendency of warm air to rise causes convection, but this ceases at the tropopause, above which is the stratosphere. It was the discovery of this boundary that led to the realisation that the atmosphere is stratified. In fact, the temperature inversion that occurs in the stratosphere is caused by the absorption of solar ultraviolet by the ozone (O_3) form of oxygen. Above the stratosphere is the mesosphere, and then the thermosphere, in which the temperature of the rarefied gas exceeds that at the surface. The thermal profile derived from Voyager 1's occultation confirmed that there is an inversion in Titan's upper atmosphere. The minimum temperature of 72K occurs at an altitude of 45 kilometres, but this is too warm for molecular nitrogen to condense. Above this, the temperature rapidly soars to 150K at 100 kilometres, and then increases somewhat more slowly to 175K by the top of the optically-thick haze at 200 kilometres. This heating results from the particles of the orangey smog absorbing shorter wavelength insolation. Titan's atmosphere is distinctly stratified, with the tropopause at 45 kilometres, the stratopause at 280 kilometres, just above the main haze, and the mesopause at 600 kilometres, in among the detached layers of thin haze. The visible surface is more properly called haze than cloud because it is composed of small particles that are well spaced. It is optically thick because it is so deep, but it is likely to be transparent over shorter distances. In the mesosphere, the dissociation of methane by solar ultraviolet has created a photochemical smog which is most evident in a concentration of aerosol in the 340- to 360-kilometre altitude range that was viewed on the limb as a distinct layer. The irradiation of the atmosphere by magnetospheric electrons when inside Saturn's magnetosphere, and energetic protons when exposed to the solar wind, will enhance the range of dissociation products and increase the number of molecular species. Methane readily polymerises, combining with itself to form chains. These complex molecules will form solid aerosols ranging in size from 0.2 to 0.8 micron. Polarisation measurements indicated that these are irregular conglomerates of smaller particles fused together by collisions, and are responsible for the atmosphere's orangey hue. Although the precise composition of the aerosols was not determined, Carl Sagan of Cornell University introduced the term 'tholins' to describe the range of solid, oily and tarry substances.[58,59] The aerosols will initially remain in suspension, but will sink as they grow. As they penetrate the chilly troposphere they will be coated by ambient moisture in the form of liquid methane, accumulate into droplets and fall as rain, suggesting a surface coated with sludges and ices of hydrocarbons. The limb observations at closest approach enabled the IRIS to determine the composition with altitude. Nitrogen forms 90 per cent of the lower atmosphere by number, and 98 per cent of the upper atmosphere. Although methane is being dissociated in the upper atmosphere, it is evidently being renewed by diffusion from the troposphere, where it is concentrated. However, the diffusion rate is controlled by the 'cold trap' at the tropopause, where most of the methane condenses and rains out.

Titan cannot hold the hydrogen released by the photochemical reactions.[60] As the rapidly rotating magnetic field sweeps past the moon, the hydrogen leaks away into the 'cavity' of the magnetospheric wake in a manner reminiscent of a plume from a smokestack streaming 'downwind'. Because it cannot escape Saturn's gravitational field, the hydrogen has formed the broad torus of neutral hydrogen that pervades the vicinity of the moon's orbit.[61,62] The magnetospheric wind is compressed against the moon's trailing hemisphere. This turbulent shock wave induces an ultraviolet glow in the thermosphere of the moon's atmosphere, extending 1,000 kilometres into space. Because the hydrogen continuously leaks away, the polymerisation reactions are irreversible on a macroscopic scale, and the falling hydrocarbons accumulate on the surface. In the 20-micron band, a wavelength at which the atmosphere is more or less transparent, the IRIS was able to measure the surface temperature as 92K – a value that applied globally to within 3 degrees. As this was near the temperature at which methane can exist in both liquid and solid phases, it was speculated that in addition to sludges and ices of hydrocarbons there might be an ocean of liquid ethane diluted with 25 per cent liquid methane and 5 per cent liquid nitrogen.[63] As methane melts at 91K and boils at 118K, it was further speculated that it might serve a similar role in Titan's weather system as water does on the Earth, following a cycle of evaporating from the sea, rising in the atmosphere on air currents, condensing, falling as rain and flowing over the surface back into the sea.

"The Titan data are very exciting," enthused R.A. Hanel, leader of the IRIS team, "and much more important than we thought they would be."

Having successfully achieved one of its main objectives in the Saturnian system, Voyager 1 now directed its instruments to the other members of the retinue.

Mimas was expected to be very heavily cratered due to its proximity to Saturn and as a result of the planet's gravitational field drawing in and accelerating material that strays too close. An early view presented the astounding sight of a crater whose 130-kilometre-wide raised rim spanned fully one-third of the moon's spheroidal form. Upon seeing it, one wag whimsically exclaimed that they had found the 'Death Star' of George Lucas's *Star Wars* movie. It also prompted gasps when later shown to the reporters thronging the von Kármán Auditorium. It was remarkable that the icy moon had survived such a proportionately massive impact: the projectile must have been at least 10 kilometres across. Although initially dubbed Arthur, after King Arthur of legend, this crater was eventually named Herschel by the International Astronomical Union. As Voyager 1 penetrated more deeply into the system, it was able to document the crater from 400,000 kilometres at 4 kilometres per pixel, but it soon rotated over the terminator and the later sequence was of the intensely cratered southern terrain. Grooves 10 kilometres in width, 100 kilometres in length, and up to several kilometres in depth were more probably opened by the shock of large impacts than by endogenic processes. Although its bulk density implies that Mimas is mostly ice, the craterforms have not been softened because the low surface gravity is evidently insufficient to overcome the integrity of the ice which, so far from the Sun, possesses the strength of rock. Still bearing the scars of its early bombardment, it lives up to the stereotype of a moon in the style of Callisto. As L.A. Soderblom put it, "Mimas is your basic unprocessed ice moon."

Imagery of Tethys from a range of 1.2 million kilometres, with a resolution of 11 kilometres per pixel, showed a curvilinear feature 100 kilometres wide on the Saturn-facing hemisphere that was suggestive of an active geological past. Unfortunately, by the time that the range had been halved, most of this feature had slipped beyond the terminator. However, the improved resolution showed the leading hemisphere to be

Voyager 1's view of Mimas. The top set were taken at ranges between 800,000 and 400,000 kilometres, and show the giant crater in the centre of the leading hemisphere. North is towards the top. The bottom set were taken at ranges between 300,000 and 127,000 kilometres as the spacecraft passed the moon, showing the heavily cratered south polar region. North is to the left in this sequence, and the south pole is on the terminator on the rightmost image.

intensely cratered. The closest approach occurred about an hour and a half before the Saturn slingshot, but at that time the moon was 416,000 kilometres away, on the far side of the planet. The trailing hemisphere of Dione was recorded from 620,000 kilometres at a resolution of 5 kilometres per pixel. The wispy and filamentary character of the streaks was shown to be rather different to the sulci of Ganymede. The network radiates out from a large dark patch and projects across the entire hemisphere. The imagery of Rhea, from 720,000 kilometres, established that the Saturn-facing hemisphere is an intensely cratered plain. By this time, most of the wispy feature was beyond the terminator, indicating that it is concentrated between 180 and 270 degrees longitude. L.A. Soderblom said of such streaks: "I don't believe that these can possibly have been produced by impact processes." Their extent and connectedness suggested that they were created by internal processes. For bodies so small, so lightweight, and so cold, this was a surprising discovery.

The encounter sequence included imaging the co-orbiting Janus and Epimetheus, which were mere specks of light in telescopes. Both are irregular with their primary axes aligned towards Saturn. Neither exceeds 200 kilometres on its longest dimension, and Janus is slightly the larger. Both have cratered surfaces and they looked to be fragments of a single progenitor. Helene ('Dione-B') is no more than 30 kilometres across. Tethys was confirmed to have a pair of co-orbiting moonlets, with 'Tethys-B' (Telesto) leading and 'Tethys-C' (Calypso) trailing.[64] However, a suspected moonlet co-orbiting with Mimas was not confirmed. The recently discovered 'A' ring and 'F' ring shepherds could not be inspected in detail because their orbits were insufficiently defined to aim the scan platform.

An image of the 'F' ring from a range of 750,000 kilometres, with a resolution of 15 kilometres per pixel, indicated that this feature actually comprised a pair of thin bright stands and a fainter diffuse band about 100 kilometres towards Saturn. The thin strands were distorted and in places intertwined, and the distribution of material was rather 'clumpy'. The 'braiding' of the strands "seemed to defy all of the laws of celestial mechanics" an amazed B.A. Smith told journalists. In fact, the braiding is the result of gravitational interactions with the shepherding moonlets Prometheus and Pandora, which are very close to the ring. The material ahead of one of the moonlets will be retarded and will descend, and the material behind will be accelerated and will rise, thereby distorting the ring. These distortions will travel with the moonlets, and may well interact with one another when the inner shepherd overtakes its outer companion.[65,66,67,68,69,70]

At 15:45 on 12 November, Voyager 1 flew 123,500 kilometres above Saturn's cloud tops. When passing Titan, it had crossed onto the far side of the ring system, so during its final approach to the planet it was able to view the non-illuminated face. The trajectory had been constrained by the requirement to inspect Titan on the way in, and then to set up the Saturn occultation so that the spacecraft would emerge (as viewed from Earth) between the limb and the ring system. The planetary occultation began at 19:08, and lasted an hour and a half. Within a few minutes of re-emerging, the spacecraft started the highly prized ring occultation. The way in which the signal fluctuated served to measure the density of the material in a 'radial' across the entire system.[71] This showed that there are many more fine ringlets than

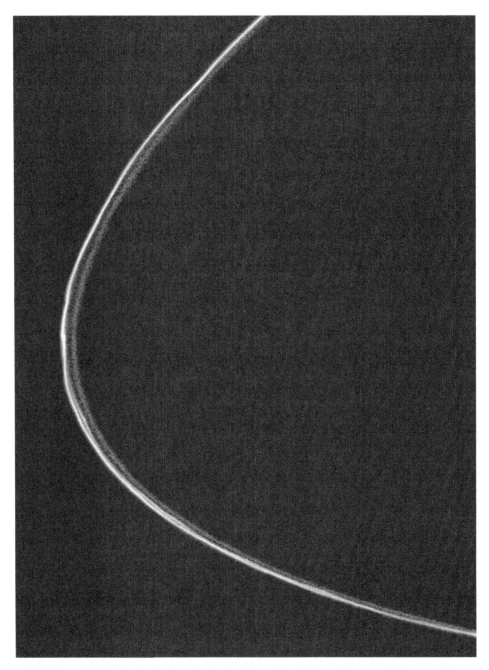

The kinked, clumped and multiple 'F' ring imaged at high resolution on 12 November 1981 by Voyager 1. Note that the faint inner band is unperturbed by the force that is distorting the main pair. It is possible that they are not physically intertwined, but in slightly different planes and the overlapping is a perspective effect.

could be seen in even the highest-resolution imagery.[72] The size of the particles ranges from a few microns to a few tens of metres. The tenuous 'C' ring evidently comprises particles which are typically 1 metre in diameter: "boulders flying around Saturn", explained G.L. Tyler, the leader of the radio science team. The data also finally established that the 'B' ring, which is the visually brightest and most opaque section of the system, is no more than 100 metres thick and comprises particles ranging in size from a few centimetres up to 1 metre. The 'A' ring material ranges from a few microns up to 10 metres. The outer edge of the 'A' ring is 10 metres thick. If the largest particles are towards its periphery, it is only one 'particle' thick.[73] Given its overall span, the ring system is the *flattest* object in the Solar System.[74]

The rings received much of Voyager 1's attention because, apart from the spin–scan longshots by Pioneer 11, this was the first opportunity to inspect their detailed structure. Some telescopic observers had reported fine structure in the 'B' ring, but others had only reported a smooth feature. Clearly, if such structure was real, it was of a transient character, and it had been speculated to be due to spiral density waves propagating through the system. The Voyager imagery established that there are indeed such waveforms. A search for moonlets embedded in the ring system proved to be frustrating, implying either that the finer structure is not induced by such rocks, or they are smaller than the camera's resolution.[75,76] Nevertheless, a detailed analysis of the density waves in the 'A' ring subsequently prompted the prediction of a moonlet within Encke's Division, and thus Pan was discovered. The historical variability of Encke's Division had prompted the belief that it was only an intermittent thinning out of the 'A' ring particles rather than a clear zone. Now the task for the theoreticians was to explain how, with a moonlet present, the gap could seem to close. Furthermore, as Voyager 1 closed in, it became evident that this division, which was just 270 kilometres wide at the time, hosted several ringlets that were both discontinuous and 'kinky'.

Once Voyager 1 was able to view the rings forward-scattering sunlight, it saw a region of tenuous material extending in from the 'C' ring to about 3,200 kilometres of the cloud tops. In 1969, when the ring system was opening up, Pierre Guerin had reported material in this zone, and it had been designated the 'D' ring. However, the feature found by the spacecraft was too tenuous to have been seen telescopically by reflected sunlight – indeed, it had not been visible even an hour before the spacecraft crossed the ring plane. As O.W. Struve had speculated in 1851, material is spiralling onto the planet. Perhaps the flow rate is variable, and Guerin had been fortunate and seen a clump spiralling in.

Although Voyager 1 could not modulate its carrier wave to transmit data during the planetary and ring occultations, it had continued to make observations and stored the data on tape for later replay. As Saturn's gravitational field bent the spacecraft's trajectory, it flew within 88,000 kilometres of Mimas an hour after closest approach to the planet, recording the illuminated part of the moon's south polar region. An hour later it viewed Enceladus's trailing hemisphere from 201,000 kilometres. This was not a particularly favourable fly-by, but it showed that much of the surface is remarkably smooth and devoid of craters, at least down to the 12-

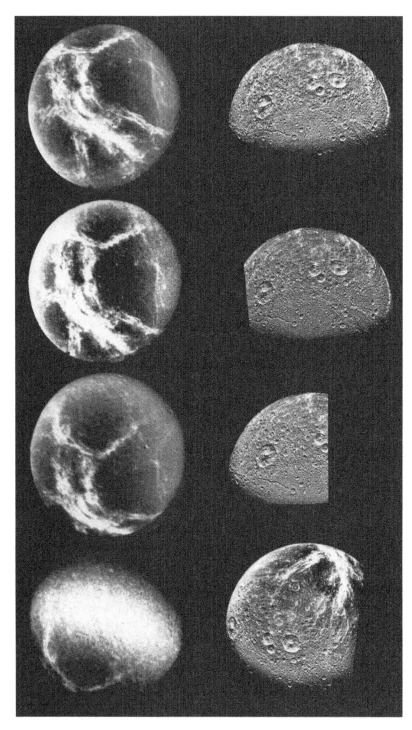

Voyager 1's view of Dione. The top set were taken at ranges between 1.7 million and 600,000 kilometres, and show the wispy streaks on the trailing hemisphere. They are recorded in finer detail in the leftmost view of the lower set, from a range of 237,000 kilometres, which also shows that they radiate out from an anomalously dark patch. As the range closed to 162,000 kilometres, the perspective transitted the heavily cratered terrain of the Saturn-facing hemisphere. North is towards the top in each case.

Rhea's trailing hemisphere is dominated by bright streaks. The first three images were taken by Voyager 1 as it closed from 1.8 to 1.3 million kilometres. The perspective is migrating around onto the Saturn-facing hemisphere. In the 720,000-kilometre image (right) one streak is seen in detail, as is the large dark region and the cratered surface. North is towards the top in each case.

kilometre-per-pixel resolution. The theoreticians promptly set to work developing models for how this moon might have come to be so extensively resurfaced. Its similarity in size to Mimas suggested to some researchers that the cause had to be exogenic, perhaps tidal heating derived from the gravitational stresses of orbital resonances with other satellites, several of which also displayed features suggestive of activity, although to a lesser extent. However, E.M. Shoemaker countered that the resurfacing may have been driven by heat liberated by early massive impacts which melted large areas of the icy surface, transforming them into smooth plains on which crevasses formed as the water froze. Other researchers cited the paucity of craters as evidence that the resurfacing occurred 'recently'. What was certain, however, was that Enceladus did *not* fit the stereotype of an ancient icy moon. An hour and a half later, it was Dione's turn. The fly-by geometry provided a view of the Saturn-facing hemisphere from a minimum range of 162,000 kilometres. The wispy streaks on the trailing hemisphere were mostly beyond the limb, but there was a paucity of craters on the section of the associated dark patch that was visible. Much of the rest of the moon is heavily cratered but there are degraded irregular valleys suggestive of troughs, and relatively smooth plains on the leading hemisphere. The Rhea fly-by was three and a half hours after the closest approach to Saturn. For a sequence of frames for a mosaic of the moon's northern hemisphere, the computer rotated the spacecraft during the shuttering action in order to compensate for the high relative velocity of the 72,000-kilometre encounter that would otherwise have smeared the image.[77] The imagery of the polar region, with a resolution of a few kilometres per pixel, showed a mixture of shallow craters having subdued shapes suggestive of great age and others whose sharp rims looked 'fresh'. Part of Rhea's surface is so saturated with craters that it is reminiscent of the surface of the planet Mercury.

At 21:00 Voyager 1 emerged from behind the 'A' ring. Forty-five minutes later, with its trajectory bent back by the slingshot, it re-crossed the ring plane just short of Rhea's orbit. Earlier in the year, when the ring system had been edge-on as viewed from Earth, William Baum at the Lowell Observatory utilised a CCD camera to establish that the exceedingly tenuous 'E' ring is a 90,000-kilometre-wide band that spans Enceladus's orbit.[78] The fact that the brightest part of the ring lies close to the moon prompted speculation that the material is derived from it.[79] The resurfaced area is so sparsely cratered that the process might still be active, and the ring might be ice crystals spewed into space from cryovolcanic geysers. As the spacecraft re-crossed the ring plane, an image looking 'back' showed these fine particles forward-scattering sunlight. The spacecraft was once again able to view the illuminated face of the ring system. Now, however, by virtue of its out-of-plane vantage point, its view of the system was more 'open' than before. A close examination of this imagery revealed that the system actually comprises many hundreds – *perhaps even 1,000* – ringlets. Indeed, in the early count for a press conference R.J. Terrile stopped counting at 300 ringlets after he grew bored. This tremendous complexity could not be explained in terms of resonances with the moons orbiting beyond.

The ring system was rather different in appearance from beyond Saturn's orbit.

As Voyager 1 re-crossed the ring-plane just short of Rhea's orbit, it aimed its camera back and took this time-exposure which (for the first time) documented the 'E' ring forward-scattering sunlight. This broad but tenuous feature is centred on the orbit of Enceladus, and may well be composed of ice crystals from the moon. The streaks are stars trailed during the exposure.

The spokes which had previously appeared dark on the bright 'B' ring were now brighter than the ring material. Clearly, if the spokes were particles suspended away from the ring plane by electrical forces, then the fact that they did not reflect sunlight efficiently meant that when they were viewed from up-Sun they were visible only by the *shadows* they cast on the ring, whereas because they forward-scattered sunlight efficiently they were directly visible from a down-Sun vantage point.[80]

About 13 hours after Saturn, Voyager 1 crossed Hyperion's orbit, but the moon was 877,000 kilometres away and was presenting its non-illuminated hemisphere. Nevertheless, it was obviously an irregularly shaped body. At mid-afternoon on 13 November, the spacecraft switched to its 'post-encounter' sequence. Over the next day, as it closed to within 2.5 million kilometres of Iapetus, it snapped pictures of the non-illuminated trailing hemisphere. Although this, the most enigmatic of moons, was not well placed for study, the geometry offered an oblique view over the north pole of a cluster of large craters situated on the intrinsically brighter terrain on the illuminated leading hemisphere.

As to Saturn itself, the data from the radio occultation as Voyager 1 crossed over the limb, and the remote sensing of its disk by the IRIS, gave a vertical profile of the semi-transparent outer envelope down to the 1 bar level, at which the temperature is 133K. By the tropopause, at the 0.07-bar level, the temperature had fallen to 93K.

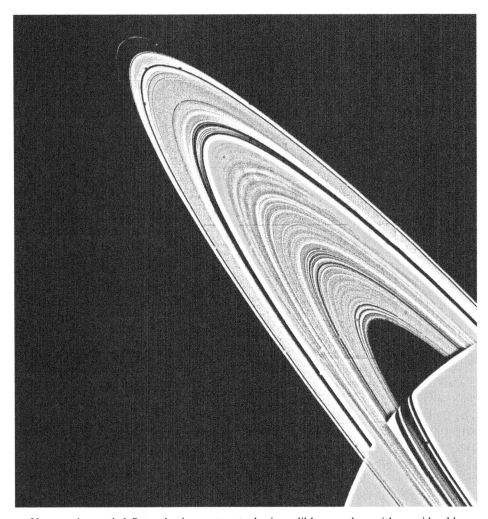

Voyager 1 revealed Saturn's ring system to be incredible complex, with considerably more structure than suspected by even the most eagle-eyed of telescopic observers. Notice the ringlets within Cassini's Division and the arc of the thin 'F' ring beyond the cusp of the 'A' ring.

There is an inversion in the stratosphere due to methane absorbing sunlight, with the temperature increasing to 143K by the 0.001-bar level. Because Saturn's magnetic field is considerably weaker than Jupiter's, the solar wind is able to enter the polar cusps to induce auroral activity in the ionosphere.[81] There is also strong ultraviolet emission in the thermosphere on the sunlit hemisphere. Orbiting farther from the Sun, Saturn's outer envelope is about 15°C colder than Jupiter's, so the ammonia-crystal layer spans a 100-kilometre range from just below the 1-bar level up to the tropopause. The cloud coverage is virtually continuous, but the infrared radiation is readily able to leak out if a short-lived hole opens.

This image of Saturn's rings was taken by Voyager 1 as it withdrew, from a range of 1.6 million kilometres. From this vantage point, the 'spokes' on the 'B' ring appear bright when forward-scattering sunlight. Notice the 'F' ring's clumpy strands, and that the planet is visible through the translucent rings.

Two Voyager 1 views of Iapetus. The early view (left) from 3.2 million kilometres is of the Saturn-facing hemisphere. The large circular feature is at longitude 330 degrees west. The illumination is actually face-on, and the dichotomy shows the boundary of the intrinsically dark Cassini Regio, on the leading hemisphere. The second image is looking over the north pole (which is just beyond and between the two large craters) onto the leading hemisphere and the winding dichotomy is a combination of Cassini Regio and the actual terminator. North is towards the top in each case.

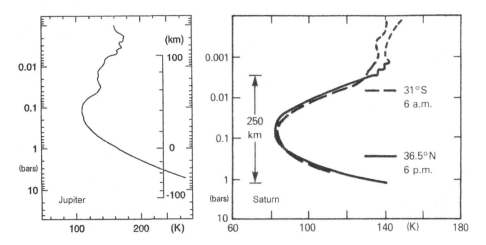

The thermal profiles of Jupiter (left) and Saturn (right). Both show a stratospheric temperature inversion. Apart from minor diurnal differences in the outermost layers, the data from Voyager 2's signal as it passed behind Saturn and then reappeared on the far side are identical.

In bulk, measured in terms of mass, Saturn is about 80 per cent hydrogen, and most of the remainder is helium. All the other elements which condensed out of the solar nebula together contribute a little over 2 per cent. The heavy elements sank into the core. Much of the oxygen combined with hydrogen as water, much of the carbon as methane, and much of the nitrogen as ammonia, all of which are present in the envelope. However, the IRIS measured the helium–to–hydrogen molecular number ratio for the cloud tops as 6 per cent.[82] The corresponding ratio for Jupiter had been 13 ($\pm$4) per cent.[83] Jupiter was comparable to a solar value of 13 ($\pm$2) per cent.[84] In terms of mass fraction, at Saturn's cloud tops helium is 11 ($\pm$3) per cent, as against 19 ($\pm$5) per cent on Jupiter. Because both planets would have started out with near-solar abundances, this implied that the helium within Saturn has, over time, migrated more deeply into its interior. The conditions for gravitational fractionation in a hydrogen–helium fluid had been computed by E.E. Salpeter, who had offered it as the source of Jupiter's excess energy.[85] Further analysis suggested that conditions in Jupiter were probably *not* conducive, but Saturn might be susceptible.[86] The depletion of helium implied that Saturn's interior is sufficiently cool for the helium to condense and fall as 'rain' towards the interior (which must be correspondingly enriched). In addition to the heat that it still derives from ongoing gravitational collapse, Saturn – unlike Jupiter – is also heated by the friction of the helium droplets falling through dense hydrogen. Saturn therefore not only radiates to space *more* energy than it receives as insolation, but at 1.8 as against 1.7 for Jupiter, it generates a proportionately *greater* 'infrared excess'. Evidently, Jupiter's interior is too hot for helium to condense.[87,88] The model for Jupiter derived from its moment of inertia, as a result of the Voyager fly-bys, has *three layers*. It has a relatively small rocky core of iron, silicates and ices at 30,000K, the incorporated superheated 'ices' being in a superfluid state. This is enclosed by a thick shell of hydrogen which transitions from metallic to molecular phase at a radius of about 46,000 kilometres, at a pressure of 3 megabars and a temperature of 11,000K. Despite Saturn being less massive, and having a much lower density, it would seem to have a similar three-layer structure, except that its core is only 15,000K, half that of Jupiter's core, and the milder thermal profile enables the helium to settle. Although the mass of Saturn's core is about three times that of the Earth, it is denser and hence only slightly larger. But how did such giant planets form? Computer modelling conducted soon after the Pioneer 10 fly-by of Jupiter prompted the idea that they formed by a two-stage process. In each case the process began with the accretion from the nebula of a rocky core. Once this attained a mass several times that of the Earth, it induced instability in the gravitationally bound portion of the nebula, which underwent hydrodynamic collapse and promptly enveloped the rocky core.[89,90,91,92,93,94,95,96,97,98,99,100,101]

Synoptic imagery revealed that Saturn's zonal winds are different to Jupiter's. The fastest wind occurs in what, in 1891, A.S. Williams dubbed "the great equatorial atmospheric current". Indeed, at 1,750 kilometres per hour, this almost-supersonic wind was four times faster than any measured in Jupiter's violent atmosphere. The eastward flow forms a broad swath that extends some 35 degrees north and south of the equator. In the mid-latitudes, the flow first stagnates and then adopts a series of contra-rotating flows, as on Jupiter, except that the maximum and minimum velocities occur in the centres of the alternating features rather than at their boundaries. At that time, there was no certain knowledge of the process driving Jupiter's atmospheric circulation, but the probe that the Galileo spacecraft dropped into the atmosphere in December 1995 established that the circulation is deep and is driven by heat leaking from the interior, rather than being shallow and driven by differential absorption of insolation.[102] Because Saturn radiates proportionately more energy than Jupiter, its winds are likely to be driven by the same process. The predominant easterly wind certainly suggests that the air flow is not confined to the outer atmosphere, and may well extend to a depth of several thousand kilometres. Voyager 1 documented considerable atmospheric structure at the surface. The fact that it is not as distinctive as on Jupiter could mean either that the condensates are chemically different and less colourful, or that the winds are so strong that the troposphere has been homogenised into a bland hue.[103] As the atmosphere turns over, the 'exotic' chemicals will form strata (working outwards) of water droplets, water ice crystals, ammonium hydrosulphide crystals and ammonia crystals, above which, in view of the fact that it is not cold enough for methane to freeze, there is a haze of methane.[104,105,106]

Voyager 1 established that isolated storm systems are common at high latitudes. Their anticyclonic character, and the fact that they were concentrated in latitudes at which the winds were slow, implied that localised high-pressure system at a deeper level were forcing their way up through the ammonia cloud deck. Most of the spots lasted for only a few days, but a few survived long enough to be studied by both Voyagers. Telescopic observers had noticed that when 'white spots' appeared in the equatorial zone they were large, indicating that the ammonia cloud deck was rent by

The zonal banding of Saturn's atmosphere (right) is the result of rising and falling air masses being 'stretched' by the planet's rapid rotation. Windspeeds were measured by monitoring the motions of features over several planetary rotations. The wind speed varies with latitude symmetrically about the equator. The eastward equatorial flow is approximately 1,100 metres per second. The Jovian-style alternating jet streams (left) are confined to higher latitudes. The dark band on Saturn's equator is the near edge-on ring system.

a *major* updraft, and the emerging blobs were soon drawn out in longitude by the strong eastward wind.

The Saturn fly-by deflected Voyager 1 north of the ecliptic. Although this ended the planetary phase of the mission, it enabled the spacecraft to investigate the solar wind above the ecliptic as it searched for the heliopause. Its total success at Saturn cleared the way for Voyager 2 to adopt a trajectory for a slingshot that would set up the Grand Tour.

As Voyager 2 pursued its interplanetary cruise, it sensed filaments of the Jovian magnetotail washing over it several times,[107,108,109] so this very likely does extend as far as Saturn's orbit, which is itself as far from Jupiter as that planet's orbit is from the Sun.[110]

On 21 June 1981, Voyager 2 began the 'observatory phase' of its programme. Saturn's atmosphere was much more active than in the previous year, with more prominent latitudinal banding and a multitude of spots, waves and storms. One early finding was that the banding was visible almost to the limb, indicating that there is very little haze, which confirmed that the very low contrast at visible wavelengths is the result of the chemicals in the clouds being either intrinsically less colourful, or thoroughly mixed.

As Saturn had moved around the Sun, the angle of illumination on the rings had

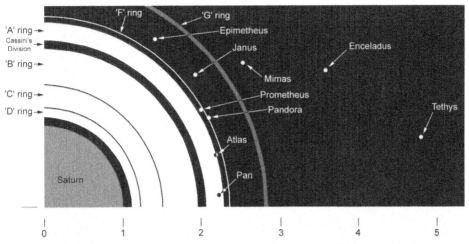

A diagram of the inner part of the Saturnian system showing the rings and the newly discovered moonlets in relation to the innermost icy moons.

increased to 8 degrees; they now appeared rather brighter, which made it easier to determine their composition by the absorption features in the reflection spectrum. Filtered imagery showed that while the ring material is predominantly water ice, the composition of the 'C' ring is different from the 'B' ring, with a distinct transition – it is not simply a case of particles from the dense 'B' ring spiralling down through the tenuous 'C' ring towards the planet. The particles of the 'B' ring are mainly icy, but with a 'redder' spectrum than the 'C' and 'D' rings. However, the material in the ringlets within Cassini's Division is similar to that of the 'C' ring.[111]

The dark spokes on the 'B' ring were more prominent, too, so during the lengthy approach it was possible to monitor the long-term character of this phenomenon. A series of movies were made, in one case a 'tight shot' which followed the motion of a specific group as it emerged from the planet's shadow.

When Voyager 1 revealed the complexity of the ring system, it was realised that, with the exception of a gap as striking as Cassini's Division, Daniel Kirkwood's idea that the fine divisions were induced by resonances with the moons orbiting beyond was untenable. As an alternative, it was suggested that there were a large number of moonlets ranging up to several dozen kilometres wide embedded in the system, each of which was responsible for forming a fine division. J.N. Cuzzi, a specialist in the rings from the Ames Research Center, had studied the 500-kilometre-wide clear zone near the inner edge of Cassini's Division, and concluded that it might be being swept by an object about 25 kilometres in diameter. As Voyager 2's resolution improved, it secured imagery of almost the entire circuit of two broad gaps in Cassini's Division, but no bodies as large as 10 kilometres wide were evident within them.[112] A detailed search of 'B' ring imagery failed to find any embedded moonlets. If they exist, they are smaller than the 10-kilometre-per-pixel resolution. Despite R.S. Ball's yearning a century ago "to see the actual texture of the rings", the cause of their fine structure defied even a spacecraft's close-up inspection.

Although taken from 34 million kilometres, this image of Saturn on 21 July 1981 by Voyager 2 shows considerably more atmospheric structure than its predecessor had been able to see from this range.

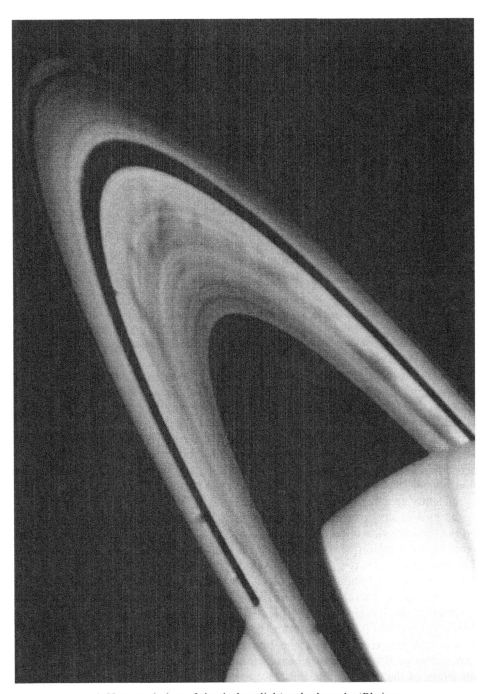

A Voyager 2 shot of the dark radial 'spokes' on the 'B' ring.

On 15 August 1981, Voyager 2 snapped this image of the 'F' ring showing the two shepherds: Prometheus orbits 1,250 kilometres inside and Pandora 1,100 kilometres outside the ring. Abiding by Kepler's laws of orbital motion, Prometheus is moving slightly faster, so we were lucky to catch them in the same vicinity.

On 19 August, at a distance of 10 million kilometres, Voyager 2 refined its trajectory to move its point of closest approach to the planet 900 kilometres closer to optimise the slingshot for the Grand Tour. Its trajectory complemented that of its predecessor. Voyager 1 had inspected Titan, Mimas, Dione and Rhea, and Voyager 2 would investigate Iapetus, Hyperion, Enceladus and Tethys. In addition to the search for new moonlets, it was to snap the co-orbital moonlets and – now that their orbits had been determined with sufficient accuracy to aim the scan platform – the ring shepherds, so its workload deep within the Saturnian system would be significantly greater than that of its predecessor.

On 21 August, Voyager 2 took its first look at Iapetus. Although the moon was on the opposite side of its orbit when inspected by Voyager 1, the view was of the same cratered region of the illuminated leading hemisphere near the north pole. The best imagery, taken late on 22 August from a range of 1 million kilometres, had a resolution of about 20 kilometres per pixel. The contrast between the light and dark terrains was striking. David Morrison of the University of Hawaii, and a member of the imaging team, explained to the reporters that the contrast was "equivalent to that between snow and asphalt". In fact, the albedo of the dark side is less than one-tenth that of the bright side. The cratered bright terrain was reminiscent of Rhea, Iapetus's near-twin in size. Some detail was apparent along the fringe of the dark terrain, but since *nothing* could be discerned on that terrain (subsequently named Cassini Regio) it was impossible to determine whether this is cratered terrain which had been coated black or a smooth plain which was intrinsically black. The imagery enabled the moon's diameter to be measured

As Voyager 2 flew past Iapetus, it took a sequence of images. The main image shows the north polar region, with the dark patch of Cassini Regio on the limb (lower left).

accurately as 1,436 kilometres, and the degree to which it deflected the trajectory enabled its mass to be refined to within 10 per cent, so the bulk density of 1.1 g/cm^3 means that there is almost no rock and it is virtually all ice. The bright terrain is therefore almost certainly 'native', and typical of the other icy moons. The dark area is the anomalous surface. However, it was not immediately evident whether it is of endogenic or exogenic origin.

Voyager 2 switched its attention to Hyperion on 23 August. The distant imagery by its predecessor had determined only that this small moon is irregularly shaped. The first image, from 1.2 million kilometres, showed an oblong shape with squared-off edges, reminiscent of a brick. Surprisingly, the primary axis was not aimed at the planet, implying it did not rotate synchronously. An image from 700,000 kilometres from a different perspective revealed it to be roughly 360 by 280 by 236 kilometres, like a fat hamburger. The final image, taken on 24 August from 500,000 kilometres, showed that the surface is cratered. An arcuate ridge suggestive of an impact crater comparable in size to the moon's longest dimension hinted that Hyperion might be a fragment of a much larger body.

Voyager 2 took these images of Hyperion as it closed in from 1.2 million kilometres (left) to 500,000 kilometres (right) showing that it is non-spherical. The prominent arcuate ridge, Bond–Lassell Dorsum, was named for the moon's co-discoverers. The ridge was suggestive of a crater comparable in size to the moon's longest dimension, and hinted that Hyperion might be a fragment of a much larger body.

Voyager 2's particles and fields instruments showed the solar wind as strong and gusty, so the bow shock was not expected until near, or maybe even inside, Titan's orbit. After being first sensed early on 24 August at a range of 32 planetary radii, the shock washed back and forth several times during the day, and the spacecraft did not finally meet the magnetopause until midnight, at 18.6 radii, just inside Titan's orbit, which confirmed that when the wind is strong the moon is exposed while on the sunward side of its orbit. Titan was not a priority for Voyager 2, and it approached no closer than 663,400 kilometres – a distance 150 times that of its predecessor. Nevertheless, this was an opportunity for remote-sensing of the north polar region. The Photopolarimeter on Voyager 1 had been crippled during the deep penetration of the Jovian magnetosphere, but Voyager 2 had kept its distance and its instrument had fared better. Despite being partially disabled, it made the surprising discovery that the reflected insolation from Titan is highly polarised, which provided valuable insight into the size of the aerosol particulates in the upper haze layer.

As Voyager 2 finally entered the magnetosphere in the early hours of 25 August, and closed to within 1 million kilometres of Saturn, it initiated its hectic 'near encounter' schedule. Images were streaming into JPL every 3 minutes. As the time for the next one to appear approached, everyone not otherwise engaged turned to the nearest TV monitor. At 04:00 Voyager 2 returned a series of images of Enceladus's intriguingly smooth surface from a range of 800,000 kilometres, then again at 06:00, by which time the range had reduced to 600,000 kilometres. At 09:00 it snapped the co-orbital moonlets of Tethys and Dione, then documented the 'F' ring which, surprisingly, was now regular and free of clumping and braiding. At noon, it returned to Enceladus, then made a start on Tethys. The scan platform was slewing from target to target at its fastest rate of 1 degree per second. At times when the remote-sensing instruments were inactive, the spacecraft rotated to enable the particles and fields instruments to sense the magnetosphere most efficiently.

On the final approach, the imaging ceased and the scan platform fixated upon the star delta Scorpii as this passed behind the rings so that the Photopolarimeter could take measurements every 10 milliseconds.[113,114] Monitoring started at 18:18, some 22 minutes prior to the star's contact with the inner edge of the 'C' ring, and ran until 20:40, at which time it cleared the 'F' ring. With such a close-in vantage point, the angular resolution on the sample was much greater than can be achieved from Earth, and the light curve was not complicated by the twinkling effect of the signal passing through the Earth's atmosphere. With a linear resolution of 100 metres, this gave the best profile of the opacity across the rings to date. Instead of the light passing unhindered through a series of gaps, as had been expected, the ceaseless 'flickering' indicated that, despite it being only 100 metres thick, there is very little 'empty' space within the main ring system, even where the highest-resolution imagery seemed to show gaps. The real shocker, however, was that the material extends beyond the 'A' ring. This made Pioneer 11's transit at 2.87 radii all the more remarkable, and elevated concern over whether Voyager 2 would survive its attempt to fly this 'proven' route. In fact, concern was already running high, because in documenting the broad but tenuous 'E' ring Voyager 1 had spotted a section of a narrower ring, later designated the 'G' ring, at 2.83 radii. A re-examination of Pioneer 11's particles and fields data uncovered a hint of it. However, with the Grand Tour reliant upon this ring-plane crossing, there was little option but to trust to luck. The 'window' for the Grand Tour had already closed, and would not reopen for an century and a half, so for this most privileged of generations of planetary scientists, with their only real prospect of reconnoitring the outer Solar System in the fickle hands of Lady Luck, the tension was electric.

At its closest point to Saturn, Voyager 1 had been behind the ring system and had thus inspected the non-illuminated face, but as Voyager 2's ring-plane crossing would be after closest approach, it was able to record the illuminated face at an increasingly 'open' perspective. "The images of the rings we receive today will be of much higher resolution than anything we've ever seen before," E.C. Stone promised. The best Voyager 1 imagery had suggested that there might be as many as a thousand individual ringlets, but the new imagery revealed that there are actually *tens of thousands*. It seemed that every time the resolution was increased, every ringlet was found to incorporate many finer ringlets. "This is about as far as imaging can take us," B.A. Smith informed the reporters, referring to the 10-kilometre-per-pixel resolution. Only the Photopolarimeter's stellar occultation light curve could identify finer detail. "We have a superb set of ring data, there is no question about it," assured A.L. Lane, that instrument's leader. It had to be recognised, however, that this data pertained only to the narrowest of lines across the system. It would not be easy to infer whether a specific 'dip' in the light curve was due to the starlight being occulted by material in a narrow ringlet or by a single object. With the 'F' ring now looming, a very-high-resolution image was secured from a range of 50,000 kilometres which showed that this comprised one bright strand, two fainter ones and one diffuse band. The brightest strand was shown by the stellar occultation to be

Voyager 2 took this shot of the 'F' ring from 50,000 kilometres with a resolution of a few kilometres per pixel (the best ever) shortly after crossing the ring plane. It shows the bright ringlet and three fainter strands, but no 'braiding'. The light curve from the stellar occultation established that the main strand is itself composed of a dozen fine ringlets.

composed of a dozen finer ringlets, each only a hundred metres wide.[115] Was there no end to the ever-finer subdivisions?

At 21:50 on 26 August, Voyager 2 passed 101,300 kilometres above Saturn's cloud tops. Fifteen minutes later, in preparation for passing behind the planet's trailing limb at 22:26, it boosted its signal strength and ceased to modulate it. The signal took about a minute to fade, and the manner in which it did so provided information on the atmosphere. Although out of sight, the spacecraft was not idle; it pursued its programmed observations, saving the data on magnetic tape for later replay.

Voyager 2 was to cross the ring plane at 22:44 at 2.86 planetary radii, some 32,000 kilometres beyond the 'F' ring. "I think we're all confident," E.C. Stone told the journalists, alluding to the fact that the crossing point was just 3,000 kilometres beyond the only vaguely defined boundary of the 'G' ring. There was wild cheering in the von Kármán Auditorium when the signal was re-acquired at midnight. The celebration was pre-empted, however, by the shocking realisation that the computer had ceased scan platform operations. Instead of returning home for a few hours of well-deserved sleep, the engineers remained to find out what had gone wrong.

When the Plasma Wave Spectrometer data was transmitted, this revealed that the ring-plane crossing had not been plain sailing. For several minutes the spacecraft had been immersed in a cloud of plasma, and several times fired its thrusters to correct a drift in its orientation. An analysis suggested that the plasma had been generated by impacts with many thousands of micron-sized dust grains striking the spacecraft at a relative velocity of 10 kilometres per second, and been instantly vaporised.

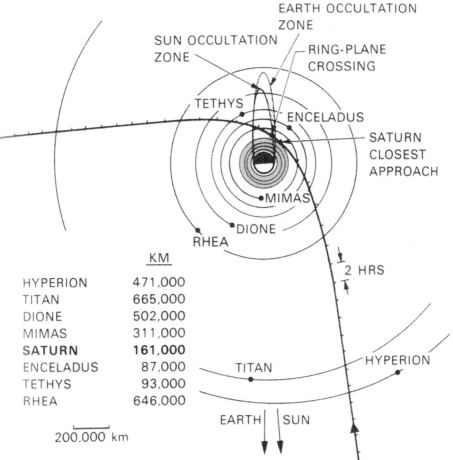

	KM
HYPERION	471,000
TITAN	665,000
DIONE	502,000
MIMAS	311,000
SATURN	**161,000**
ENCELADUS	87,000
TETHYS	93,000
RHEA	646,000

Voyager 2's route through the Saturnian system.

Immediately thereafter, the spacecraft was to document the non-illuminated side of the ring system and secure close-in mosaics of the terminators of Enceladus and Tethys. Unfortunately, as the scan platform was slewed, its azimuth actuator began to fail, and at 23:40 it seized, so this section of the encounter sequence was lost. The fact that the spacecraft had been in Saturn's shadow at the time prompted some suspicion that the gear train's lubricant may have frozen, so a command was sent to activate a heater, but to no avail. In fact, it took almost three days of thermal control to coax the mechanism back to life. After a detailed analysis, the engineers concluded that the sheer intensity of the slewing had driven the lubricant from the gear train, and it was not released until the heat enabled the components to separate sufficiently for the lubricant to re-enter the mechanism. While some data had been lost, E.C. Stone assured the disappointed reporters that science was measured in terms of *increased understanding* rather than the number of data bits, and when considered in this way Voyager 2 had achieved "200 per cent" of its science programme.

"We found so many things to do with Voyager 2 after the Voyager 1 fly-by; that may have led to the scan-platform problem, because we had the thing swinging back and forth in the sky so fast and so furiously that we think that's what drove the lubricant out of the gears and caused the gear train to seize," reflected E.D. Miner of JPL, Voyager's assistant project scientist.

Voyager 1 had shown little *detail* on Enceladus, but it had appeared tantalisingly smooth. Voyager 2's trajectory enabled it to document the northern hemisphere at a resolution of a few kilometres per pixel. Imagery from 120,000 kilometres shortly before the ring-plane crossing showed a section of grooved plain transecting the cratered terrain, a landform strongly reminiscent of the sulci on Ganymede. The hint of geological activity on Dione had stirred up interest, but Enceladus was amazing. "The oldest terrains on Enceladus are similar in crater density to the least cratered plains of Dione," L.A. Soderblom told reporters. The sparse cratering of the smooth plains meant that they are even younger, perhaps forming within the last 100 million years, or as Soderblom expressed it, "at a time when dinosaurs roamed the Earth". There was evidence of *repeated* episodes of fluid flooding across its surface. Even with the possibility of cryovolcanic activity having been predicted by analogy with Io,[116] and with the broad hints provided by Voyager 1, E.M. Shoemaker joyfully admitted that Enceladus had turned out to be "far beyond my wildest expectations".

Voyager 2 dramatically improved on its predecessor's coverage of Tethys. There are areas that are slightly darker than the norm. The 'strange' feature on the trailing hemisphere is indeed a crater (named Penelope). Its western rim is juxtaposed with a dark band and so, despite the full illumination, had given the appearance of shadow. As the range reduced and the resolution improved, the feature which had previously been the focus of attention paled into insignificance as, firstly, a 400-kilometre-diameter crater (Odysseus) was revealed on the leading hemisphere, one so vast in proportion that its floor has 'risen' to match the moon's radius of curvature, and then the moon was found to be girdled by a system of canyons (Ithaca Chasma). The fact that these canyons trace out a 'great circle', one of whose 'poles' coincides with

As Voyager 2 closed in on Enceladus from 440,000 to 268,000 kilometres it took a series of pictures in which the perspective shifts from the anti-Saturn to the trailing hemisphere and north migrates from the upper right to the right, with the north pole on the terminator.

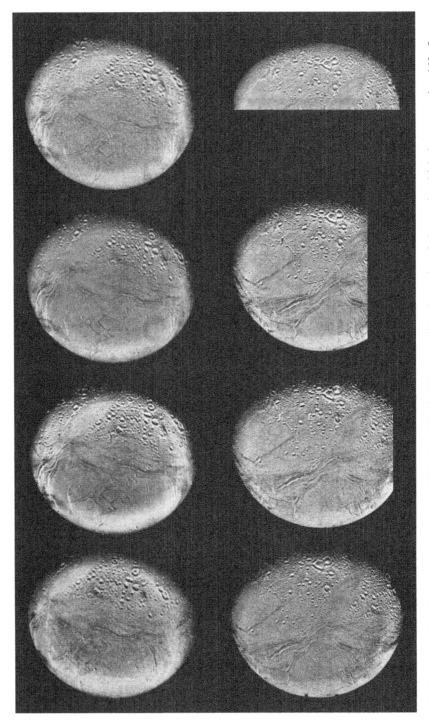

As Voyager 2 closed in on Enceladus from 240,000 to 108,000 kilometres it took a series of pictures in which the perspective shifts from the trailing to the Saturn-facing hemisphere and north migrates from the right to lower right, with the north pole on the terminator in Samarkand Sulcus.

As it closed between 1.5 million and 800,000 kilometres, Voyager 2 took a series of three images of Tethys showing the vast crater Odysseus whose central peak lies at 130 degrees west. Later, with the crater now beyond the terminator, it recorded the shifting perspective from the anti-Saturn hemisphere at 680,000 kilometres, around the trailing hemisphere onto the Saturn-facing hemisphere at 282,000 kilometres. In each case North is towards the top. The later sequence revealed the global extent of Ithaca Chasma. The prominent crater Penelope on the trailing hemisphere is the strange albedo feature seen by Voyager 1.

the vast crater, suggested that they were opened by the shock of that impact. Tethys has a bright, heavily cratered terrain and a less intensely cratered, somewhat darker, plain which could be the result of geological activity. Unfortunately, some 30,000 kilometres short of the 120,000-kilometre point of closest approach – at which time the best terminator view would have been afforded – the imaging sequence was pre-empted by the fault in the scan platform, and the best view was found to be that from 600,000 kilometres.

The scan platform was coaxed back to life on 27 August. However, the departure trajectory was at a fairly steep angle to the system's equatorial plane, so once it was through the ring plane it rapidly left the moons behind. Only distant Phoebe, whose orbit is more or less in the plane of the ecliptic, was favourably presented, and as the 2.2-million-kilometre fly-by on 4 September would not require the scan platform to slew in its high rate this unusual object was investigated. It was found to be more or less spherical and about 220 kilometres in diameter, with a very dark cratered surface showing a few relatively bright patches. Although at that time no main belt asteroids had been inspected by spacecraft to provide a basis for comparison, it seemed likely from spectroscopic considerations that it is similar to the reddish asteroids whose spectra suggest that they are 'primitive' objects made of condensates left over from the solar nebula – either that or a giant cometary nucleus – and hence that it was indeed captured by Saturn.

Three views of Phoebe taken by Voyager 2 on 4 September 1981 as it flew past at a range of 2.2 million kilometres. It is spheroidal and about 220 kilometres in diameter. At 6 per cent, its albedo is darker than the icy moons orbiting closer in. The 'bright' patches may be the sites of impacts. Because it is in an elliptical 550-day retrograde orbit, its rotation has not been captured; the imagery indicated that its period is 9 to 10 hours.

PLANETS BEYOND

The Saturnian slingshot put Voyager 2 on course for Uranus, as the first step on its Grand Tour of the outer Solar System. Up to this point it had been serving in a backup capacity, but now it took the lead in exploring beyond the farthest planet, as Saturn had been when Galileo Galilei had first aimed his primitive telescope into the sky in 1609. C.E. Kohlhase, in charge of mission planning, estimated a 65 per cent chance of the spacecraft surviving to reach Uranus.

MAPS

A Lambert azimuthal equal-area projection of Mimas showing its anti-Saturn (left) and Saturn-facing (right) hemispheres.

A Mercator projection of the equatorial zone of Mimas. (USGS map I-1489.)

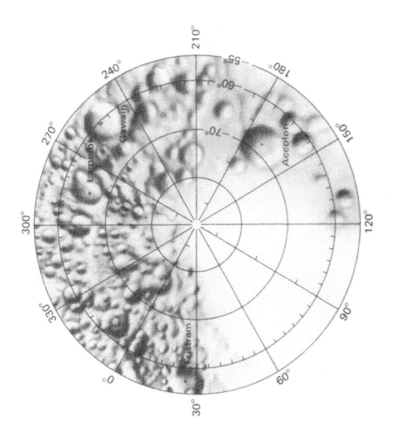

A polar stereographic projection of the south polar region of Mimas.

A Lambert azimuthal equal-area projection of Enceladus showing its anti-Saturn (left) and Saturn-facing (right) hemispheres.

A Mercator projection of the equatorial zone of Enceladus. (USGS map I-1485.)

A polar stereographic projection of the north polar region of Enceladus.

A Lambert azimuthal equal-area projection of Tethys showing its anti-Saturn (left) and Saturn-facing (right) hemispheres.

A Mercator projection of the equatorial zone of Tethys. (USGS map I-1487.)

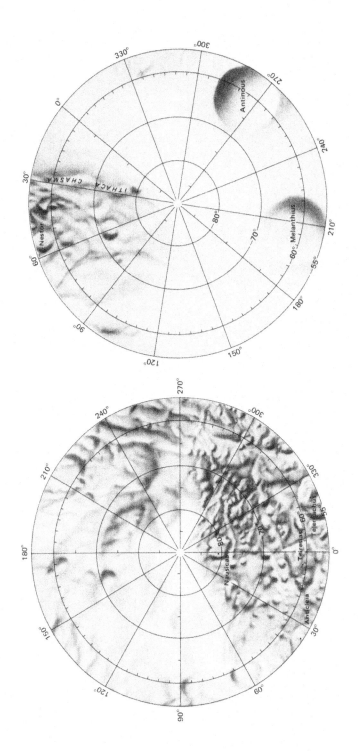

A polar stereographic projection of the north (left) and south (right) polar regions of Tethys.

A Lambert azimuthal equal-area projection of Dione showing its anti-Saturn (left) and Saturn-facing (right) hemispheres.

A Mercator projection of the equatorial zone of Dione. (USGS map I-1488.)

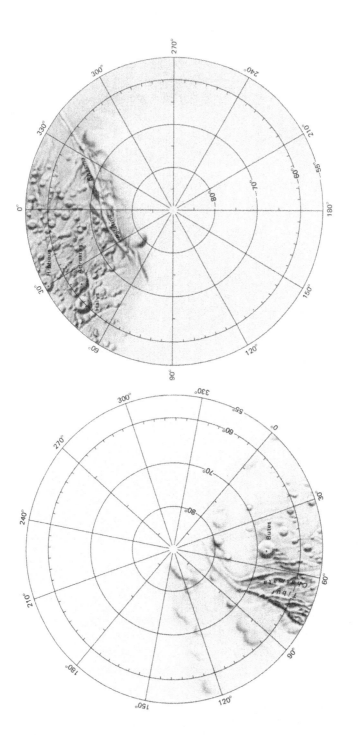

A polar stereographic projection of the north (left) and south (right) polar regions of Dione.

A Lambert azimuthal equal-area projection of **Rhea** showing its anti-Saturn (left) and Saturn-facing (right) hemispheres.

A Mercator projection of the equatorial zone of Rhea. (USGS map I-1484.)

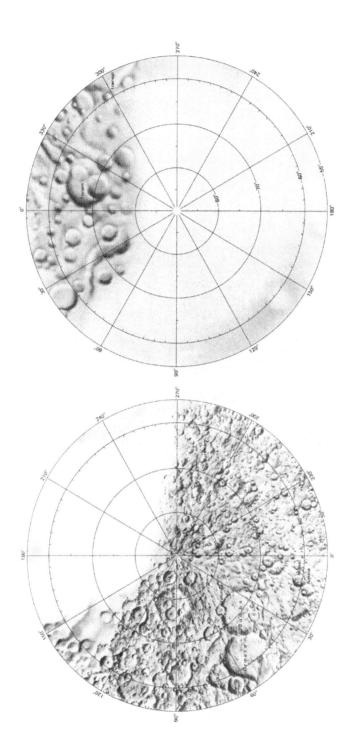

A polar stereographic projection of the north (left) and south (right) polar regions of Rhea.

A Lambert azimuthal equal-area projection of Iapetus showing its anti-Saturn (left) and Saturn-facing (right) hemispheres.

A Mercator projection of the equatorial zone of Iapetus. (USGS map I-1486.)

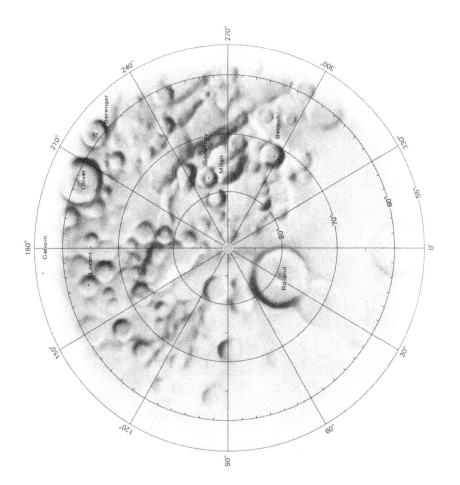

A polar stereographic projection of the north polar region of Iapetus.

4

The Titans

A NOMENCLATURE

After five Saturnian satellites had been discovered, a scheme was introduced in which they were referred to numerically in accordance with their distance from the planet. This was satisfactory until William Herschel discovered two additional satellites orbiting closer in. In 1847, John Herschel, who had just returned from South Africa where he had made a thorough study of the southern sky to extend his father's study in the north, decided to rectify the situation by naming Saturn's moons.

In mythology, Uranus had sired Saturn who, after siring Jupiter, devoured his subsequent offspring. Herschel therefore decided to name the satellites after Saturn's siblings, a group known as the Titans. Since the satellite discovered by Huygens was the largest, he named it Titan. Then, working inwards, he assigned the next three to Rhea, Dione and Tethys, sisters of Saturn. In order to protect them from Saturn, he interposed Mimas and Enceladus, two giants "from a younger and inferior (though still superhuman) brood". The enigmatic outermost satellite was named Iapetus. Fortunately, the clan of Titans was sufficiently populous to be able to accommodate subsequent discoveries.

GENERAL CONSIDERATIONS

Using masses of the Saturnian moons derived from a study of their orbits, and an estimate of Titan's diameter, in 1931 Georg Struve estimated the sizes of the others by assuming that they all had the same albedo. Given the masses and the sizes of the moons, he was able to estimate their densities (Table 4.1).

Not surprisingly, however, Struve's assumption that all the satellites had similar albedos was eventually proved false. As spectroscopic studies identified water frost on their surfaces,[1,2,3] they were evidently more reflective than Struve had assumed, and he had overestimated their diameters and underestimated their densities. With improved infrared-sensing, their visual reflection and thermal emission spectra could

Table 4.1   Satellite diameters and densities (pre- and post-Voyager*)

Satellite	Diameter (km)		Mass (kg)		Density (g/cm^3)		Albedo
	Pre	Post	Pre	Post	Pre	Post	
Mimas	640	398	3.5×10^{19}	3.76×10^{19}	0.4	1.44	0.7
Enceladus	800	498	1.4×10^{20}	7.4×10^{19}	0.4	1.16	0.95
Tethys	1,280	1,046	6.2×10^{20}	6.2×10^{20}	0.6	1.21	0.8
Dione	1,200	1,120	1.1×10^{21}	1.05×10^{21}	1.3	1.43	0.5
Rhea	1,760	1,528	2.3×10^{21}	2.3×10^{21}	1.0	1.33	0.6
Titan	4,160	5,150	1.4×10^{23}	1.36×10^{23}	3.3	1.88	0.2
Hyperion[†]	500	–	1.1×10^{20}	1.1×10^{20}	4.0	–	0.3
Iapetus[‡]	1,584	1,436	5.7×10^{21}	1.9×10^{21}	2.6	1.16	–

*   Pre-Voyager data from W. Sandner,[4] using K.H. Struve's mass estimates and his son Georg's densities. Post-Voyager data from Moore and Hunt,[5] and Morrison et al.[6]
†   Hyperion is irregularly shaped, measuring $360 \times 280 \times 236$ kilometres, and its density had not yet been reliably determined.
‡   Iapetus's albedo varies depending upon its orbital position; one side is between 40 and 50 per cent and the other side is between 2 and 5 per cent.

be interpreted on the basis of a theoretical model of their surface characteristics to calculate their diameters.[7,8] When the Saturnian system was occulted by the Moon in 1974, high-speed photometry enabled their diameters to be directly measured.[9] In fact, Titan is the least reflective of the retinue. The others are relatively small bodies whose densities indicate that they are predominantly composed of ices, and whose high reflectivities show that their surfaces are ice with varying admixtures of a dark non-ice component, probably dust.[10,11] With their diameters accurately measured from Voyager imagery, and their masses refined by how they influenced the trajectories of spacecraft,[12] it was possible to refine their bulk densities. With densities in the range 1.2 to 1.4 g/cm^3, Mimas, Enceladus, Tethys, Dione, Rhea and Iapetus are 60 to 70 per cent ices. Apart from the anomalously dark hemisphere of Iapetus, their albedos are in excess of 40 per cent. Excluding Titan, which is unique, there is a trend of increasing size with distance from Saturn, but there is no trend of decreasing density corresponding to that of the Galilean satellites of Jupiter. Computer models suggest that there was a compositional trend in the coalescing disk of gravitationally bound nebula, with water ice close in to the planet (Mimas and Enceladus), then ammonia monohydrate (Tethys and Dione) and possibly a mixture of clathrate of methane and argon in the outer reaches (Rhea, Titan and Iapetus).

Comparing the pre- and post-Voyager data is illuminating. Firstly, the mass estimates stood up quite well, confirming the analytical power of Newton's law of Universal Gravitation. Enceladus has a remarkable albedo. In fact, it is the most reflective body in the Solar System – far exceeding even brilliant cloud enshrouded Venus. The density of 'dark' Titan is somewhat lower than predicted, placing it in the same category as Jupiter's Ganymede and Callisto. Iapetus's 'intermediate' density was found to be illusory, showing clearly that no matter how much a planetary system is studied telescopically, there is no substitute for sending a probe to take a closer look. Voyager 1 gave close views of Mimas, Dione and Rhea, and Voyager 2 was able to investigate Iapetus, Hyperion, Enceladus and Tethys. The best imagery had a resolution of a few kilometres per pixel, but inferences on geological histories derived from comparisons of their surfaces are necessarily tentative.[13,14,15,16] Considered together, the cratering of the icy moons indicates that they have been subjected to (at least) two quite distinct impactor populations. In general, craters with diameters exceeding about 100 kilometres are ancient, evidently dating to the end of the post-accretional bombardment.[17,18] The heavily cratered terrains of Rhea, Dione and Tethys, and the bright hemisphere of Iapetus, seem to date similarly. However, the plains regions of Dione and Tethys have been disfigured by a population of younger craters with diameters less than about 20 kilometres, and most of Mimas appears to be of this population. The resurfacing of Enceladus has been so comprehensive that the post-accretional craters have been totally erased. The post-accretional impacts occurred mainly as the newly formed accretions swept up the residual material of the gravitationally bound part of the solar nebula. The later population may derive from the fragmentation of satellites by planetesimals that strayed into the system. As a result of gravitational focusing, the disruption would have been greatest in the system's inner region, and the innermost moons may have been shattered and re-accreted several times before the environment was finally cleared. There are orbital resonances involving Mimas and Tethys, Dione and Enceladus, Titan and Hyperion.[19] The interlocking resonances could only have been formed after a lengthy period during which the moons would have inflicted pulses of tidal stress upon one another.[20] The expectation was that the larger satellites would have undergone thermal differentiation, but the interiors of the smaller ones would be homogeneous.[21] However, the chilly state of the solar nebula from which the retinue condensed suggested that the ice may have had a significant fraction of ammonia, and even for an object of such diminutive size a modest amount of radiogenically produced heat might have stimulated significant internal melting, possibly prompting surface activity.

MIMAS

As measured from Saturn's centre, the outer edge of the main ring system is 2.3 planetary radii, the slender 'G' ring lies at 2.9 and Mimas is just beyond, at 3.1, which is sufficiently far outside the Roche radius to be safe from tidal disruption. At 390 kilometres in diameter, it is the smallest of the historically known satellites. Its rotation is synchronous. One study of gravitational focusing suggested that Mimas should be the most heavily cratered member of Saturn's retinue, with a preference for strikes on its leading hemisphere.[22]

Voyager 1 was able to view most of the side that faces away from the planet and

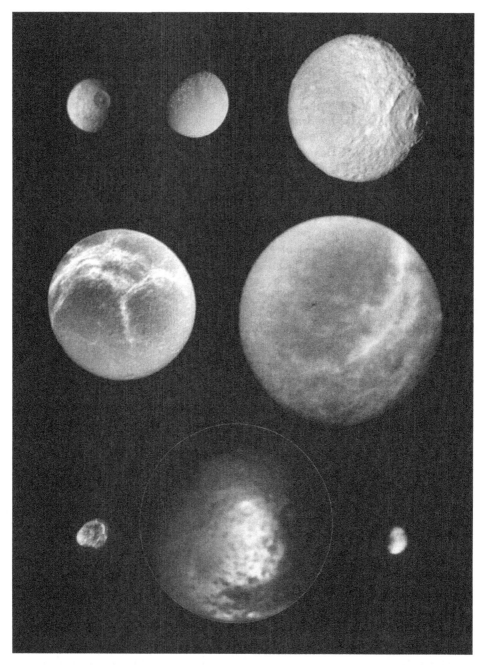

The Voyagers provided our first views of the major Saturnian satellites. In order of distance from the planet, and to scale, they are Mimas (top left), Enceladus, Tethys, Dione, Rhea, Hyperion, Iapetus and Phoebe. Notice that Mimas would fit into the giant crater Odysseus on Tethys.

Two views of Mimas, showing (right) the crater Herschel on the leading hemisphere and
(left) the south polar region with the pole on the terminator. North is towards the top in
each case. Notice the curvilinear feature crossing the southern hemisphere.

much of the southern part of the near side with a resolution of a few kilometres per
pixel. It certainly lived up to the stereotype of an ancient battered icy moon, in that it
is heavily cratered. The cratering is not uniform, though. Most of the craters form
deep cavities. The larger craters tend to have central peaks, and their ejecta blankets
are indistinct, which suggests that on a small body with weak gravity the material
was distributed far and wide. The most prominent crater, appropriately named
Herschel, is on the leading hemisphere. At about 130 kilometres across, it spans fully
one-third of Mimas's diameter. Its wall rises about 5 kilometres from the
surrounding terrain, its floor is depressed 10 kilometres, and its well-defined central
peak rises 6 kilometres. It is remarkable, therefore, that the moon survived the
impact. Mimas's cratered surface has been transected by several prominent chasms
up to 10 kilometres across, several kilometres in depth and 100 kilometres in length,

Saturn's moonlets as seen by Voyager 2. They are shown (working left to right) in order
of distance from the planet and to scale. Ranging from a few dozen to a few hundred
kilometres in size, the irregularly-shaped moonlets are either loose accretions of ice or
fragments of bodies which were shattered by collisions.

whose origin may be related to the Herschel impact.[23] In terms of the impactor populations, none of its craters, not even Herschel – which is large only in proportion – dates back to the post-accretional bombardment. This suggests that the impacts by the interlopers were devastating. E.M. Shoemaker concluded that Mimas must have been shattered and re-accreted several times.[24] The inner moonlets have albedos and spectra that resemble those on Mimas, so they would seem to be water ice, and their irregular shapes suggest that they are either loose accretions of ice or fragments from shattered bodies. A few of them might even be fragments of an earlier 'Mimas' that failed to re-accrete. The presence of craters on some of these inner moonlets indicates that their surfaces have been exposed for at least several billion years.

ENCELADUS

Although Voyager 1 did not approach Enceladus closely, it confirmed the telescopic inference that the moon has a *very* highly reflective surface. In fact, with a geometric albedo of 95 per cent, it is more reflective than a field of fresh snow. Voyager 2 was able to provide imagery of the northern portion of the trailing hemisphere at a resolution of several kilometres per pixel.

At 500 kilometres in diameter, Enceladus is only slightly larger than Mimas, and it had been thought that as they were both deep in the Saturnian system they would be physically similar. However, Enceladus was found to have undergone extensive resurfacing. The absence of large craters indicates that none of the surface dates back to the post-accretional bombardment.[25] A physiographic analysis identified cratered, smooth and ridged plains. The cratered plains have an abundance of craters in the 10- to 20-kilometre size range. However, in some areas these craters are sharply defined and in other areas they are softened, possibly due to viscous flow. The degree of relaxation of the craters differs markedly on the various terrains, suggesting differences in viscosity due to their composition and varying heat flow. Studies of the crater forms concluded that the moon's lithosphere is a mixture of ammonia ice and water ice.[26,27]

In large areas the early craters have been 'replaced' by plains whose sparsity of impacts implies either a very low cratering rate or the relative youth of the surface. In some places, these smooth plains are criss-crossed by rectilinear fractures which appeared to be grabens formed as subsurface ice froze, expanded and cracked the brittle outer shell. There are also ridges rising a kilometre or so, with smooth plains lying between. These ridged plains, which predominate on the trailing hemisphere, could be fluids that oozed from fractures and solidified in place. In fact, the extent of the resurfacing and tectonic activity on this small moon is surprising. One analysis concluded that Enceladus would rapidly have lost its accretional heat (due to its high ratio of surface area to volume) and it is probably too small to have facilitated significant radiogenic heating. It must therefore have endured an exogenic mode of heating at a later stage.[28] Indeed, it appears to have undergone *several episodes* of resurfacing over an extended period by lava consisting of a water–ammonia eutectic

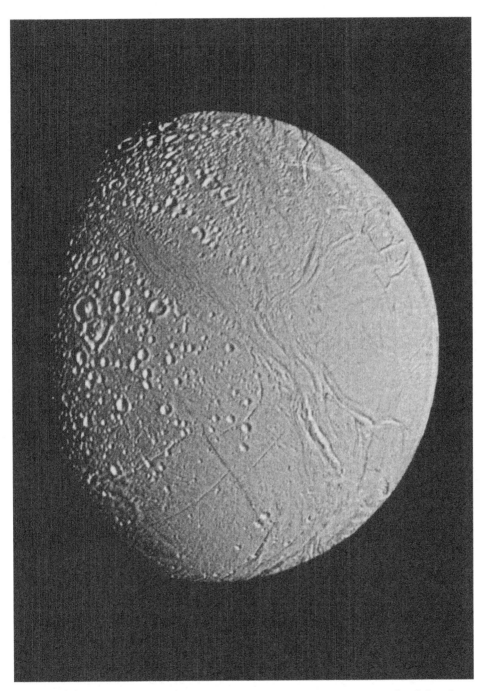

The remote Voyager 1 imagery suggested that Enceladus had been resurfaced, but the variety of activity evident in this Voyager 2 view was astonishing.

that would melt at around 173K and, being of a lower density, would tend to rise through the water ice to flood the surface. The most plausible cause of heating is tidal stress.[29] The moon's free eccentricity is very small, indicating that it must have suffered significant tidal dissipation in the past. Although the eccentricities of the current orbits of the moons are too small to induce significant tidal stress, they were probably greater in the past, as the satellites settled into their present locations. One suggestion is that since the co-orbital pair of Janus and Epimetheus are in an almost 2:1 resonance with Enceladus, if these bodies are pieces of a progenitor that was in the same orbit prior to being shattered by a catastrophic impact, the resonance with *that* object might have driven much of Enceladus's activity.[30] However, if the substantial cratering observed on these moonlets indicates that they date back to the end of the post-accretional bombardment,[31] then this could not have driven the most recent phase of activity, which would appear to have occurred at some time within the last billion years. As an alternative, it has been suggested that the most intense activity was driven by a temporary resonance with Mimas as that moon migrated outwards due to its interaction with the ring system.[32] Enceladus is currently in a 2:1 resonance with Dione, farther out, and this will provide an ongoing source of heat.[33] While this may be *just* sufficient to melt the interior if it still contains a significant fraction of either methane clathrate or ammonia monohydrate, it was not clear whether this would be able to drive cryovolcanism.

The densest part of the tenuous 'E' ring is coincident with Enceladus's orbit.[34] It is thought that the moon's surface is highly reflective because it is coated with a layer of fresh ice crystals. The 'E' ring would appear to be composed of very fine ice particles.[35] The sparse cratering on Enceladus argues against this material having

An enlargement of Samarkand Sulcus cutting across the cratered terrain on Enceladus. Note that the resurfacing process breached the craters Ahmad and Peri-Banu at its margin.

An artist's impression of the surface of Enceladus, depicting a hypothetical geyser venting water vapour into space, where it replenishes the ice crystals in the 'E' ring. (Courtesy of David Seal.)

been ejected by large impacts. It may be being ejected by micrometeoroid impacts, but if so why do the other icy moons not form rings? The most intriguing possibility is that water is being explosively vented by cryogenic geysers to freeze into a myriad of ice crystals, many of which fall back to coat the surface, but some of which are able to escape to space.[36] Without replenishment, the 'E' ring would long since have faded away. Its continued presence is therefore strong evidence that Enceladus is undergoing geyser activity.

TETHYS

At 1,060 kilometres in diameter, Tethys is twice the size of Enceladus, but its bulk density is comparable. Its surface is uniformly bright, but it is not quite as reflective as Enceladus. It received fairly comprehensive imaging coverage, with only a section of the southern hemisphere being missed. A physiographic study identified several terrains.[37] The oldest region is a hilly cratered terrain that is characterised by rugged topography. It is densely cratered, but most of the larger craters are degraded. They would appear to date from the post-accretional bombardment. As with Mimas, there is a vast impact crater on the leading hemisphere. However, while Odysseus, at 440 kilometres in diameter, is in similar proportion to Herschel, the fact that the moon is more than twice the diameter means that Mimas itself would fit comfortably within Odysseus's rim. Although Odysseus is large, it is nevertheless very shallow, and its floor actually follows the curvature of the moon. This may be the result of isostatic

On a body the size of Tethys, the 440-kilometre-diameter crater Odysseus is better described as a 'basin'. Over time, its floor has risen to restore the moon's spherical surface. Ithaca Chasma follows a 'great circle' corresponding to a 'pole' centered on Odysseus. The view (left) shows the region antipodal to that massive impact.

adjustment of the icy lithosphere soon after the crater was formed. The rim has been softened by a series of terraces and the central peak has degraded into a ring-like structure. The antipodal area could reasonably be expected to have been torn up by the focusing of seismic shock waves, but it is *less* rugged, forming a plain displaying faint lineations and albedo variations. It is possible that the eruption of material that formed the plain was prompted by the impact.[38] Any antipodal relationship is speculative, however.

Tethys's next most spectacular feature is Ithaca Chasma, a vast canyonland that winds almost all the way around the globe. This narrow branching terraced trough system is at least 1,000 kilometres long, 100 kilometres wide, and up to 3 kilometres deep.[39] Sections of its rim are raised 500 metres above the adjacent terrain. While the canyon system may be related to the Odysseus impact, it does not actually encroach upon the crater. In fact, if the crater's central peak is considered to be a 'pole', the chasm tends to trace the great circle of the associated 'equator'. It has been suggested that the Odysseus impact occurred when the moon's interior was still 'soft' and the force of the impact temporarily deformed the satellite into an ellipsoidal shape, inducing transient tensional stresses sufficient to crack the surface almost all the way around.[40] Because the chasm is so wide, it is possible to compare the cratering inside it with that on the adjacent terrain.[41] This showed that the chasm's formation was contemporary with the post-accretional bombardment, and may indeed have been formed as a result of the Odysseus impact.

The objects in the Lagrangian points of the inner moons may be fragments from 'parent' bodies that were locked in position as their shattered parents re-accreted most of their material. If so, then Dione and Tethys, which have these companions, must have been totally disrupted. However, since some areas of these moons seem to date to the post-accretional bombardment, this would imply that the companions, if they were created in this way, must have been present since that time.[42]

DIONE

In terms of size, Dione is the near-twin of Tethys. A bulk density of 1.43 g/cm^3 and an overall albedo of about 50 per cent indicated a predominantly icy body with an exposed icy surface. Telescopic studies noted a 0.6-magnitude variation around its orbit.[43] The Voyager imagery revealed why: there is dark mottling with an albedo of about 20 per cent on the trailing hemisphere, with a pattern of bright streaks with 70 per cent albedo superimposed upon it making a striking contrast.[44,45]

A physiographic study identified several associated terrain types.[46] Predominant is a rugged 'highland' terrain with many craters up to 100 kilometres in diameter. In general, these craters are shallower than on Tethys. Most of the larger craters have terraced walls and central peaks. Scarps up to 100 kilometres in length run over this terrain. In addition to a less rugged cratered terrain, defined as cratered plains, there are also smooth plains. There are some troughs on the smooth plains, and although these are typically less than 100 kilometres in length, a few exceed 500 kilometres, and in most cases there is a 'pit' at either end. In fact, there may be a global network of fractures. The largest craters – about 200 kilometres across – are on the trailing hemisphere; however, Dione has no crater to match Tethys's Odysseus. Evidently, early resurfacing erased the record of the post-accretional bombardment and the extant cratering derives from the later population of impactors. Dione's relatively high density indicates that it contains a significant proportion of rock, so radiogenic heating would have kept its interior warm. It has been suggested that after the formation of the brittle lithosphere, internal heat prompted fluid ammonia–water ice to erupt from fractures to form the plains.[47] Indeed, a series of ridges that rise only a few hundred metres but extend up to 100 kilometres in length may be the fronts of low-viscosity flows. The fact that both Enceladus and Dione, whose orbits are in mutual resonance, have undergone sufficient internal melting to prompt surface flows is evidence for tidal stresses being the main heating agent. In the middle of the radiating pattern of broad wispy streaks on the trailing hemisphere there is an elliptical feature (named Amata) several hundred kilometres in diameter with a dark central patch which might mark an impact, but the streaks are *not* rays of bright ejecta splashed out from it. In higher resolution imagery, narrow linear troughs and ridges are evident emerging from some of the bright streaks, projecting around onto the leading hemisphere.[48] The fact that the streaks seem to be associated with lineaments hints they might be pyroclastic deposits of a clathrate that was warmed at shallow depth by a pocket of radioactive elements and vented under pressure through fissures.[49] However, volcanic models are problematic due to an absence of expected visible evidence, such as overwhelmed or flooded craters. Nevertheless, Dione has undergone significant endogenic activity. As for the dark mottling, the 'middle' icy moons orbit in the most intense region of the rapidly rotating magnetosphere, and their trailing hemispheres are irradiated as they synchronously

The largest crater in this Voyager mosaic of part of the Saturn-facing hemisphere of Dione is Aeneas, and the prominent curvilinear feature near the terminator is Latium Chasma. The brighter feature towards the south pole is Palatine Chasma. Notice the extremities of the wispy streaks radiating over the limb from the trailing hemisphere.

rotate. Perhaps the dark material is a modification of the surficial material by this plasma sputtering out ions and inducing chemical reactions.[50]

RHEA

At 1,530 kilometres across, Rhea is second in size among Saturn's moons only to Titan, and it is the near-twin of Iapetus. The bulk density of 1.34 g/cm^3 and albedo of about 60 per cent implied a predominantly icy body with an exposed icy surface. As Voyager 1 flew within 60,000 kilometres of Rhea, it secured excellent imagery of its north polar zone. As in the case of Dione, telescopic monitoring had shown that Rhea's trailing hemisphere was slightly darker, there being a 0.2-magnitude variation across its orbit,[51] and the spacecraft revealed this to be due to dark mottling with a superimposed pattern of bright streaks, similar to Dione's, but not as prominent. They seem to be uncorrelated with cratering, and there is no radial pattern to suggest that they are rays of ejecta. The dark mottling upon which they are superimposed may be due to irradiation by the magnetospheric wind, which is most intense at Rhea's range from the planet.[52]

In terms of impactor populations, parts of Rhea's surface would seem to date to the post-accretional bombardment, while other areas show signs of resurfacing. Like Tethys, Rhea has had at least one massive impact. This left a double-ringed structure of the order of 450 kilometres in diameter. Unfortunately, this structure was on the terminator in the Voyager 1 imagery, and most of it was in darkness. Nevertheless, the inner ring seems to comprise ridges and the outer ring is a pair of closely spaced concentric inward-facing scarps (the inner one is more prominent, and the outer one is discontinuous).[53] There is also a hint of an even larger impact on the trailing hemisphere whose cavity has been almost completely resurfaced. Its only relic is a ring of subtle ridges protruding through the in-fill.[54] The degraded state of these larger craters indicates that they are very old, and Rhea was struck when its interior was still warm and its lithosphere was sufficiently pliable to enable the surface deformation to relax. There are areas that are saturated with smaller, younger craters with sharper features, central peaks and, in some cases, bright patches on their walls that could be 'clean' ice which has been exposed by slumping.[55] A paucity of craters with diameters of less than 10 kilometres in the equatorial zone on the leading hemisphere suggests that a mantling deposit several kilometres thick is masking the underlying cratered topography.[56]

TITAN

Voyager 1 revealed Titan's atmosphere to be predominantly molecular nitrogen with methane as the secondary constituent. The atmosphere is both dense and deep. The surface pressure is 1.5 bars. As a result of reactions in the upper atmosphere driven by solar ultraviolet and magnetospheric charged particles, a wide variety of complex hydrocarbons are produced. It is these that bestow the orangey hue on the optically

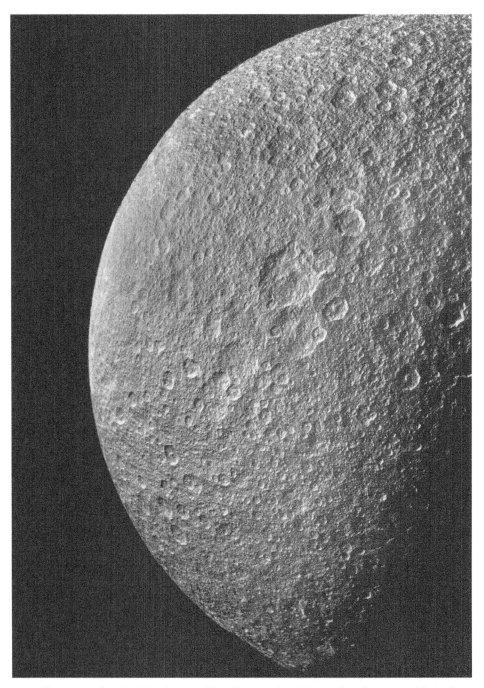

As Voyager 1 flew by Rhea it assembled this mosaic of the north polar region (the pole in on right of frame, on the terminator) displaying a variety of terrain types.

As Voyager 1 left Rhea behind, it snapped this view showing what appears to be a multiple-ringed impact structure on the terminator. The north pole is located on the terminator at the top.

thick haze that forms the visible 'surface' at an altitude of 200 kilometres. A number of thin layers of semi-transparent haze exist above this on a global basis, at altitudes ranging up to about 750 kilometres.[57] While this was fascinating for the atmospheric specialists,[58] the fact that the surface was obscured made the fly-by rather frustrating for the geologists. Nevertheless, the atmospheric composition prompted speculation as to the nature of the hidden surface.

Carl Sagan and Stanley Dermott at Cornell suggested in 1982 that Titan might be completely covered with an ocean of methane.[59] As methane is continuously being dissociated in the upper atmosphere, it must be being replenished from the surface, and it was argued that this source must be in a liquid state. Because Titan's orbit is slightly eccentric, the resulting tide in this methane ocean would generate friction on the sea floor, which would tend to slow the moon's rotation. By the conservation of angular momentum, the moon's orbit would become perfectly circular. However, the fact that the orbit is still eccentric implied that there could not be *shallow* seas; there must either be *no* ocean, or a *very deep* one (a depth of 300 metres was suggested) in which the surface tides are decoupled from the ocean floor.[60]

With a diameter of 5,150 kilometres and a density of 1.88 g/cm^3, Titan is broadly comparable with Ganymede. It should have inherited enough radioactive elements

Voyager 1's infrared spectrometer was able to identify a variety of hydrocarbons in Titan's upper atmospheric temperature inversion by virtue of their emission lines.

from the solar nebula to have kept its interior sufficiently warm for differentiation to have occurred.[61] This would have driven the early devolatisation which formed the atmosphere. When the Infrared Space Observatory accurately measured the ratio of deuterium to hydrogen in Titan's stratosphere, it found it to be four times less than in comets. This implied that the atmosphere is not of cometary origin, and therefore had to be derived from outgassing.[62] Recent millimetre-wavelength measurements of the ratio of nitrogen isotopes in Titan's atmosphere suggest that the 'air' was once 30 times thicker than it is now. The lighter isotope has been depleted.[63,64,65,66] The fact that such an imbalance is not present in the carbon isotopes of methane implies that the methane content of the atmosphere is ephemeral. As the rate at which methane is currently being lost to photochemical reactions will erode the observed concentration in 10 million years, the atmosphere must be being replenished – if a global ocean is present this would be by evaporation, and if not then as a result of cryovolcanism. If this is done only on an intermittent basis, the atmosphere may occasionally thin and chill as methane's greenhouse effect is depleted.[67] The cryovolcanically released methane might be vented directly into the atmosphere, but it might also be extruded as liquid and then drain into low-lying areas and become dissolved in *ethane* and higher hydrocarbons such as propane and acetylene, whose vapour pressure is too low for them to evaporate, and so must be accumulating on the surface in liquid form. Because the tropospheric methane concentration is close to its saturation point (it cannot condense in the lower troposphere, it can do so only by rising to the 'cold trap' of the tropopause), it will evaporate slowly.[68]

Surface structures will have been physically eroded by the hydrocarbon rainfall, and chemically eroded as this drained off into the low-lying areas. If there is a near-global ocean, the several-metre-amplitude tides will erode the icy shoreline. In the absence of ongoing tectonism to rebuild topography the icy surface might have been more or less levelled, leaving a predominantly oceanic environment. Such a reservoir of hydrocarbons on the surface would draw a significant amount of nitrogen from

Once the radio data from Voyager 1's Titan occultation had been analysed, it gave a profile of the atmosphere. Above the surface, the temperature decreases towards the tropopause at an altitude of about 45 kilometres. There is a temperature inversion in the stratosphere. The ultraviolet and infrared instruments detected several 'detached' layers of haze far above the optically-thick orangey haze that forms the visible limb, and a layer of cold methane cloud near the surface.

the atmosphere.[69] If the atmosphere is controlled by its being in thermodynamic equilibrium with the condensed volatiles on the surface, then the climate may be subjected to strong positive feedback mechanisms, and the atmosphere's density and temperature might be extremely changeable over geologically brief timescales. But is Titan really a vast ocean?

A glimpse of the surface

In the aftermath of the Voyager fly-by, it was presumed that Titan's orangey haze was optically thick across the visual and infrared, and that we would therefore be denied a direct view of the surface until a microwave imaging radar mission could be mounted. However, it was found that methane absorption is weak in parts of the near-infrared, and in the narrow band in the region between the Voyagers' ISS and IRIS instruments the atmosphere is semi-transparent.

In 1994, a team led by P.H. Smith of the Lunar and Planetary Laboratory at the University of Arizona observed Titan in the 0.94-micron 'window' using the Hubble Space Telescope's Wide Field Planetary Camera, and characterised its surface in terms of reflectivity.[70] In fact, they had been hoping to see tropospheric

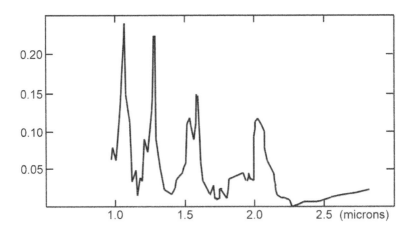

A plot of the transparency of Titan's atmosphere as a function of wavelength in the near-infrared region of the spectrum. There are 'windows' at 0.83, 0.94, 1.075, 1.28 and 1.59 microns through which it is possible to see tropospheric clouds and measure the surface reflectivity. (Based on data in 'Windows through Titan's atmosphere?', W. Grundy, M. Lemmon, U. Fink, P. Smith and M. Tomasko. *Bull. Amer. Astron. Soc.*, vol. 23. p. 1186, 1992.)

clouds to enable them to track their motions. "Try as we might, looking at the data we couldn't see anything that was a significant brightening on the surface and moved. It looked like there were no clouds or, if clouds were there, they were right at the noise level and we couldn't tell the difference," Smith reflected. He had looked for clouds precisely because there was no guarantee of being able to see the surface. "When I wrote my proposal, I said that we intended to map the surface features, but the reviewers said: 'That's impossible.' Of course, we saw no clouds, and we mapped the surface!" The result was crude, but it was the first glimpse of the mysterious surface.[71] The entire disk was barely 20 pixels across, which gave a resolution of 300 kilometres per pixel. Geometrical constraints meant that only the equatorial and mid-latitudes could be studied in this way. In fact, variation in reflectance can arise from compositional, textural or topographical differences. A large bright patch on the leading hemisphere explained an earlier study which had shown a variation in brightness that was correlated with the moon's rotation. The data was sufficient to create a preliminary map with a prominent infrared-bright feature comparable in size to Australia stretching one-sixth of the way around the equatorial zone.

Although this feature was dubbed a 'continent', it could not be confirmed to be a topographic structure – let alone that it rose from an ocean. In fact, there was no definitive proof of *any* liquid hydrocarbons on the surface, and terrestrial microwave radar studies were inconclusive on this question.[72] Infrared observations using 'adaptive optics' on the Canada–France–Hawaii Telescope noted several even

By repeatedly imaging Titan during its 16-day axial rotation using the Hubble Space Telescope the near-infrared reflectivity of the moon's surface was able to be charted. The low-albedo areas might be seas of ethane. The prominent bright area on Titan's leading hemisphere may be a 'continent'. (Courtesy of P.H. Smith, M.T. Lemmon, R.D. Lorenz, J.J. Caldwell, L.A. Sromovsky and M.D. Allison in STScI-PR94-55, December 1994.)

brighter spots dotted along its length.[73,74,75] "I think this could be a plateau with peaks," speculated Athéna Coustenis of the Meudon Observatory in Paris, venturing that the brightest features might represent a frost of methane ice on their summits. "I'm a big fan of mountains on Titan," admitted Ralph Lorenz of the University of Arizona, but he dismissed the possibility of their being tall enough to have become coated with frost. Indeed, by analogy with Ganymede, it has been argued that the range of elevation on Titan should rise no more than 3,000 metres from the mean;[76,77] considerably lower than Coustenis presumed. Lorenz argued that the peaks might appear bright because they stimulate rainfall that has washed them 'clean' of a mantling deposit of dark material.[78] As Peter Smith summed up: "I think what we see is a very large range of ice mountains. You've got a constant wind that blows the wet air from the methane ocean. This freezes out and clouds form on the top of these mountains. The methane rain erodes these hills and exposes fresh ice." But why is the mountainous feature so expansive? "There has to be something different about this place," agreed Smith. "People have suggested that Titan was hit by a gigantic asteroid which exposed fresh ice, but that would have had to have happened very recently, in light of the rate at which the haze falls out from the atmosphere. And presently an impact by a body large enough to leave a crater of that size is very unlikely."

What of the weather system? After years of fruitless observations by terrestrial telescopes, in September 1995 the UK Infrared Telescope on Hawaii made a lucky

The first images of Titan at 1. 3 microns were secured using 'adaptive optics' on the Canada–France–Hawaii Telescope in 1997. They showed several particularly bright spots on the 'continent'. (Courtesy 'Adaptive optics images of Titan at 1.3 and 1.6 microns at the CFHT', A. Coustenis, E. Gendron, O. Lai, J.-P. Veran, M. Combes, J. Woillez, Th. Fusca and L. Mugnier. *Icarus*, vol. 154, p. 501, 2002.)

sighting of tropospheric methane clouds. The first clue was a rapid increase in the 2-micron brightness. Caitlin Griffith of the Northern Arizona University in Flagstaff, leading the team, explained this as a thick methane cloud at an altitude of about 15 kilometres which gradually rotated over the limb into view during the two days available for their observations, eventually covering 10 per cent of the disk.[79] It was strongly suggestive of a hurricane-sized system. Unfortunately, an immediate follow-on study was impracticable, and when this site was inspected in 1997 this feature was absent. Later observations, using more sensitive instruments, detected small clouds, each covering no more than 1 per cent of the disk, with rapid temporal variations that offered the prospect of using their variability to infer the tropospheric winds.[80] "Standing on the surface of Titan, we would see a very dimly lit world, as bright as Earth under a 'full' Moon," Griffith predicted. "The Sun would appear as a diffuse light source through Titan's high smog. At night, we wouldn't see stars through this veil. On the ground, the atmosphere would be clear and the visibility unobscured, temperatures would be uniform and winds quiescent. Every week, sparse clouds would appear below the orangey haze, high in the sky, barely visible. They would quickly produce rain and disappear. Perhaps once a year, clouds would blanket the sky for a day or two."

As the resolution of mapping improved, the case for a global ocean diminished. Infrared 'speckle interferometry' by the 10-metre Keck telescope in Hawaii was able to produce a sharper image than that of the Hubble Space Telescope.[81] This showed a complex surface comprising bright areas that might be ice-and-rock continents and a large number of extremely dark areas, each several hundreds of kilometres across. This material is "one of the darkest things in the Solar system", pointed out Bruce Macintosh of the Lawrence Livermore National Laboratory, a member of the team. A thick hydrocarbon would have an albedo of 2 per cent.[82] The rain of photolytic

This map of Titan depicts the 'continent' in terms of surface reflectance at 2.1 microns. (Adapted from 'Titan: high-resolution speckle images from the Keck Telescope', S.G. Gibbard, B. Macintosh, D. Gavel, C.E. Max, I. de Pater, A.M. Ghez, E.F. Young and C.P. McKay. *Icarus,* vol. 139, p. 189, 1999. Courtesy of Seran Gibbard of the Institute of Geophysics and Planetary Physics at the Lawrence Livermore National Laboratory.)

particulates formed in the upper atmosphere must be accumulating on the surface. It has been estimated that over the moon's history this 'fall out' would correspond to a global blanket 1 kilometre thick. The organics may form of a semi-solid sludge.[83] Their areal extent would seem to indicate that they have pooled in impact craters.[84] Later observations in the windows from 1 to 5 microns gave a crude spectrum of the surface and, surprisingly, implied that water ice is predominant on a global basis,[85] in turn suggesting that, contrary to expectation, much of Titan's surface is *free* of hydrocarbons.

Musings about Titan's surface have therefore undergone a turnaround since the model of a global ocean was prompted by the Voyager atmospheric data, with the result that many researchers now expect there to be no more than shallow seas in craters. The action of the tides in isolated seas would *not* affect the ellipticity of the moon's orbit.[86] In their early analysis, Sagan and Dermott had argued for either a deep ocean or no ocean, and it would now seem that there is no ocean. If so, where is the reservoir that replenishes the atmospheric methane? Perhaps the icy lithosphere is porous and the fluid is underground in an 'oil field',[87] or perhaps it is soaked into the regolith[88] and methane is slowly diffusing into the atmosphere.

Life on Titan?

The Infrared Space Observatory that was launched in late 1995 made spectroscopic observations of Titan. In addition to verifying and refining the abundances reported by Voyager 1, its wider wavelength coverage enabled it to penetrate more deeply into the atmosphere. Its vertical profiles of the abundances of some of the more complex hydrocarbons enabled the models developed from the Voyager data to be refined. It also established the presence of carbon dioxide. Observing in the far-infrared in late 1997, the ISO's Short Wavelength Spectrometer discovered two lines at 40 microns indicating water vapour in the upper stratosphere.[89,90] "Water vapour makes Titan much richer," pointed out Athéna Coustenis, the leader of the team. "We knew there was carbon monoxide and carbon dioxide, so we expected water vapour. Now that we've found evidence for it, we can better understand the organic chemistry taking place on Titan, and also the sources of oxygen in the Saturnian system." In fact, the surface is too cold for ice to release water vapour. One suggestion is that the water comes from the rain of interplanetary dust, which in turn contains grains released by comets. "We are seeing a mix of elaborate organic molecules closely resembling the chemical soup out of which life emerged, so it will help us to understand the organic chemistry that took place also in the young Earth," Coustenis observed.

H.C. Urey, the Nobel prize-winning chemist, argued that the Earth formed with a gaseous envelope drawn directly from the solar nebula. If so, this must have been dominated by hydrogen-containing gases. Any molecular hydrogen would promptly have been lost to space, but prodigious amounts would have remained in the form of methane, ammonia and water vapour. In such an environment, the free oxygen would have been 'reduced' by being bound up chemically. The chemical reactions that gave rise to life could not have occurred in an oxygenated environment, as the

In 1997 the Infrared Space Observatory's Short Wavelength Spectrometer (SWS) detected emission lines at 39.37 and 43.89 microns corresponding to pure rotational transitions of water vapour that is evidently present in Titan's upper stratosphere. (Adapted from 'Evidence for water vapor in Titan's atmosphere from ISO/SWS data', A. Coustenis, E. Lellouch, Th. Encrenaz and A. Salama. *Astron. & Astrophys.*, vol. 336, L.85, 1998.)

extremely reactive oxygen would have readily broken down the organic molecules. Voyager 1's IRIS measured the abundance of molecular hydrogen in Titan's troposphere as being 0.2 per cent, and it is therefore a reducing atmosphere in which there can be no free oxygen or oxygen-rich compounds – the oxygen Titan inherited from the solar nebula is bound up in the refractory oxides in the rock in its interior and in the water ice in its crust. Titan's atmosphere is therefore believed to be similar to the oxygen-free envelope of early Earth. On the fair assumption that intense volcanism on the early Earth would have generated powerful lightning discharges, S.L. Miller, one of Urey's graduate students, performed an experiment in 1952 in which he put a spark generator into a jar containing water, ammonia, methane and hydrogen. After only a week he found that a significant fraction of the methane had been drawn from the atmosphere by reactions which had produced surprisingly large amounts of various amino acids, the 'building blocks' of life.[91] Wilhelm Groth and H. von Weyssenhoff in Germany re-ran the experiment in 1959 by illuminating the jar with ultraviolet light rather than simulated lightning, with the same result. The Earth's surface is currently shielded from solar ultraviolet by the ozone layer in the stratosphere, but the ozone could not have formed until the atmosphere became oxygenated, and this is believed to have occurred subsequent to – and as a side-effect of – the development of life, so in the earliest of times the surface would have been irradiated by solar ultraviolet. Although Titan's low temperature means that water and ammonia must be frozen, and form part of its surface, the atmosphere is similar to that of these experiments. Could life develop on Titan?

The IRIS's detection of hydrogen cyanide was significant because it was the first molecule to be confirmed to be present that is not a straightforward C–H bond. It is formed from radicals left by the dissociation by ultraviolet of nitrogen and methane molecules in the upper atmosphere. It is a precursor to biotic chemistry.[92] It plays a critical role in the chemical synthesis of amino acids and the bases present in nucleic acids.[93] Apart from the Earth, Titan is unique in the Solar System in its abundance of the building blocks of complex organic chemistry.[94] If there are pools of 'primordial soup' on its surface, it could be thought of as a cryogenic version of the prebiotic Earth.[95] Although Carol Stoker of the Ames Research Center established that bacteria can grow on tholin material in the laboratory, conditions on Titan are – probably – too chilly for bacteria to survive. Nevertheless, as Carl Sagan and Reid Thompson at Cornell suggested in 1992, meteor impacts could melt the crust and create temporary pools of liquid water.[96] Although the atmosphere lacks oxygen, it was calculated that for several million years organics in the crust around an impact site might be able to react with the oxygen in liquid water; sufficient time perhaps for simple amino acids to develop. If it could be established that self-replicating nucleic acid molecules exist on the moon's surface, it would have profound implications for the mystery of the origin of life. Finding the basis of life elsewhere, even if it has not yet developed a resilient cellular structure, would be a discovery of the first magnitude because, as Philip Morrison of the Massachusetts Institute of Technology once observed, this would "transform the origin of life from the status of a miracle to that of a statistic".

It has been speculated that indigenous biology may develop on Titan in 6 billion years, when the Sun evolves into a 'red giant'.[97] Initially, the increased insolation will be absorbed by the haze-laden stratosphere, which will 'inflate', so the temperature at the surface will rise only to about 200K, but the enhanced greenhouse may raise this sufficiently to melt the icy surface. Arthur C. Clarke has pointed out that Titan will act as a convenient 'lifeboat' for our descendants as, fleeing the destruction of the inner Solar System, they migrate outwards with the 'habitable zone'. After a few hundred million years, however, Titan's atmosphere will leak away, so it will end up as an inhospitable rock.

HYPERION

Hyperion turned out to be an irregular body of $360 \times 280 \times 236$ kilometres with a passing resemblance to a thick hamburger. As the largest of the irregularly shaped satellites, it is uniquely located, being neither deep within the system nor out beyond Phoebe.[98] If it is the largest fragment of a progenitor, and if that object occupied the same orbit, which is in resonance with Titan, then the other fragments ejected into nearby but non-resonant orbits may well have been so perturbed as to have been unable to re-accrete into a single body. Close encounters would have sent some fragments deeper into the Saturnian system, where they may have collided with the inner satellites, and others would have been ejected from the system. Most would have been accreted by Titan itself.[99] W.H. Pickering's Themis, if it had existed, may well have been another fragment. Hyperion has an icy surface, suggesting that its composition is the same as the other moons.[100,101] If the arcuate ridges are relics of large features on the surface of the predecessor, their size implies that the body was subjected to the post-accretional bombardment. Although parts of Hyperion's current surface must have been on the progenitor's interior, the cratering indicates that these faces have been exposed to the ongoing flux for a considerable time. For much of its time, Hyperion is outside the magnetosphere. The admixture of dark material that gives its spectrum a distinctly reddish hue might be the result of chemical reactions induced by long-term solar irradiation, as a 'space weathering' effect.

Initially, Hyperion's rotation was puzzling because the Voyager imagery did not identify its spin axis, but a photometric variation suggested a 13-day periodicity.[102,103,104] It orbits in 21 days, so this indicated that it rotated non-synchronously. In 1983, it was realised that Hyperion has a *chaotic* rotational state.[105,106] This was proved in 1988 by nearly continuous observations of the light curve over a period of three months.[107] Hyperion is in resonance with Titan: for every three orbits it makes, Titan makes four. Hyperion's orbit is also slightly eccentric, and its opposition with respect to Titan occurs when it is at apoapsis. Perturbations on the asymmetric mass appear to have set it rotating on several axes in a pendulum-like motion.

IAPETUS

Iapetus was the most enigmatic of the Saturnian moons for the early telescopic astronomers. Soon after discovering it, G.D. Cassini realised that the moon's brightness was varying by fully two visual magnitudes in a systematic manner as it moved around its orbit. It was brightest at western elongation and faintest at eastern elongation. The fact that this pattern remained fixed implied that the moon's rotation was synchronised with its orbital period. For some reason, the leading hemisphere was considerably darker than the trailing hemisphere. At first, there was debate as to whether part of the surface was really darker, or whether the moon was of a peculiar (that is, non-spherical) shape. William Herschel concluded that one hemisphere was partially obscured. Iapetus does not show a measurable disk, so early estimates of its diameter were made on the basis of assumptions about the albedo of its brighter hemisphere. Although there was considerable variation in the various estimates, the fact that they ranged up to 3,200 kilometres prompted speculation that in the chilly Saturnian environment this moon might have a rarefied atmosphere of dense gases. However, Herschel had pointed out that if it possessed an envelope, it was not thick enough for major obscurations to show up in the light curve. In fact, if Iapetus does have a residual atmosphere, it may well freeze onto the surface during the almost-40-day-long night as the moon pursues its remote orbit, at which time the temperature must chill down to near absolute zero. Photometric studies concluded that the bright terrain extended over both polar regions, and that the dark material formed a large spot centred on the apex of orbital motion.[108,109,110] This strongly implied that Iapetus has an intrinsically bright surface disfigured by a dark patch, rather than being a dark object marred by an anomalously bright patch.

Iapetus's public profile was immeasurably boosted when Arthur C. Clarke, in his 'extended' novel of Stanley Kubrick's movie *2001: A Space Odyssey*, set a 'star gate' black monolith in a white circle in the centre of the dark hemisphere.[111] There was therefore a sense of *déjà vu* as Voyager 1 approached. Although Iapetus was not particularly well placed for viewing by the Voyagers, and their best imagery had a resolution of 10 kilometres per pixel, this was sufficient to show that the dark patch, which has been appropriately named Cassini Regio, is featureless.[112,113,114] Its boundary is well defined, with a 'transition zone' several hundred kilometres wide. The telescopic inference that the dark material does not encroach upon the poles was confirmed. Dark streaks beyond the transition zone suggest that the dark material is a mantling deposit whose source was on the leading face. The albedo of the dark material varies from 2 per cent near its centre to 5 per cent around the periphery. It is as black as tar. In fact, it may be a tarry substance. Such dark material elsewhere in the Solar System is believed to be carbonaceous compounds. A spectroscopic study revealed that Cassini Regio has a reddish hue,[115,116] which argues for a carbonaceous material such as that in meteorites derived from the 'primordial' material left over from the formation of the Solar System. As little as 1 per cent of a carbonaceous impurity can reduce to 7 per cent the normally 40–50 per cent reflectivity of water ice. But how might carbonaceous material have 'painted' Iapetus? The outermost member of the retinue, Phoebe, is also dark.[117] Perhaps Iapetus's leading hemisphere is sweeping up fine material ejected from Phoebe by micrometeoroid impacts.[118] Significantly, Phoebe orbits in a retrograde manner, so any motes blasted off its surface will 'fall' towards Saturn. As they spiral down in a retrograde manner, they will meet Iapetus, striking its leading hemisphere with their combined velocities, producing particularly energetic impacts that might produce unusual reactions in the ices on the surface. Hyperion, the next moon in towards the planet, does not appear to have been affected, but its chaotic rotation prevents it from maintaining one side facing 'forward'. Perhaps its entire surface has been uniformly darkened to some degree. On the other hand, Hyperion may have been darkened by some other process. As Phoebe's orbit is inclined at 30 degrees to the plane of the Saturnian system, its motes will form a tenuous but deep annular disk. Iapetus's orbit is inclined at 14.75 degrees. Perhaps its gravity draws in most of the dust, preventing it from penetrating more deeply into the system. Some of the inner moons do have dark patches, but these are on their *trailing* hemispheres and are very likely induced by the charged particles that circulate deeper within the planet's magnetosphere. Although Phoebe, Hyperion and the dark side of Iapetus all have a dark character, a preliminary spectrum derived from multicolour data indicated that Phoebe's spectrum is 'flatter' than the 'reddish' hue of Cassini Regio,[119,120] casting doubt on Phoebe being the source of Iapetus's anomaly.

If the dark material of Cassini Regio was not picked up progressively, perhaps it was acquired by a single violent event. The most probable intruder into the system is a comet. When the Giotto spacecraft flew by Halley's Comet in 1986, it found the nucleus to be extremely dark.[121] Perhaps Iapetus took a head-on strike and its leading hemisphere is coated with carbonaceous residue. The material absorbs much of the incident sunlight, but reflects just enough light for a modern instrument to provide a hint of spectral features. Initial laboratory tests suggested a polymerised hydrogen cyanide called poly-HCN. However, later research revealed that a blend of water ice and a nitrogen-rich solid tholin could account for the strong 2.9-micron absorption. Adding amorphous carbon (soot) provided an excellent match for the spectrum from 0.3 to 3.8 microns.[122] It has even been argued that the dark material originates from Titan, having been ejected by a major impact involving Hyperion's progenitor.[123]

Could the dark material be of endogenic origin? One proposal has interpreted the dark-floored craters adjacent to Cassini Regio as suggesting that the material erupted from fractures in their floors as a carbonaceous icy slurry in cryovolcanic activity.[124] Although Iapetus is similar in size to Rhea,[125] its lower density means that it is more ice than rock. Nevertheless, it is large enough to have accreted sufficient rock to have undergone mild radiogenic heating,[126] which may have stimulated chemical reactions between the ices of methane and ammonia, with convection beneath the frozen lithosphere prompting the extrusion of fluid. But why should Iapetus have been so active? Why on only one hemisphere? Why the leading hemisphere? Centred on that hemisphere? Was there no high-standing terrain? Was the surface flow accompanied by venting of volatiles whose fall-out painted the high ground? The absence of bright ray craters on Cassini Regio implies either that this deposit is too thick to have been penetrated by recent small impacts, or that the process of venting is still active and the bright ejecta is rapidly masked. Initially, no detail was apparent in Cassini Regio but a recent re-examination of the Voyager imagery identified craters, suggesting that the material is either ancient and too thick for impacts to excavate the underlying ice, or it is so recent that it has painted the pre-existing terrain without burying it.[127] This study also spotted mountain-like structures about 50 kilometres across and rising to heights of 25 kilometres – an astonishing elevation – near the boundary of Cassini Regio and the brighter terrain on the anti-Saturn hemisphere. Iapetus is ellipsoidal ($1,534 \times 1,484 \times 1,426$ kilometres in diameter) and the distribution of albedos on the flanks of these peaks suggest that the source of the dark material – irrespective of its origin – reached the surface *from the direction of* the apex. Even if the source was internal, the symmetry implies an external *control function*. Although Iapetus travels mostly outside the magnetosphere, surface darkening by the charged particles of the solar wind would not have produced an asymmetric distribution.

Irrespective of how Cassini Regio originated, the profusion of large craters in the polar region implies that at least this area dates to the tail end of the post-accretional bombardment.[128] In fact, the cratering density is comparable to the lunar highlands, the most intensely cratered areas of Mercury, and virtually all of Callisto, confirming that the Saturnian system was invaded by a population of external objects at an early stage in its development.

PHOEBE

With the exception of Iapetus and Phoebe, Saturn's satellites all travel in circular prograde orbits coplanar with both the planet's equator and the ring system. Phoebe has an elliptical, inclined and retrograde orbit. Voyager 2's best imagery from a range of 2.2 million kilometres had a resolution of 20 kilometres per pixel, just enough to establish that this moon is spheroidal with a diameter of about 220 kilometres, rotates in 9.5 hours in a prograde manner, is cratered, and that while its mean albedo is 6 per cent there are several relatively 'bright' patches.[129] Neither Voyager flew sufficiently close for the perturbation of its trajectory to yield an estimate of the moon's mass. The fact that its orbit is almost in the plane of the ecliptic strongly implies that it is a captured asteroid or comet, and as such it is a respectable size. Its surface, which is much darker than any of the icy Saturnian satellites,[130] has spectral traits indicative of 'primitive' carbonaceous material; hence, if it is an asteroid, it is a 'C' type relic of the solar nebula. Apart from short periods spent in Saturn's magnetotail, Phoebe spends almost all of its time in the solar environment.

As the Space Age dawned, it was reasoned that stray asteroids were unlikely to be able to reach Saturn without being intercepted by Jupiter. It was thought that this was why several families of moonlets only a few dozen kilometres in diameter had been spotted in irregular orbits around Jupiter whereas none had been found orbiting Saturn.[131] Nevertheless, it was acknowledged that Saturn was farther from

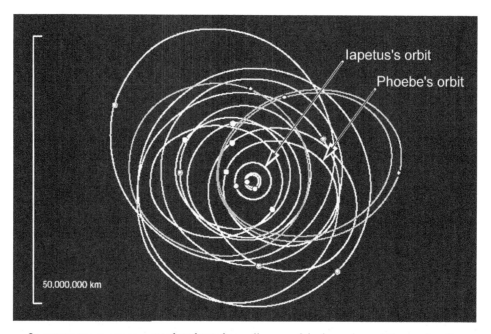

In recent years, many moonlets have been discovered in irregular orbits ranging far from Saturn, several of which pursue retrograde orbits which suggest that they are fragments chipped off Phoebe by a major impact. (Adapted from 'Saturn saturated with satellites', D.P. Hamilton. *Nature*, vol. 142, p. 132, 2001.)

the Sun and if such tiny objects were present they would be difficult to identify. In recent years, however, advances in detector technology and automated analysis have prompted a rash of such discoveries.[132] In one year, a dozen tiny moons were spotted in irregular orbits ranging far from Saturn. However, the fact that these split into three or four groups implied that they are fragments of a small number of objects that strayed into the Saturnian realm and were subsequently disrupted. Interestingly, one group has retrograde orbits which suggest that they are fragments chipped off Phoebe by a major impact. The 'bright' areas on Phoebe may mark the sites of these impacts.

If ever humans explore the Saturnian system, Phoebe will make a superb vista point, its inclined orbit offering the most spectacular views of the ring system. In the event that its retrograde motion proves inconvenient, then Hyperion will serve almost as well. A base on Titan would not offer much of a view, at least not in the sky. Although bland, Saturn would be overwhelmingly large in the sky from one of the inner satellites, and because they travel in the ring plane they would not offer much of a view of the ring system, which would form a bright line crossing in front of and projecting out to each side of the globe. However, until the first humans settle the Saturnian system, we will have to rely upon robotic probes to explore it by proxy.

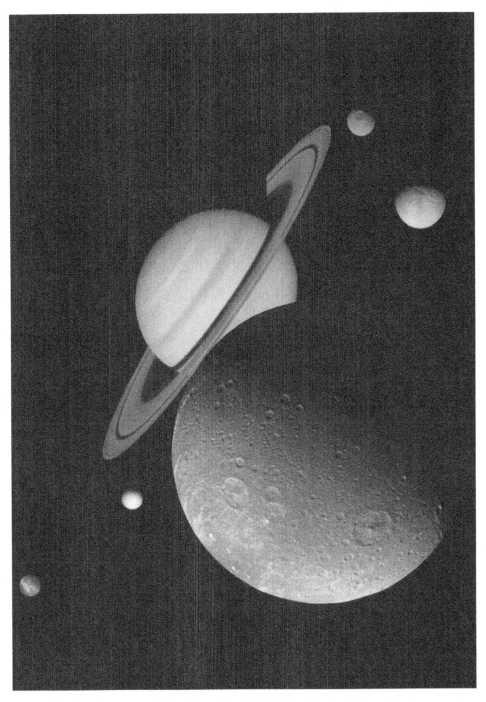

A montage of the Saturnian system using Voyager imagery.

5

Cassini–Huygens

PLANNING

The Saturnian system was visited by robotic explorers from the planet Earth in three successive years: 1979, 1980 and 1981. In contrast to the expected bland frozen realm, the system has turned out to be incredibly rich in diversity: Enceladus has been comprehensively resurfaced, and might still be undergoing cryovolcanic activity with geysers venting jets of water into space; Titan has a dense reducing atmosphere that may well be in a prebiotic state; Iapetus has its enigmatic dark hemisphere; and Phoebe is almost certainly a captured comet or asteroid – maybe even an object that strayed in from the Kuiper Belt. It was inevitable, therefore, that within a few years planning would get underway to dispatch a follow-up mission. As in the case of the Galileo mission to Jupiter, the next spacecraft to Saturn would enter orbit to conduct an *in-depth* study. In 1982, a Joint Working Group of the US National Academy of Sciences and the European Science Foundation recommended a 'Saturn Orbiter and Titan Probe'. The following year, NASA's Solar System Exploration Committee, in recognising Titan's uniqueness, assigned its top priority in the outer Solar System to a 'Titan Probe and Radar Mapper'.

In June 1985, following a joint NASA/ESA study, it was agreed that the Cassini spacecraft to enter orbit around Saturn would be built by NASA, and that the Titan Probe would be supplied by ESA.[1] The integrated spacecraft would be mated with a Centaur 'escape' stage and be ferried into space by the Space Shuttle in May 1994. Upon entering Saturn orbit in January 2002, Cassini would release the Titan Probe and then undertake a four-year orbital tour. However, the loss of the Challenger in January 1986 prompted NASA to cancel the version of the Centaur for the Shuttle, and this decision effectively sent both the Galileo and Cassini teams back to their drawing boards.[2] In an effort to provide time for the redesign, in 1987 NASA slipped Cassini's launch date to March 1995 and added a Jovian fly-by to the interplanetary cruise to pick up some energy from this slingshot. JPL finished its Phase 'A' studies for the main vehicle in September 1987, and a month later, with this information at hand, ESA started the corresponding phase of its programme. In September 1988, upon completing its study, ESA named its aspect of the mission after Christiaan Huygens.[3]

Outlines of the Comet Rendezvous and Asteroid Fly-by (left) and Cassini–Huygens (right) spacecraft as envisaged using the generic Mariner Mark 2 configuration

In 1988, in order to overcome a $1.6 billion limit set by Congress for the Cassini mission, JPL decided to develop the spacecraft within a broader programme in which several deep space missions would share systems, as doing so was expected to reduce the overall cost by $500 million. In the early 1970s, when starting to plan the Grand Tour of the outer Solar System, JPL had proposed the development of a new class of spacecraft optimised for deep space missions. As the successor to the Mariner series with which it had explored the inner Solar System, JPL dubbed the new design 'Mariner Mark 2'. It was to be a generic design utilising radioisotopic power, high data rates, scan platforms for state-of-the-art instruments, sophisticated computers for autonomous operations with the flexibility of in-flight reprogramming, and highly redundant and fault-tolerant systems to undertake missions lasting a decade or more. However, when Congress refused the development funding, JPL had been obliged to adapt its existing design. As Voyager 2 neared Neptune, JPL reintroduced the idea of developing generic systems for deep space missions, arguing that it would eventually save money. Costs would be further minimised by re-using spare parts from the Voyager and Galileo spacecraft. In 1989 Congress approved the development of the Cassini spacecraft in parallel with the Comet Rendezvous and Asteroid Fly-by (CRAF) mission. At this time, Cassini was set for launch on 6 April 1996, with arrival at Saturn on 6 December 2002. In 1989, with the project finally approved, the space agencies issued requests for proposals for science projects. In October 1990 ESA announced the experiments selected for the Huygens probe, and the next month NASA announced the scientific themes and the list of principal investigators for the orbiter's instruments. In early 1991, Cassini and CRAF swapped launch dates, with Cassini moving forward to 25 November 1995. Furthermore, a Venus slingshot was added to enable the spacecraft's mass to be increased (adding a third RTG power cell meant that the storage batteries that had been included to assure sufficient power at times of heavy demand could now be eliminated). The new trajectory permitted a 40-kilometre fly-by of asteroid 302 Clarissa on 18 November 1998 *en route* to the Jovian slingshot, but with the resulting diversion slipping Cassini's arrival at Saturn to 15 May 2004.

The path that Cassini would have flown if it had been launched in 1995, employing one gravitational assist from Venus, one from the Earth and a fly-by of asteroid 302 Clarissa *en route* to the Jupiter for the final slingshot for Saturn arrival in May 2004. Although Cassini was not launched until 1997, the fact that it picked up more energy by virtue of an extra Venus slingshot enabled it to make up time.

In January 1992 Congress cancelled CRAF, and Cassini's launch date was slipped to October 1997 to ease its annual funding, even though this would push the total cost over the $1.6 billion cap. By eliminating the scan platforms and cancelling pre-Saturn science activities such as asteroid encounters, JPL was able to pull the mission back within budget. Time was now of the essence, because if the mission failed to launch in 1997 it would be in serious trouble as this was the last launch window that would facilitate a Jovian slingshot. In 1995, the House Appropriations Committee, unable to postpone Cassini any further, raised the prospect of outright cancellation, but lobbying by the science community, and the international aspect of the mission, overcame the threat.

CASSINI SPACECRAFT

The Galileo spacecraft had incorporated a spin-bearing assembly so that one part of the vehicle could be three-axis stabilised while the remainder rotated like the Pioneers to optimise particles and fields measurements. Spinning communication satellites in geostationary orbit use similar spin bearings to hold their antennas facing the Earth. However, it was decided not to utilise such a complex mechanism on Cassini, therefore the spacecraft will have to stabilise itself to perform aimed observations and spin for particles and fields sensing. It will spend most of the time with its high-gain antenna aimed at the Earth for ideal communications. Galileo (like the Voyagers) had a scan platform to enable its aimed instruments to track a target during a fly-by without the vehicle needing to manoeuvre. Unfortunately, financial constraints forced the deletion of Cassini's scan platform (indeed, at one time it was to have *two* such platforms). With its instruments bolted onto its framework, Cassini will have to turn towards a target to make observations. Since its high-gain antenna will not be able to be held facing the Earth at such times, the spacecraft will have to 'buffer' the data from its instruments until their observations are complete, then turn its high-gain antenna back towards Earth for downloading.

The integrated propulsion system built by Lockheed–Martin incorporates a pair of redundant 445-newton engines that will make the major manoeuvres such as the Saturn Orbit Insertion and later periapsis-raising burns, and four 'quads' of thrusters to control the spacecraft's orientation. The bi-propellant system burns hydrazine in nitrogen tetroxide. Because these reactants are hypergolic, the engine does not need an 'igniter'. The redundant engines are mounted side-by-side, and the engine that fires must be aimed to ensure that its thrust is directed through the vehicle's centre of mass. The JPL-developed engine gimbal actuators that fine-tune the alignment of the main engine to compensate for the randomly shifting propellant were derived from those of the Viking orbiters.

Cassini has a system of gyroscopic reaction wheels with which to stabilise itself and make small attitude changes without firing its thrusters. Three of the quartet of wheels are mounted orthogonally to one another, one in each of the Cartesian axes; the fourth wheel is in standby, as a spare. Each wheel is driven by an electric motor. As a result of Isaac Newton's famous law that every action imparts an equal and opposite reaction, changing the rate at which a wheel rotates applies a torque to the spacecraft and causes it to rotate in space around the appropriate axis. The reaction wheels will serve as the primary attitude control system for normal operations. The

The structure of the Radioisotope Thermo-electric Generator (RTG).

primary source of inertial reference is a star tracker. Between star fixes Cassini will monitor its attitude employing gyroscopes.

Cassini is powered from a trio of RTGs, mounted singly around the base of the propulsion system. Each RTG has 18 iridium-clad high-strength carbon composite modules, each of which contains a pair of golf-ball-sized spheres of plutonium-238 dioxide in dense ceramic form. Plutonium dioxide is also present in 117 radioisotope heater units situated to keep the electronics systems on Cassini and Huygens at their operating temperatures. Plutonium-238 is a non-fissile alpha-emitting isotope having a half-life of 88 years. As the power output falls in line with the expenditure of radioactive isotope, Cassini's power supply will slide from 815 watts at the beginning of the mission to 638 watts at the conclusion of the four-year primary mission in the Saturnian system. In total, there are 32 kilograms of radioactive material. The units were built by General Electric. The Power and Pyro Subsystem (PPS) regulates the 30-volt supply, and distributes it to the core systems and instrumentation through 10 kilometres of cabling.

The 4-metre-diameter solid dish-shaped high-gain antenna is affixed to the 'front' of the vehicle – opposite the propulsion system. It was supplied by Alenia Spazio in Italy. The 20-watt X-Band transmitter can vary its rate between 20 bits and 169 kilobits per second. A low-gain antenna is mounted on the end of the tower in the dish's centre, and another projects from the body of the vehicle, beneath the Probe Support Equipment, so Cassini will be able to communicate irrespective of its orientation, although at a much lower bit-rate when on these other antennas. Whereas Galileo had a reel-to-reel tape recorder for storage of data being buffered for transmission to Earth, it was decided that Cassini should utilise a 3.6-gigabit solid-state memory subsystem instead. The wisdom of this decision was reinforced when Galileo's recorder developed a fault shortly before the spacecraft entered the Jovian system. Although the radiation in the Saturnian magnetosphere is less intense than Galileo had to endure in close to Jupiter, Cassini's solid-state memory has nevertheless been heavily shielded in order to protect it from the kind of random damage suffered by the computer memories on board Galileo.

Instrumentation
Cassini has 335 kilograms of scientific instruments, divided between aimed remote-sensing and particles and fields studies.

Aimed imaging systems
Cassini has four optical remote-sensing instruments, collectively operating from the ultraviolet to the far-infrared, together with a radio-frequency imaging radar. There is always a trade-off in optical instrumentation between high spatial and spectral resolution. Whereas the main imaging system has high spatial resolution and a number of broad filters in specifically selected spectral bands, the spectrometers emphasise spectral resolution at the expense of spatial resolution, so these are more accurately described as 'mappers' than 'imagers'.

The *Imaging Science System* (ISS) is a high-spatial-resolution imaging system. In effect, it is Cassini's 'eye'. It comprises two subsystems, each with a $1,024 \times 1,024$

The major components of the Cassini spacecraft in its cruise configuration, with the Huygens probe in-carriage on its far side.

Stellar reference units

Visible and Infrared
Mapping Spectrometer
(IR channel)

Visible and Infrared
Mapping Spectrometer
(visible channel)

Ultraviolet Imaging
Spectrograph

Imaging Science Subsystem
(narrow-angle camera)

Imaging Science Subsystem
(wide-angle camera)

Composite Infrared Spectrometer

Cassini's Remote Sensing Pallet.

pixel array CCD camera. Although referred to as 'wide angle', the 200-millimetre-focal-length f/3.5 refractor yields a square image that spans only 3.5×3.5 degrees, so by any popular measure it is actually a telephoto. The 'narrow angle' camera uses a 2,000-millimetre-focal-length f/10.5 folded-optics reflecting telescope with a field of view that is ten times narrower. The hardware was supplied by JPL, but C.C. Porco of the University of Arizona is the principal investigator for imaging science.

The spectral range of the CCD detector is much broader than the vidicon tube of the Voyager's system, extending shortward into the ultraviolet and longward into the near-infrared. The wide-angle camera is sensitive from 0.38 to 1.1 microns, and the narrow-angle camera extends this shortward to 0.2 micron. Each camera has a wheel with filters for a variety of specific purposes. The wide-angle camera is fitted with 18 filters and the narrow-angle camera with 24 filters. CCDs offer other advantages over vidicons. As the dynamic range is greater, the image will not be so readily 'washed out' when an object is subjected to a wide range of illumination. Furthermore, the output of a CCD is 'linear' across its dynamic range, so it is more readily calibrated and will be better suited to photometry measurements for albedo studies. The ISS will map the three-dimensional structures of the atmospheres of Saturn and Titan, determining the composition and distribution of clouds and aerosols and measuring their scattering, absorption and solar heating properties. It will also seek lightning, auroral displays and airglow phenomena. In the case of Titan, the filters for the 0.94- and 1.1-micron 'windows' will enable Cassini to image tropospheric methane clouds and, where the weather is clear, map the reflectivity of the surface at high resolution to complement the topographic data from radar mapping.

The *Visual and Infrared Mapping Spectrometer* (VIMS) is a improved version of the Galileo spacecraft's Near-Infrared Mapping Spectrometer (NIMS) incorporating a Visual Subsystem covering the range 0.35 to 1.07 microns in 96 channels and an Infrared Subsystem from 0.85 to 5.1 microns in 256 channels. The Visual Subsystem consists of a Shafer telescope, a holographic spectrometer grating, and a silicon CCD area array focal-plane detector. Because it is configured as a 'pushbroom' imager, the instantaneous field of view (IFOV) is an entire line of pixels. This is scanned across the scene with a single-axis scanning mirror to produce a series of contiguous rows, which together produce a two-dimensional image. The Infrared Subsystem employs a Cassegrain telescope, a conventionally ruled spectrometer grating, and a 256-element linear array focal-plane assembly. It is a 'whiskbroom' scanning imager, so the IFOV is a single 32×32 milliradian square pixel. A two-dimensional image is assembled by raster scanning along a row of pixels, dropping down a row, scanning that row, etc., utilising a two-axis scanning mirror. The VIMS therefore operates differently to the NIMS. Both of its spectrometers use fixed gratings. The timing is such that the Visual Subsystem integrates as the Infrared Subsystem pixels are being scanned out one-by-one (cross-track) on the same piece of the scene. The Visual Subsystem then steps one slit width (down-track) and the Infrared Subsystem does a fly-back-and-step, after which it scans the next piece of the scene. Thus, both spectrometers can scan a two-dimensional scene using the same pixel scale.[4] In contrast, the NIMS employed a single channel and a moving grating, and a scan mirror which nodded up and down in one direction, with the second dimension being built up either by virtue of the relative motion between the spacecraft and the target or by slewing the scan platform. The VIMS has twice the spectral resolution of the NIMS. The sensitivity of the instrument derives from its radiative cooling system. The Infrared Subsystem was supplied by JPL, and the Visual Subsystem by Officine

A diagram of the structure of Cassini's Visual and Infrared Mapping Spectrometer.

The Visual Subsystem of Cassini's Visual and Infrared Mapping Spectrometer was supplied by Officine Galileo in Italy, and the Infrared Subsystem was built by JPL, where the instrument was integrated by engineers Paul Kirchoff (on the left), Ed Miller (VIMS systems engineer), Michael Brenner and Dave Rosing. (Courtesy of Robert H. Brown of the Departments of Planetary Sciences and Astronomy at the University of Arizona.)

Galileo in Italy. R.H. Brown of the University of Arizona's Lunar and Planetary Laboratory is the overall principal investigator.

By measuring reflected and emitted radiation, the VIMS will be able to determine the composition and temperature of Saturn's atmosphere, charting the temporal behaviour of the surface winds, eddies and other features, and the cycling of the deep atmosphere. It will map the distribution of gases, haze and cloud species in Titan's atmosphere, and observe the surface to search for active cryovolcanism. It will also determine the composition and distribution of materials on the icy satellites and rings.

The *Composite Infrared Spectrometer* (CIRS) was originally named the Infrared Fourier Spectrometer. It serves the same role as the IRIS used by the Voyagers, but it is more technologically advanced with a new chilled detector providing an order of magnitude improvement in spectral resolution. Its three interferometers are fed by a single 500-millimetre-diameter telescope. As the far-infrared spectrometer senses from 17 microns to 1,000 microns, it detects primarily thermal emission. It has a 4.3-milliradian circular field of view. A pair of mid-infrared spectrometers draw the range shortward to 7 microns. These use 1×10 linear array detectors, each element having a 0.273-milliradian square field of view. Thus, despite the trade-off of spatial versus spectral resolution, the spatial resolution of these mid-infrared detectors is 15 times narrower than that of the IRIS, so the CIRS will be able to make a series of measurements along a line that would have been entirely contained within its predecessor's field of view. The great sensitivity of the instrument derives from its radiative cooling system.

The CIRS will measure infrared emission from vibrating molecules in the atmospheres of Saturn and Titan and provide three-dimensional maps of the temperature, clouds and hazes, and chemical composition. Observing on the limb will not only eliminate surface emissions from the field of view, but also optimally investigate the vertical structure. At Voyager 1's fly-by distance, IRIS did not have the resolution to make detailed limb profiles, but Cassini's approach will be much closer and its linear detector array will be able to secure detailed profiles. By repeatedly doing this as the viewing angle changes, it will be able to assemble regional maps, and by looking 'straight down', it will be able to measure the temperature of Titan's surface.[5] The CIRS will also measure the amount of energy reflected from, and radiated by, the rings at infrared and millimetre wavelengths to determine the size, composition, texture, shape, and rotation of the constituent particles. Its measurements of the thermal properties and composition of the surfaces of the icy moons will complement such observations by the other remote-sensing instruments. The principal investigator is V.G. Kunde of NASA's Goddard Space Flight Center in Greenbelt, Maryland.

The *Ultraviolet Imaging Spectrograph* (UVIS) integrates four instruments with a common microprocessor control system.[6] Between them, the Far-Ultraviolet Spectrometer (FUV) and the Extreme-Ultraviolet Spectrometer (EUV) span from 55 to 190 nanometres with a resolution of 0.2 to 0.5 nanometre. The High-Speed Photometer (HSP) will trace the light curves of stellar occultations. The Hydrogen Deuterium Absorption Cell (HDAC) will measure the ratio of deuterium to hydrogen. Apart from their grating ruling density, optical coatings and detector details the three spectrometers are similar, having a telescope with a three-position slit changer, a baffle system, and a spectrograph with a microchannel plate detector employing a coded anode converter. Each telescope consists of an off-axis parabolic section with a focal length of 100 millimetres, a 22×30 millimetre aperture, and a baffle with a field of view of 3.67×0.34 degrees. A mechanism places one of the three entrance slits at the focal plane of the telescope, each corresponding to a different spectral resolution. The HSP measures undispersed light from its own parabolic mirror using a photomultiplier tube detector. The HDAC has a hydrogen cell, a deuterium cell, and a Channel Electron Multiplier (CEM) detector to record photons that are not absorbed in the cells. The resonance absorption cells are filled with pure molecular hydrogen and deuterium, as appropriate, and are located between an objective lens and a detector. Both cells are made of stainless steel coated with teflon and are sealed at each end with magnesium fluoride (MgF_2) windows.[7]

By observing stars as they are occulted by the limbs of Saturn or Titan, the UVIS will provide thermal and compositional profiles of the thermosphere to complement those by the CIRS more deeply within the atmospheres. It will be able to monitor the auroral displays that form as solar wind particles flood into the magnetosphere and

Cassini's Ultraviolet Imaging Spectrograph. (Courtesy of Ione Caley, administrative assistant to L.W. Esposito of the Laboratory for Atmospheric and Space Physics at the University of Colorado.)

excite the ionosphere, and correlate the diurnal variations of the charged particles with the state of the magnetosphere and the solar wind. It should also be able to detect any exospheres of the icy satellites comprising species sputtered from the surface by the charged particles in the magnetospheric wind. The photometer's light curves of stars that the spacecraft sees passing behind Saturn's ring system will yield profiles of the distribution of ring material.[8] The principal investigator is L.W. Esposito of the Laboratory for Atmospheric and Space Physics of the University of Colorado at Boulder.

The Italian Space Agency provided Cassini's sophisticated telecommunications system, which has been designed to double as the *Titan Radar Mapper* (RADAR). In this role, it will utilise the five-beam feed assembly of the spacecraft's high-gain antenna as a 'synthetic aperture' radar. Once the spacecraft has oriented itself to aim its antenna at Titan, it will transmit a radar pulse and then monitor how the energy is reflected by the surface. It will operate for approximately an hour before closest approach. On a typical fly-by with a minimum altitude of 1,000 kilometres, the spatial resolution will vary between 0.35 and 1.7 kilometres along the imaging strip. A short wavelength has been selected to document the surface relief accurately, so it will not be able to penetrate water ice to 'sound' the subsurface. A radar altimeter will determine the mean elevation of each sample point to within 90 to 150 metres vertical resolution. Varying incidence angle and polarisation will allow simultaneous retrieval of the temperature and dielectric constant of the surface, and thereby yield compositional information. Although the radiometer's sample spot will span at least 7 kilometres, this will facilitate thermal mapping. Michael Janssen of JPL is leading the radiometer project. The radar data will be stored until Cassini can reorient itself to turn the dish towards the Earth. The Magellan spacecraft used a similar mode of operation, with its high-gain antenna doubling as a radar to map Venus through its cloud cover. Magellan's orbital plane was fixed relative to the stars so that the radar produced a series of strips (dubbed 'noodles' since they were long and thin) as the planet rotated on its axis, and these were subsequently integrated into a global map by sophisticated software. By being limited to a series of fly-bys of Titan, the radar strips from Cassini will not overlap, so rather than assemble a global map it will be able to survey only isolated swaths of the surface at high resolution, with lower resolution data providing the context in which to interpret the fine detail. The radar data will characterise the moon's surface by its reflectance, which depends upon composition, slope and degree of roughness. It will be possible to distinguish between compositional and topographic features using the altimeter data. With knowledge of the surface morphology, geologists should finally be able to infer the processes which formed and modified Titan's surface. It should also resolve the question of whether there are fluid hydrocarbons on the surface. Despite its distinctive chemistry, Titan's morphology might turn out to be remarkably familiar. Later, the radar may also be used to study the ring system. Charles Elachi of JPL is the principal investigator for the radar imaging experiment.[9]

Voyager 1's single fly-by of Titan took six months of detailed planning. Cassini will make 44 Titan fly-bys during its four-year primary mission. "We have to be many times more efficient at planning the observations," noted R.D. Lorenz of the University of Arizona's Lunar and Planetary Laboratory, and a member of the team. "There is a lot of work in setting up the software and deciding which places to look at. This is a very difficult decision-making process, because there are lots of good instruments on Cassini and you have to think about what you can learn from each observation." The plan is to map the entire surface at medium resolution, and at least 25 per cent at high resolution. The initial focus, of course, will be to investigate the site where the Huygens probe sets down.

Particles and fields sensors
Many of the instruments in Cassini's particles and fields suite are technologically updated and functionally enhanced versions of those of the Pioneers, Voyagers and Galileo spacecraft. To an extent, these instruments address general themes, and each provides a particular part of the 'big picture'. When dedicated to particles and fields observations, Cassini will slowly rotate in order to optimise the spatial resolution of its sensors.

The *Ion Neutral Mass Spectrometer* (INMS) has an 'open' ion source, a 'closed' ion source, a quadropole deflector and lens system, a quadropole mass analyser and a dual detector system. The open ion source produces ions by ionising neutral gases. It includes an ion trap-deflector that forms trapped ions into a beam. This minimises interaction effects between the gas environment and the open source surface as the source directly samples the gaseous species. The closed ion source also makes ions by ionising neutral gases but uses ram density enhancement to make measurements of higher accuracy and sensitivity for the more inert atomic and molecular species than is possible with the open ion source. This is achieved by maintaining a high input flux to an enclosed antechamber and then limiting the gas conductance by the use of an orifice. Ions are directed to the mass analyser from the selected ion source by changing the potentials on a 90-degree quadropole deflector, an electrostatic device which allows both ion sources to be sequentially switched into a common exit lens system. The quadropole mass analyser has four precision-ground hyperbolic rods mounted in a rigid mechanism. The transmitted mass, the resolution, and the ion transmission are controlled by varying the radio-frequency and direct-current electric fields between adjacent rod pairs, while opposite rod pairs are kept at the same potential. The ion dual detector system amplifies and measures the input from the mass analyser using two continuous dynode multipliers. As with most of the particles and fields suite, the INMS senses the spacecraft's immediate environment. Its quadropole mass analyser will measure the distribution of positive ions and neutral species of up to 99 atomic mass units. In several Titan fly-bys, Cassini will pass close enough to sample the moon's thermosphere and investigate its interaction with the planet's magnetosphere or, if the magnetopause is compressed within the moon's orbit, its interaction with the solar wind.[10] The principal investigator is Hunter Waite of the University of Michigan.

The *Cosmic Dust Analyser* (CDA) will directly measure dust and ice particles in interplanetary space and in the Jovian and Saturnian systems. It will measure the physical, chemical and dynamical properties of particulate matter as functions of the range to the Sun, to Jupiter, and to Saturn and its satellites and rings. It consists of a Dust Analyser (DA), a High-Rate Detector (HRD) and an Articulation Mechanism (AM). The DA has a quartet of charge pick-up grids, a hemispherical target, an ion collector, an electron multiplier and associated sensor electronics. The charge pick-up grids mounted at the entrance of the sensor collect the initial impact particles. The hemispherical target is divided into a ring-shaped impact ionisation target and a chemical analyser target set in the middle of the ionisation target. The chemical analyser target has an acceleration grid placed 3 millimetres in front of it. The ion collector has a grid that is negatively biased in order to collect the positively charged plasma ions produced at the impact ionisation target. The electron multiplier is located in the centre of the hemispherical ion collector target. It amplifies the signal produced by ions capable of penetrating the ion collector grid. These ions originate from plasma produced by particle impacts either on the impact ionisation target or on the chemical analyser target, and the output signal from the multiplier differs depending upon the target from which the impacts are being measured. The sensor electronics are in an electronics box attached to the DA sensor chassis. Among other components, this has Charge-Sensitive Amplifiers (CSAs) that measure the signals from all of the grids in the DA. The main electronics has amplifiers and transient recorders, a control and timing unit, a microprocessor, a bus interface unit, a power input circuit, a low-voltage converter and a housekeeping unit. All CSA and electron multiplier signals are separately amplified by logarithmic amplifiers, digitised by an analogue-to-digital converter and stored on transient recorders. Only the recorder connected to the pick-up grids runs continuously. All the others are activated only when a signal is detected at a target or the acceleration grid. The control and timing unit stores and decodes the information provided by the microprocessor and produces all timing and synchronisation signals for operating the instrument. The microprocessor samples and collects the buffered data, coordinates the subsystem measurement cycle, controls the operating modes, processes the data according to a programme loaded in its memory, and outputs data to the spacecraft upon request. The housekeeping unit is a data system that multiplexes, digitises and stores information on the instrument current, the low voltages, the high voltages and the temperature measurements. The AM allows the entire package, including the HRDs, the DA, the main electronics, and the AM electronics, to be oriented relative to the spacecraft's coordinate system. The HRDs are two redundant independent sensors. The electronics for the sensors are contained in the HRD electronics box, and each sensor has its own electronics, independent of its partner. The HRD will be operated in two modes: 'normal' and 'calibrate'. In the normal mode, the operational HRD continuously collects dust particle data. In the calibrated mode, a sequence of pulses is issued to the HRD by the In-Flight Calibrator (IFC) to verify the stability of the electronics.

The CDA is similar to instruments sent into deep space aboard the Ulysses and Galileo spacecraft. Its high-rate impact detector and dust analyser will measure the flux of particulate matter in the space through which the spacecraft passes, determining the mass, velocity, flight direction, electric charge and composition of each individual mote. The multi-coincidence detector has a resolution of 10,000 impacts per second. During the interplanetary cruise, the instrument will follow up earlier studies of the material in the asteroid belt and will then monitor the 'streams' of dust in the space beyond. Once in the Saturnian system, it will measure the interactions of particles with the rings, satellites and magnetosphere. This should settle the question of how much material there is in the ring plane beyond the 'A' ring. The principal investigator is Ralf Srama of the Max Planck Institute for Astrophysics in Heidelberg, Germany.

The *Cassini Plasma Spectrometer* (CAPS) measures the flux of ions as a function of mass per charge, and of ions and electrons as functions of energy per charge in the spacecraft's environment.[11,12] It incorporates an Ion Mass Spectrometer (IMS),[13,14] an Ion Beam Spectrometer (IBS)[15,16] and an Electron Spectrometer (ELS).[17,18] As Cassini conducts its orbital tour of the Saturnian system, CAPS will progressively map the planet's magnetic field, investigate the interaction between the solar wind and the magnetosphere, identify the sources and sinks in the ionospheric plasma and study the aurorae. During fly-bys of Titan, CAPS will study how the moon's ionosphere interacts with the planet's magnetosphere, or, when it is exposed to it,

Cassini's Cosmic Dust Analyser. (Courtesy of Ralf Srama of the Max Planck Institute for Astrophysics in Heidelberg.)

Cassini's Plasma Spectrometer.

with the solar wind. There is a torus coincident with Titan's orbit which is mainly hydrogen, but there may be more complex species leaking from the moon. CAPS will characterise the ion flow around Titan to investigate the processes by which the material leaking from the atmosphere enters the torus.[19] The principal investigator is D.T. Young of the Southwest Research Institute in San Antonio, Texas.

The *Radio and Plasma Wave Spectrometer* (RPWS) is an improved version of an instrument carried on the Galileo spacecraft.[20] It has a Langmuir probe to measure the temperature and electron density of plasma in the immediate vicinity, and separate electric and magnetic field sensors with high-, medium- and wide-band receivers. The electric field sensor has three antennas composed of interlocking beryllium–copper sections, and a preamplifier to boost the antenna signals. Individual electric motors will deploy the antennas to a length of 10 metres. The magnetic search coil sensor assembly comprises a tri-axial sensor assembly and preamplifier. The tri-axial sensor consists of three mutually orthogonal metallic alloy cores with two sets of windings each, one to produce flux in the core and another to detect the flux. The Langmuir probe has a sensor, preamplifier and associated control electronics.[21] The sensor is a 5-centimetre-diameter sphere located at the end of a 1-metre-long rod which is folded in a stowed state for launch and then deployed in space. On the interplanetary cruise RPWS will monitor the solar wind. Once within Saturn's magnetosphere, it will investigate the configuration of the magnetic field to determine the planet's rotation rate; diurnal variations in the ionosphere; electrical discharges between storms in the atmosphere, and perhaps to the ionosphere; the kilometric radio emissions; and the long-term variability of the planet's interaction with the solar wind. During the Titan fly-bys, RPWS will listen for evidence of discharges to suggest lightning in the troposphere and investigate the processes by which the material leaking from the atmosphere enters the torus.[22,23,24,25] The principal investigator is D.A. Gurnett of the University of Iowa.

The *Magnetospheric Imaging Instrument* (MIMI) used three sensors. The Low-Energy Magnetospheric Measurement System (LEMMS) measures the angular and spectral distributions of low- and high-energy protons, ions and electrons in the

Cassini's Magnetospheric Imaging Instrument.

energy range 20 keV to 20 MeV. It is an improved version of part of the Energetic Particle Detector on the Galileo spacecraft The Charge-Energy Mass Spectrometer (CHEMS) is similar to an instrument provided for the Geotail spacecraft as part of the International Solar–Terrestrial Physics Programme. It will measure the charge and composition of the ions in the most energetically important portion of a magnetospheric plasma. The Ion and Neutral Camera (INCA) can undertake two different types of measurements: firstly, the camera has a two-dimensional field of view covering 90 × 120 degrees and it can remotely image the global distribution of the energetic neutral emission of hot plasmas in a magnetosphere and determine the composition and velocities of those energetic neutrals for each image pixel; secondly, it can measure the three-dimensional distribution, velocities and composition of ions in the regions of interplanetary space and planetary magnetospheres in which the energetic ion fluxes are very low.[26,27]

In addition to remotely imaging magnetospheres, the MIMI will directly measure the composition, charge state and energy distribution of the fast neutral species and the energetic ions and electrons in order to investigate how magnetospheres interact with the solar wind. When at the apoapsis of its Saturn orbit, Cassini will be outside the magnetosphere, and the MIMI will be able to image Titan's torus as a whole, and measure the composition, charge and energy distribution of fast neutral species and energetic ions and electrons in order to determine how this interacts with the magnetosphere, or, when it is exposed to it, with the solar wind. The principal investigator is S.M. Krimigis of the Applied Physics Laboratory, part of Johns Hopkins University in Maryland. "Every new spacecraft has instruments which expand our ability to see things," he pointed out, "and with MIMI, we are able to visualise the invisible." The CHEMS team is headed by D.C. Hamilton of the University of Maryland, the INCA team by D.G Mitchell and the LEMMS team by Stefano Livi, both of the Applied Physics Laboratory.

The *Dual-Technique Magnetometer* (MAG), the latest in a series of ever more sophisticated magnetometers, will measure the strength of the magnetic field in the vicinity of the spacecraft. It consists of a vector/scalar helium magnetometer sensor, a fluxgate magnetometer sensor and a Data Processing Unit (DPU). The Vector/ Scalar Helium Magnetometer (V/SHM) sensor, supplied by JPL, will make both vector (magnitude and direction) and scalar (magnitude only) measurements of magnetic fields. The Fluxgate Magnetometer (FGM) sensor that will make vector field measurements was provided by the University of London's Imperial College. The DPU provided by the Technical University of Braunschweig will interface with Cassini's computer. Because magnetometers are sensitive to electric currents and ferrous components on the spacecraft, they are mounted on an extended boom away from the vehicle. The FGM sensor is halfway along the magnetometer boom and the V/SHM sensor is at its end.[28] The boom comprises thin non-metallic rods that are compressed compactly prior to launch and then deployed *en route*. In addition to monitoring the solar environment during the interplanetary cruise, MAG will map the magnetic fields in the Jovian and Saturnian magnetospheres in order to study how they interact with the solar wind. During the Saturnian tour, it will also investigate how that planet's magnetosphere interacts with the ring system, the icy satellites and Titan's atmosphere.[29,30] D.J. Southwood of the University of London's Imperial College is the principal investigator.

In addition to the instruments carried for specific experiments, the spacecraft's communications link to Earth doubles as the *Radio Science Subsystem* (RSS). When Cassini is undergoing superior conjunction, the way in which its signal is degraded by passing close to the Sun provides information concerning the solar corona. In an occulation, the signal's refraction can profile an atmosphere to reveal its physical state and chemical composition. When in interplanetary space, 'anomalous' Doppler shifts could indicate the passage of a gravitational wave through the Solar System. During fly-bys, tracking of how the spacecraft is deflected by a body's gravity will measure its moment of inertia, the key parameter from which internal structure can be inferred. A.J. Kliore of JPL is the principal investigator for the radio science investigations.

HUYGENS PROBE

The Huygens Descent Module consists of a Fore Dome, Experiment Platform, Top Platform and Aft Cone. The 2.7-metre-diameter conical Front Shield will protect it from the heat of the plasma generated during the deceleration phase of penetrating Titan's atmosphere. The Descent Module and its parachute systems are enclosed by the Back Cover. Huygens is attached to the side of the Cassini spacecraft by the ring of the Probe Support Equipment (PSE). Apart from its bi-annual functionality checks, the probe will remain dormant during the interplanetary cruise. As Titan looms, Cassini will adopt the proper orientation, charge the probe's batteries, perform a final check, load its 'coast timer' with the time at which it should activate the sequencer, and release it.

The Entry Subsystem works as the probe is released, and during the subsequent atmospheric entry. It consists of the Spin-Eject Device, the Front Shield and the

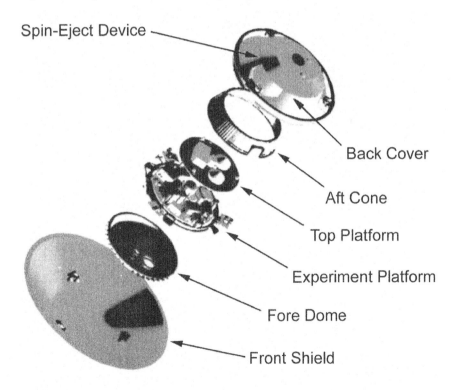

Spin-Eject Device

Back Cover

Aft Cone

Top Platform

Experiment Platform

Fore Dome

Front Shield

The structural components of the Huygens probe.

Back Cover. Activating the Spin-Eject Device severs the bolts attaching the probe to the PSE. As a trio of springs ease the probe away, a curved track and roller system will spin it up to 7 revolutions per minute in order to stabilise it during its several-week ballistic 'fall' to Titan. The spacecraft's attitude at the moment of release will ensure that the probe will contact the moon's atmosphere at the correct angle to penetrate safely. The Thermal Regulation Subsystem will utilise a variety of passive controls to protect the probe from the extremes of temperature to which it will be subjected on the protracted interplanetary cruise. As the probe will chill down upon being released by Cassini, it will be protected by a multiple-layer insulation blanket whose low emissivity will retard radiative cooling, and 1-watt radioisotope heater units strategically situated to ensure that its internal equipment does not freeze. The insulation blanket will be burned off early in the deceleration phase of entering Titan's atmosphere. Aerospatiale in France supplied the Front Shield, which uses Space Shuttle-style silica tiles. As the thermal stress on the Back Cover will be less intense, its shielding is a spray-on silica foam. Nevertheless, at just over 100 kilograms the thermal protection system contributes almost one-third of the probe's mass. The Inner Structure Subsystem comprises two aluminium–honeycombed platforms and an aluminium shell. It is connected to the Front Shield and Back

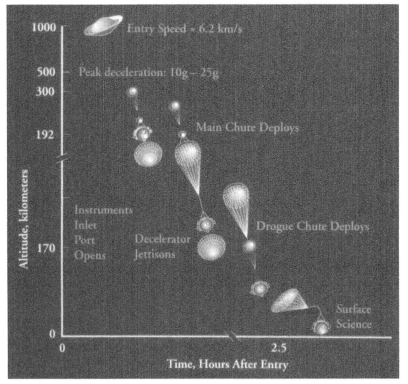

The Huygens probe's descent profile through Titan's atmosphere.

Cover by fiberglass struts with pyrotechnic mechanisms. The parachutes are carried on the Top Platform. The upper and lower surfaces of the Equipment Platform accommodate the five lithium sulphide batteries of the Electrical Power Subsystem and the six science experiments.

As Titan's atmosphere is so deep, the first contact will be at an altitude of about 1,250 kilometres, travelling at 6,150 metres per second. The timer will awaken the probe's sequencer 15 minutes beforehand. Atmospheric penetration will be over the day hemisphere in order to facilitate imaging during the descent. Although the probe will be subjected to a peak deceleration of 16 times that of the Earth's gravity, this is an order of magnitude less than that endured by the Galileo probe in penetrating the Jovian atmosphere. Simulations indicate that the temperature of the shockwave that will form in front of the probe will peak at about 12,000°C as it passes from 300 to 200 kilometres altitude, and this heat may stimulate reactions in the detached layers of hydrocarbons in this altitude range.[31] The deceleration phase will last about three minutes, during which the probe's velocity will fall to about 1,400 kilometres per hour. At an altitude of about 170 kilometres, towards the end of the deceleration phase, as the onboard accelerometers sense that the probe has slowed to Mach 1.5, a mortar will deploy a 2.6-metre-diameter drogue parachute, which will pull off the Back Cover in order to permit the deployment of the 8.3-metre-diameter braking parachute. About 30 seconds later, having slowed to Mach 0.6, the probe will jettison its Front Shield and expose the inlets for the sampling instruments, which will start to report data at this point. Fifteen minutes later, at an altitude of 125 kilometres, and now falling at 360 kilometres per hour, the main parachute will be released, in the process drawing out the 3-metre-diameter parachute that will be used for the rest of the descent. If the probe were to remain on its large braking parachute, it would not be able to reach the surface before its batteries expired. This smaller parachute has been scaled so that the probe will rapidly descend through the chilly tropopause without freezing, and yet reach the surface sufficiently slowly to survive the impact. On Titan's chilly surface, water ice will be as hard as steel, so the probe will hopefully strike the surface at no more than 6 metres per second. The first part of the probe to make contact with the surface should be the spear-like penetrometer on the base, which was built by Ralph Lorenz while working for his doctorate at the University of Kent. On the surface, the probe's accelerometers will indicate whether it is stationary on dry land, or has splashed down and is riding the waves.

Overall, the parachute descent is expected to take just over two hours (even with the benefit of the Voyager data it is difficult to predict the timing accurately), so the probe will have at most half an hour on the surface before its batteries expire. They are rated for a total of 350 watts, which ought to be more than enough for the descent. The probe will report the data from its instruments over an 8-kilobit S-Band radio link using a pair of redundant radio transmitters. On board Cassini, systems in the PSE will recover the data and feed this to Cassini for storage in solid-state memory for subsequent replay to Earth, and monitor the Doppler shift on the probe's relay link in order to track its motions within the atmosphere. The Deep Space Network will forward the probe's data to ESA's operations facility in Darmstadt, Germany, for analysis. The Data Relay Subsystem was built by Alenia Spazio in Italy.

Parachute Compartment

Back Cover

Fore Dome

Descent Module

Front Shield

A cutaway of the Descent Module of the Huygens probe snuggled within the shield that will protect it during the deceleration phase of entering Titan's atmosphere.

DASA undertook the assembly, integration and testing of the Huygens probe for shipment to NASA. Once the 2,150-kilogram 'dry' mass of the main spacecraft was augmented with 3,132 kilograms of propellant, the 373-kilogram Huygens probe and the 165-kilogram launch vehicle adapter, the total payload was 5,820 kilograms. Like all spacecraft designed to enter orbit around a giant planet in the outer Solar System, Cassini's mass is dominated by its propulsion system and propellants. The finished vehicle stood 6.8 metres tall in the Kennedy Space Center's preparation facility.

Instrumentation

Huygens carries 48 kilograms of science instruments to perform the first-ever *in situ* study of Titan's atmosphere and surface. In the parachute descent, the atmospheric instruments will profile the temperature, pressure, density, composition and wind as functions of altitude. As sampling will start at the 170-kilometre level, this 'sample column' will be entirely within the 200-kilometre-deep optically thick orangey haze. After landing, or perhaps splashing down – the Descent Module is designed to float in a hydrocarbon fluid – the physical properties of the surface will be measured and pictures taken of what is sure to be an intriguing out-of-this-world land(sea)scape.[32,33]

The *Huygens Atmospheric Structure Instrument* (HASI) is the counterpart of an instrument carried by the atmospheric probe dropped by the Galileo spacecraft into Jupiter's atmosphere. As the only one of the Huygens suite to be activated during the deceleration phase, it will profile the temperature and pressure. Its measurements of the state of the atmosphere at altitudes at which the detached layers of thin haze form should help to identify their composition.[34] While Voyager 1 found no evidence of lightning in the atmosphere, HASI will assess this possibility. A thermodynamic study suggested that the formation and evolution of clouds in Titan's troposphere are governed by convection, in which case it will resemble the Earth's in being driven by latent heat released from the primary condensable species.[35] Electrical breakdown may occur in regions of strong vertical motion, resulting in high-current discharges. Even if HASI fails to detect anything, this will not rule out the possibility. Lightning on Titan ought to be concentrated near the subsolar point where the energy input is greatest. The probe's brief *in situ* study will be supplemented during a succession of fly-bys by Cassini's RPWS.[36] HASI's principal investigator is Marcello Fulchignoni, initially of the University of Rome but now at the Meudon Observatory in Paris.

The *Gas Chromatograph Neutral Mass Spectrometer* (GCMS) will draw in gas, mix it with the hydrogen 'carrier', and feed this into the mass spectrometer which will first ionise the sample then separate the ions using a magnetic field to enable them to be counted. It will be able to identify elements up to 146 atomic mass units with a sensitivity of one part in 10^{12}. The results will profile atmospheric composition. The principal investigator is Hasso Niemann of the Goddard Space Flight Center. The *Aerosol Collector and Pyrolyser* (ACP) will sample the stratosphere (above 45 kilometres) and the 'middle' troposphere (in the 15- to 30-kilometre altitude range), evaporate each sample, thermally dissociate its molecular species, and feed its output into the GCMS in order to measure the aerosol composition as a function of altitude. Specifically, it will determine the proportions (a) of carbon, hydrogen, nitrogen and oxygen in aerosols, (b) of condensed hydrocarbons and nitriles in the stratosphere, and (c) of condensed methane in the troposphere. Guy Israel of Service d'Astronomie du Centre National de la Recherche Scientifique (CNES) in France is the principal investigator. About 20 per cent of the GCMS's operating time will be devoted to the analysis of the ACP's samples.

The *Descent Imager and Spectral Radiometer* (DISR) is a composite instrument. A set of fibre optics viewing upwards, to the side, and downwards will illuminate a silicon photodiode, a CCD array, and a pair of near-infrared linear arrays in order to

The layout of the apparatus on the upper (left) and lower (right) faces of the main platform in the Descent Module of the Huygens probe.

undertake spectroscopy right across the ultraviolet to near-infrared spectral range to measure the insolation-scattering properties of the aerosols in the orangey haze. It will profile how solar energy is transmitted through the atmosphere. Throughout the descent, the upward-looking spectrometer will peer at the bright patch in the sky as the haze forward-scatters sunlight, and it will measure the absorption spectrum in order to profile the methane mixing ratio. Meanwhile, other sensors will monitor the descent by imaging the surface repeatedly. In fact, DISR incorporates *three* imagers with low, medium, and high resolution. "With these three cameras and a spinning spacecraft, we'll time our images in such a way that we'll take an entire hemispheric panorama from above the horizon down to the surface," explained P.H. Smith of the University of Arizona's Lunar and Planetary Laboratory, who helped to develop the instrument. Transmitting the panoramas during the final 20 kilometres of the descent will require fully two-thirds of the relay bandwidth. "The information we'll get out of these images will be absolutely stupendous," Smith promised.

Imaging will start soon after the probe sheds its thermal shielding, at an altitude of 170 kilometres, at which time the narrow field of view will provide a resolution of about 1 kilometre. If the probe is drifting in a strong wind, this should be evident in the imagery, and the imagery will serve to map the swath of the ground track. Titan, being ten times farther from the Sun, receives 1 per cent of the insolation received by the Earth, a few per cent of which will filter through the optically thick orangey haze to the surface. At the subsolar point, the illumination should be equivalent to a clear terrestrial day shortly after sunset. Elsewhere on the illuminated hemisphere it will be rather gloomier, but not dark. Unless there is a thick overcast of cloud, there ought to be a broad solar aureole. Although many wavelengths will penetrate to the surface, the methane will absorb strongly, and in the final phase the probe will turn on a lamp to enable it to take a surface spectrum. The field of view of the final look-down image should span about 75 metres, so it ought to be possible to determine precisely where the probe settles and put the measurements of the surface characteristics into a proper context. The principal investigator is M.G. Tomasko of the University of Arizona.

In the same way as the Galileo spacecraft made high-resolution measurements of the Doppler on the radio relay from the probe that it sent into Jupiter's atmosphere, Cassini will monitor its probe. The *Doppler Wind Experiment* (DWE) will measure the direction and strength of the winds with altitude with a resolution of 1 metre per second, starting when the probe's transmitter switches on and continuing throughout the parachute descent. Zonal (latitudinal) winds are expected to be much faster than the meridional (equator-to-pole) winds. If the data is 'clean', it should be possible to measure the rates at which the probe swings and rotates beneath its parachute. The principal investigator is Michael Bird of the University of Bonn in Germany.

At the end of the descent, the *Surface Science Package* (SSP) will employ a suite of independent subsystems to determine the physical and chemical properties of the surface. As seven of the sensors require intimate contact with the surface, they are mounted inside, or on, the lower rim of a 100 × 100 millimetre square cross-section cavity (dubbed the 'Top Hat') in the Fore Dome, this having been deemed preferable to either developing a deployment mechanism, or competing with the inlets for the ACP and GCMS instruments and the radar altimeter's antenna for access to the Fore Dome's surface. The sensors not requiring exposure to the surface are mounted internally. In the event that the probe splashes down, many of the sensors are designed to characterise the fluid. Some of the sensors will also be able to make measurements during the descent. The principal investigator is J.C. Zarnecki of the Open University in England. The scientific objectives of the SSP are:

- to determine the physical nature and condition of Titan's surface;
- to measure the abundances of the major constituents to place constraints on theories of atmospheric and oceanic evolution;
- to measure the thermal, optical, acoustic and electrical properties and density of any ocean, providing data to validate physical and chemical models;
- to determine wave properties and ocean–atmosphere interaction; and
- to provide ground-truth for interpreting the main spacecraft's radar mapping and other remote-sensing data.

The Accelerometer (ACC) subsystem comprises two piezoelectric sensors, one of which (ACC-E) is a penetrometer mounted on a spear that will come into contact with the surface. Being internally mounted, the other sensor (ACC-I) will be able to measure descent and surface accelerations. The Tilt (TIL) sensor subsystem works on an electrolytic principle. It uses platinum terminals and a methanol-based liquid enclosed in a sealed glass housing; it also has fluid damping to improve its operation in moderately dynamic environments. All the sensors are mounted internally. One individual sensor element senses the local vertical about a single axis, and pairs of elements determine the tilt angle in any plane. The Thermal Properties (THP) sensor assembly in the Top Hat will measure the temperature and thermal conductivity of the lower atmosphere and any surface fluid. An electrical current passed through platinum wires 5 centimetres in length and 10 and 25 microns in diameter will act to heat the surrounding medium, and a series of resistance measurements will measure the rate of heating of the element at 0.1-second intervals in order to detect the onset of convection. The Acoustic Properties (API) sensors employ piezoelectric ceramic devices similar to those used in marine applications. Two of the three transducers will face each other across the diameter of the Top Hat, and will alternate between transmit and receive modes in order to measure the velocity of sound (API-V). The third transducer is an array of elements pointing vertically down that will operate as an acoustic sounder (API-S) by issuing pulses and then listening for returns from the surface during descent or, following splashdown, from the ocean floor. The Fluid Permittivity (PER) sensor in the Top Hat uses simple electrodes whose capacitance varies with the permittivity of the environment. In the case of a splashdown, a single conductivity measurement (CON) will test for polar molecules in the fluid, and the displacement of a buoyant float will be measured by four strain gauges in a bridge arrangement for the Fluid Density (DEN) measurement. The Refractometer (REF) in the Top Hat is a specially shaped prism fitted with a pair of light-emitting diodes to yield internal or external illumination of its curved surface through light guides. Light passing through its top surface will be fed to a linear diode array, and the refractive index measured from the position of the light/dark transition on the array. Together, these various measurements should yield at least a partial characterisation of either a solid or a liquid surface.

Although the stratospheric zonal winds rage at about 200 knots, the troposphere is expected to be milder. In fact, as a thermal gradient is required to induce the flow of air, and Titan's surface temperature is the same from pole to pole to within a few degrees, conditions at the surface are likely to be stagnant. It will be "cold, dark and quiescent" forecasts Caitlin Griffith of Northern Arizona University in Flagstaff. A near-infrared study by the UK Infrared Telescope in Hawaii found that the amount of methane in Titan's troposphere waxes and wanes on a timescale of hours.[37] The terrestrial weather system is very active because the Earth receives a great deal of insolation energy, and because it rotates sufficiently rapidly for the Coriolis effect to stimulate vorticity. However, Titan's atmosphere is chillier, much more massive and sluggish, and the moon rotates very slowly. A different force is thought to drive its weather system. Perhaps the latent heat that is released when a gas condenses plays a significant role. Clouds on Titan are rare and sparse, but at least one large-scale tropospheric storm system has been noted. In part, this is due to the fact that Titan's upper troposphere is super saturated with methane, and so a cloud particle that forms around a smog particulate at that altitude will rapidly grow into a large raindrop and fall.[38] Although sporadic drops will evaporate as they fall, an occasional storm may well dump a large amount of precipitation onto the surface. If Huygens is (un)lucky, it will descend into what passes for the local monsoon.

FINAL PREPARATIONS

By the start of 1996, the subsystems for the Cassini spacecraft were being assembled in the 'clean room' of JPL's Spacecraft Assembly Facility.

The Attitude and Articulation Control Subsystem (AACS) that will enable the spacecraft to maintain its bearings in space was connected to the Command and Data Subsystem (CDS) that will serve as its 'brain', and to the Power and Pyro Subsystem that will regulate its electrical power supply. Meanwhile, at White Sands in New Mexico, one of the main engines built by Lockheed-Martin in Denver, Colorado, was run for 200 minutes in order to certify its design.

In February, the first scientific instrument – the University of Iowa's Radio and Plasma Wave Spectrometer – was fitted, and the others followed over the next few months. Although each instrument had been thoroughly tested by its builder, it was necessary to re-run the tests in order to verify that they all worked in conjunction with the spacecraft's systems. As each instrument was added, tests had to be run to make sure that it did not upset the existing suite. To facilitate testing, a new Science Operations and Planning Computer in the Distributed Operations Interface Element enabled the science team to remotely operate their instruments and receive its data at their home institutions. In addition to providing each team with unprecedented access to the engineering and science data from its hardware, this 'distributed' strategy significantly reduced costs and saved time. Meanwhile, experts on planetary rings met to discuss the estimates of ring particle distribution to be used by Cassini trajectory designers in identifying a safe flight path for the Saturnian orbital tour. Recent studies of Saturn's rings helped to ensure that the flight path would avoid areas that might pose a hazard. It was also decided that the minimum altitude for a Titan fly-by should be 950 kilometres. As the detached layers of haze in the moon's upper atmosphere extend to an altitude of 700 kilometres, the spacecraft will actually pass through the tenuous thermosphere. If it were to penetrate any more deeply, the 'drag' would impair its ability to control its orientation.

In March, in Germany, the integration of the Huygens probe started, and in Italy the Radio Frequency Electronic Subsystem was assembled. At the Kennedy Space Center, a full-scale mock up of the Cassini spacecraft (dubbed the 'Trailblazer') was run through the procedures for testing and mating with the launch vehicle adapter in order to identify likely handling problems and to develop alternatives. In April, JPL took receipt of a fully functional engineering model of the Huygens probe. This was mated with the spacecraft and subjected to a full range of tests in order to verify that Cassini could communicate with a probe. Meanwhile, in Europe, the real probe was being subjected to thermal, vibrational and shock testing. Once the Trailblazer was returned to JPL, it served as a model for tailoring the many individual sections of the thermal blanket that would protect the vehicle from the space environment. Once the propulsion module had been certified leak-free by Lockheed-Martin, it was shipped to JPL in July.

By this point, science planning by more than 250 scientists from 17 countries had reached an 'intermediate' level of detail, having established the general objectives for the study of Saturn, its magnetosphere, its rings, its icy companions and Titan. As a large number of individual observations would have to be made, the task was to plan the orbital tour so that the spacecraft would be in position to make the observations required by each science investigation. At a series of meetings in the USA and Europe, the scientists, engineers and mission planners evaluated different orbital tour options to determine which offered the best opportunities for most of the instruments. Enceladus has only a few impact craters disfiguring its remarkably smooth surface, and Cassini is to try to ascertain whether Enceladus has an internal heat source that occasionally melts the ice sufficiently to erase its craters. A search will be conducted for geysers venting water ice crystals into space where they form the tenuous 'E' ring. If this can be established, then Enceladus will have to be considered in the same class as the ice-enshrouded Jovian moon, Europa. Another high-priority target is Iapetus, which has a very large dark patch on its leading hemisphere. Cassini is to determine the composition of this material, and reveal whether it derives from the interior or was deposited by some external process. Another high priority is the ring system and, hopefully, Cassini will identify the reason for its tremendous complexity. The physical structure of the magnetosphere will be charted to establish how it interacts with the moons, rings and solar wind. Observations of Saturn will characterise its atmosphere at different altitudes, identify the forces driving the zonal winds, and monitor the behaviour of the weather system over an extended period. At the science group's meeting at Caltech in the final week of July, it was decided to develop an orbital tour providing about 40 Titan fly-bys and a total of up to half a dozen of the most interesting icy moons. The integration of Cassini's main components was completed on 25 September, and by 11 October the engineering model of the Huygens probe had also been integrated. At that point, all manner of people were drawn to the visitors' gallery to admire the three-storey-tall robotic explorer. Engineers and scientists who had previously seen only their individual subsystems, and mission planners who had a broader perspective but no 'hands-on' involvement with the hardware, came to see the results of their endeavours. At long last, Cassini–Huygens was a *spacecraft*. As R.J. Spehalski, the Program Manager, observed, "People seeing it for the first time are saying 'Holy smokes!'..."

A few days later the package was moved to the Solar-Thermal Vacuum Chamber for the 'shake and bake' tests. Over the next few months, enormous speakers blasted the vehicle with acoustic vibrations in order to subject it to the random shocks that it would experience during launch. The customised thermal blankets were then fitted, the chamber was reduced to vacuum, and the spacecraft was subjected to the extreme temperatures that it would have to endure in space. The thermal blankets comprised a multitude of segments. Although the fabric looked like gold foil, the shiny golden hue was due to the placement of a transparent layer of amber-coloured material on top of the extremely reflective aluminised fabric. The segments were finely sewn together as a strong but extremely lightweight protective conformal shield. "Our blankets are built unlike any others," said Mark Duran, supervisor of the 'shield shop' that had made the protection for JPL's previous vehicles. On such a long mission, the durability of the shield is a major factor. "Our goal in blanketing Cassini is to keep temperatures on board the spacecraft at room temperature," noted thermal-requirements engineer Pamela Hoffman. In space, temperatures on exposed parts of the spacecraft will range from about –220°C to +250°C. If the spacecraft maintains a fixed orientation relative to the Sun, the illuminated exposed structure will bake and the shaded parts will freeze, therefore the blanket fabrics must withstand this extreme radiation environment. For Cassini, the blankets employ as many as 24 layers of different fabrics, including aluminised kapton, mylar, dacron and other special materials. Some of the blankets were sewn with layers of beta cloth, which is a canvas-like carbon-coated fabric that is especially effective in protecting against micrometeoroids. The blankets must also satisfy electrical standards. The spacecraft will be flying through environments full of charged particles that could cause an electrical arc to form across the blankets, so thin accordion-like strips of aluminium were sewn into every layer of each blanket to ensure that an arc would be electrically grounded.

To ensure that Cassini would be ready for transfer to the Kennedy Space Center in April 1997, testing continued right through the end-of-year festivities. In the past, JPL had put spacecraft in environmentally stable containers and driven them across the country. However, 'anti-nuke' activists had started registering their protests against the use of RTG power units by taking pot shots at the containers – a practice hardly likely to ensure safe passage of what they regarded as a hazardous material! With Cassini being so large, it was decided that the Air Force should transport it in a C-17 Globemaster-II on 21 April. Once the high-gain antenna and the propulsion module had been integrated, and Huygens mounted on its side, the entire vehicle was thoroughly checked-out to verify its functionality. Once the spacecraft was ready, one final item was added. "The 'signature disk' idea began as a very small in-house effort to let team members place their signatures on the spacecraft that they had helped to build, but then we realised we could accommodate many signatures, and we opened it up to everyone," mused Spehalski. As soon as the announcement was made in November 1995, cards had begun arriving from around the globe with as many as 35,000 signatures being received during peak weeks. In excess of 600,000 signatures were eventually submitted by citizens of 81 countries. "It has blossomed into a 'message in a bottle' to Saturn," Spehalski noted. The project-within-a-project was managed by Charles Kohlhase, the science and mission design manager. "We received signatures from individuals young and old, whole families, hundreds of classes of students and indeed whole schools," he said. Some of the names were in solemn memory of deceased family members. For completeness, the signatures of the two astronomers after whom the mission had been named were retrieved from archives. At least one person claimed to be a descendant of Giovanni Domenico Cassini. Every submitted signature was electronically scanned by Planetary Society volunteers and the entire set was then stored utilising the recently developed Digital Versatile Disk (DVD) technology. The artwork on the disk was designed by Kohlhase to celebrate various elements of the Cassini mission, its scientific targets, and the flags of the countries from which the largest numbers of signatures were received. The disk was inserted into a shallow cavity on the side of the spacecraft, beneath the Remote Sensing Pallet, in between two pieces of aluminium where it will be safe from micrometeoroid impacts. The completed spacecraft was then mounted on its launch vehicle adapter and driven to Launch Complex 40 for mating with the Centaur rocket stage. The overall package was so heavy that a Titan IV-B, the most powerful rocket in the Air Force's fleet, would be required to dispatch it.

On 3 October, after persistent attempts by anti-nuke activists to have the launch prohibited, the White House's Office of Science and Technology Policy granted the space agency approval to proceed. In calling for "an end to nuclear proliferation in space" the protesters were rather naive, as interplanetary space is flooded with solar and cosmic radiation. Their immediate concern, however, was their assessment of the environmental threat from the plutonium dioxide in the RTGs and the radioisotopic heaters in the event of a catastrophic failure. The White House deemed the risk to be acceptable. "NASA and its interagency partners have done an extremely thorough job of evaluating and documenting the safety of the Cassini mission. I have carefully reviewed these assessments and have concluded that the important benefits of this scientific mission outweigh the potential risks," said J.H. Gibbons, the director of the Office of Science and Technology Policy, having signed the approval document. "The launch of spacecraft cannot be made completely risk-free," pointed out Louis Friedman, the executive director of the Planetary Society, "but the public can take satisfaction in knowing that NASA is being careful, prudent, and smart."

The Cassini–Huygens spacecraft hanging from a crane during its integration at the Kennedy Space Center.

LAUNCH

The 'launch window' that opened on 6 October 1997 would last until the end of the month. The hope of dispatching the mission on the first day was foiled when an air conditioner damaged a piece of the Huygens probe's thermal insulation, so the launch was rescheduled for 13 October, on which date the window ran from 04:55 to 07:15 local time. However, unacceptably high winds at altitude prompted a 'scrub' and a two-day recycle. Watched by a large number of project members, including several hundred Europeans, Cassini–Huygens lifted off at 04:43 on 15 October from the Air Force's Cape Canaveral facility, its plume lighting up the pre-dawn sky.

Even the Centaur stage was incapable of sending the heavy spacecraft straight to Jupiter, whose gravitational field serves as the 'doorway' to the outer Solar System, so, like the Galileo mission, Cassini was to fly a complex interplanetary cruise which began with the spacecraft heading *sunward*. Consequently, the Centaur's escape burn acted *against* the energy bestowed by the Earth's orbital motion around the Sun and when Cassini separated from the expired stage 43 minutes after launch, it headed inside the Earth's orbit. Ten minutes later, the spacecraft made contact with the Deep Space Network facility at Tidbinbilla, near Canberra in Australia, and reported that its systems were operating normally. "I can't recall a launch as perfect as this one," said Chris Jones, the Cassini spacecraft development manager. The launch vehicle's performance was "right on the money", announced a delighted Spehalski. Indeed, the energy imparted was accurate to within 1 part in 5,000 and, at better than 0.004 degree, the deviation in the trajectory was "insignificant". The flight plan allowed for an early trajectory correction of up to 26 metres per second, but tracking indicated that a mere 1-metre-per-second adjustment would be required. Propellant saved by not making corrective manoeuvres would be put towards extending the spacecraft's Saturnian tour, so the mission was already ahead of nominal. Once safely on its way, Cassini transmitted the telemetry that it had recorded in its solid-state memory during the launch, so that the engineers could assess whether any of its systems had been damaged. There were no anomalies. "The spacecraft is extremely clean," noted Ronald Draper, the deputy program manager, "and mission operations are proceeding in an excellent manner".

Over the next few days, Cassini turned to face its 4-metre-diameter dish antenna towards the Sun, so that it would act as a parasol to protect the rest of the structure from the increased insolation of the inner Solar System; it also released latches that secured instrument covers and other deployable devices. On 23 October, power was directed to the Huygens probe, which the Operations Center in Darmstadt declared healthy. Several days later, the Langmuir probe of the Radio and Plasma Wave Spectrometer was deployed to measure the electron density in the space in which the spacecraft was travelling, and the 10-metre antennas were deployed to monitor the electric and magnetic fields. It was not often that a deep space mission got off to such a flawless start. "The spacecraft is operating beautifully," confirmed Draper on 29 October.

On 15 October 1997, the Cassini–Huygens spacecraft was successfully launched from the US Air Force's Launch Complex 40 by a Titan IV-B/Centaur.

VENUS

The first trajectory correction manoeuvre was scheduled for 9 November 1997. The spacecraft terminated its solar-inertial orientation and adopted the proper attitude to make the burn. The 34.6-second main engine firing was monitored by real-time telemetry, and Draper reported that the engine had performed "very, very nicely". Afterwards, the solar-inertial orientation was automatically resumed. The 2.7-metre-per-second burn fine-tuned the trajectory for the first Venus fly-by. The rest of the year was routine, with Cassini periodically reporting its excellent state of health.[39] Throughout January and February 1998 the Deep Space Network refined the spacecraft's location to enable JPL's navigators to work out the trajectory correction manoeuvre for 3 March. The change in velocity was so small that the hydrazine thrusters were used instead of the main engine. In late March, the spacecraft survived its scorching perihelion at 0.676 AU, slightly within Venus's orbit. The plan included an option for a final trajectory adjustment in early April, but this proved to be unnecessary.[40]

Monitored by the Deep Space Network antennas at Goldstone in California and Madrid in Spain, Cassini performed its first Venus fly-by, skimming 284 kilometres above the planet's surface on 26 April. The closest approach was at 06:45 Pacific Daylight Time, but even at light-speed the signal took 7 minutes to reach the Earth. As the spacecraft crossed the limb, as viewed from the Earth, the radio occultation provided information on the planet's atmosphere. For several hours, the Radio and Plasma Wave Spectrometer listened for evidence of lightning in Venus's atmosphere. The excellent sensitivity of the negative result placed a strong constraint on lightning caused by rapid vertical convection in the dense atmosphere.[41,42,43] "If lightning exists in the Venusian atmosphere, then it is either extremely rare or is very different

Cassini's trajectory during its first fly-by of Venus.

from terrestrial lightning," reflected team leader D.A. Gurnett. "If terrestrial-like lightning were occurring in the atmosphere of Venus within the region viewed by Cassini, it would have been easily detectable." Indeed, in its subsequent fly-by of the Earth, the instrument detected lightning continuously while Cassini was within 14 Earth radii, at rates of up to 70 impulses per second. In the case of Venus, the clouds are at very high altitude (above 40 kilometres) and so there is unlikely to be electrical discharges to the surface. If there is lightning, it will therefore be either from cloud to cloud or up into the ionosphere. Nevertheless, dust rising in volcanic plumes in the stagnant dry lower atmosphere may produce discharges to the surface. If so, then as Cassini flew by there were no active volcanoes.[44] When close to Venus, the radar illuminated the surface as an engineering test of the instrument's functionality, but the geometry was not conducive to receiving the 'bounce'. Operational limitations ruled out conducting a comprehensive study of the planet using the spacecraft's full suite of instruments.

The fly-by gave Cassini an 'assist' of 7,055 metres per second, sufficient to ease its aphelion out beyond the Earth's orbit. "The accuracy achieved by our navigators is roughly equivalent to shooting a basket from Los Angeles to London for a *swish* shot," proclaimed Spehalski proudly. With Cassini firmly on track, Spehalski retired on 5 June. As a veteran of four decades at JPL, during which he had played crucial management roles in both the Galileo and Cassini missions – which JPL director E.C. Stone described as "the last great *flagship* planetary flights of discovery of the twentieth century" – Spehalski received NASA's highest honour, the Distinguished Service Medal. Robert T. Mitchell, who was at that time running Galileo's 'extended mission', was reassigned to manage the Cassini mission.

On 3 December 1998, four days before the 1.58-AU aphelion point of the orbit formed by the first Venus encounter, and out near Mars's orbit, Cassini performed a 88-minute Deep Space Manoeuvre to reduce its speed by 450 metres per second and slip the ensuing perihelion within Venus's orbit. "The performance of the spacecraft and the team in performing this manoeuvre, was just perfect," Mitchell reported. "We couldn't have asked for anything better." The refinement would maximise the effect of the slingshot during the second encounter.[45]

A depiction of Cassini's Deep Space Manoeuvre, firing its main propulsion system.

On 11 January 1999, while performing instrument tests, the spacecraft sensed a potential error in its orientation and 'safed' itself by adopting a minimum-power configuration, with its big dish aimed at the Sun for a thermally benign environment. An analysis of the telemetry established that the cause was a problematic geometry experienced during a very slow roll. The actual orientation is constantly compared to the Attitude and Articulation Control System's model, and the 'safing' was triggered when a reference star very near the edge of the sensor's field of view was lost. After resuming normal operations on 15 January, Cassini made an 11.6-metre-per-second manoeuvre on 4 February to refine its aim for Venus and on 24 March it activated its Cosmic Dust Analyser, which was to remain operational through most of the interplanetary cruise.[46]

In early March, 60 members of the science community met at JPL to consider opportunities to study Venus, Earth and Jupiter *en route* to Saturn, and to consider the optimal plan for the orbital tour of the Saturnian system.

A 0.24-metre-per-second trajectory correction manoeuvre was made on 18 May to fine-tune the fly-by of Venus, which took place at 13:30 Pacific Daylight Time on 24 June. The closest point of approach was at an altitude of 603 kilometres – some two seconds earlier and 4 kilometres higher than planned. This time, most of the scientific instruments were active to make observations for the final few hours of the approach and for two days thereafter, with the data being stored for later replay. On 29 June, Cassini made its 0.7166 AU perihelion passage.

Cassini's trajectory during its second fly-by of Venus.

EARTH

The second Venus encounter had accelerated Cassini by 6,690 metres per second and deflected it towards the Earth. Trajectory correction manoeuvres (of 43.5 metres per second on 6 July; 5.1 metres per second on 19 July; 36.3 metres per second on 2 August; and 12.26 metres per second on 11 August) refined the Earth encounter. Despite JPL's confidence in its interplanetary navigators, the anti-nuke community was vocal in expressing its concern that the spacecraft would somehow "spin out of control" and enter the atmosphere, "scattering its radioactive fuel" – even though the plutonium dioxide is in the form of a ceramic and a RTG case is designed to survive re-entry. The closest point of approach was at an altitude of 1,166 kilometres above the eastern South Pacific at 20:28 local time on 17 August (03:28 Universal Time on 18 August). Travelling at almost 20 kilometres per second, it presented a fleeting target for observers on Pitcairn and Easter Islands. The arrival time was within 0.6 second of the schedule, and the altitude was within 5 kilometres of nominal.

The magnetometer started sampling on 13 August, and its 11-metre boom was deployed on 15 August. The magnetopause was encountered at a range of about 10 Earth radii. The calibration of the mode in which the instrument will operate once in the Saturnian magnetosphere was verified,[47] and useful data on the solar–terrestrial relationship was secured.[48] The Radio and Plasma Wave Spectrometer was activated a month prior to the fly-by, while the spacecraft was some 600 Earth radii upstream in the solar wind.[49,50,51,52] Particles and fields studies were coordinated with satellites in Earth orbit. These included the Polar spacecraft, in an elliptical orbit that enabled it to spend most of its time high over the north pole. Cassini's trajectory penetrated the afternoon plasmasphere, cutting through the outer reaches of the ionosphere, out

Cassini's trajectory during its fly-by of the Earth.

through the plasma sheet on the dawn side at 5.2 radii and on into the magnetotail. Cassini's Plasma Spectrometer determined the electron populations in this region on an unprecedented timescale, defining the boundaries of the plasma regions.[53,54,55] The magnetometer detected the emergence from the magnetotail at a range of about 60 radii, after which there were multiple re-entries as the tail flapped in the gusty solar wind. As the spacecraft withdrew, it flew alongside the magnetotail.[56] The geometry provided an opportunity for the Magnetospheric Imaging Instrument to investigate the magnetotail to 6,000 radii.[57,58,59] The INCA imagery provided a significantly better spatial resolution than the 'picture' inferred from the Geotail spacecraft's data.[60] The particles and fields phase of the Earth fly-by was concluded on 19 September.

While in the Earth's vicinity, the Radio and Plasma Wave Spectrometer made a coordinated study of the Jovian decametric radio emissions with the Wind spacecraft that was orbiting the Earth.[61,62,63] The small but finite difference in viewing angles (a few degrees) meant that their data sets could be correlated to investigate 'beaming' characteristics. The beam width was 1.5 ($\pm$0.5) degrees and the instantaneous widths of the walls of the hollow conical radiation beams could be measured. The emissions from Io were confirmed to sweep around with the moon's rotation rate, rather than with the planet.

This fly-by provided an opportunity to test the imaging instruments, and verify their calibrations. C.C. Porco was delighted by the Imaging Science System, which was functioning "beautifully". A few scans of the Moon provided a useful check on the Visual and Infrared Mapping Spectrometer.[64] And while passing over South America, the radar made scatterometry and radiometry measurements in coordination with remote-sensing satellites.[65]

The Earth slingshot accelerated Cassini by 5,477 metres per second, giving it sufficient energy to reach Jupiter, and the trajectory was refined on 31 August by a 6.7-metre-per-second burn.[66] Once beyond the Earth's orbit, the Ultraviolet Imaging Spectrograph's absorption cell started a systematic study of hydrogen and helium in the interplanetary medium, its data placing a strong constraint on the distribution of the interstellar hydrogen and helium flowing through the Solar System.[67] Heading swiftly out, Cassini crossed Mars's orbit on 25 September, and in November it adopted an off-Sun orientation to place the Cosmic Dust Analyser's aperture in the plane of the ecliptic, preparatory to entering the asteroid belt a few weeks later.

Meanwhile, one team of planners had developed the operational concept for the science observations to be undertaken during the fly-by through the Jovian system, and another team was making progress towards allocating time to the various instruments for the 'near encounter' period (that is, 30 minutes either side of the point of closest approach) of the Titan fly-bys. Additionally, the first meeting was held to define the Probe Relay Critical Sequence. The team's task was to review the interfaces between Cassini and Huygens, and an end-to-end test of the relay system was scheduled for February, at which time a Deep Space Network antenna would mimic the probe by transmitting to the relay receiver on Cassini. On 15 December 1999, 21 scientific papers were presented to an American Geophysical Union meeting

While passing the Earth, Cassini tested its Imaging Science System by snapping the crescent Moon, with perfect results. The Sun has just risen at the site on the Sea of Tranquillity where Apollo 11 landed thirty years previously.

in San Francisco, reporting the preliminary results of Cassini's time spent in the inner Solar System.

DESIGNING THE ORBITAL TOUR

Much had been learned from the process by which the Galileo spacecraft's orbits of Jupiter had been planned on a just-in-time basis.[68] In the case of the Jovian system, there were four moons sufficiently massive for gravity-assists to enable a spacecraft to contrive an evolving trajectory for a specific orbital tour. Cassini would be able to use only Titan. An early decision was to set up a fly-by of the enigmatic outer moon Phoebe on the way into the system. For the 1997 launch window, this required the spacecraft to perform its Saturn Orbit Insertion manoeuvre on 1 July 2004.

Given the number of mission-critical events concentrated in the post-arrival part of the primary mission, the first 15 months of all 'serious' orbital tour options were the same. The Saturn Orbit Insertion burn was to put Cassini into a highly elliptical 'capture orbit' so that when the spacecraft was near apoapsis, about 3 months later, it could lift its periapsis above its initial point of closest approach to Saturn. On the inbound leg of this orbit, Cassini would have to refine its trajectory so that when it

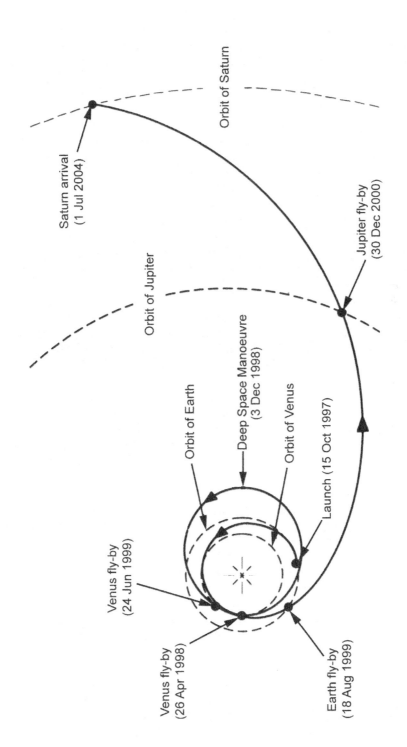

Highlights of Cassini's interplanetary cruise to Saturn.

released the Huygens probe this would enter Titan's atmosphere, and the spacecraft would have to deflect its trajectory to allow it to pass close by the moon as the probe was transmitting its data. The timing was constrained by the requirement for the probe to enter on the dayside hemisphere to enable its optical instruments to function. For a specific approach trajectory and entry angle, it would be possible to land only along a certain arc. By sheer luck, it seemed as if the probe would be able to set down on the northwestern fringe of the 'continent' that appeared bright in the infrared map. In the division of responsibilities in the international mission, JPL was to deliver the probe to the 'entry interface', this being defined as an altitude of 1,270 kilometres, and ESA would be responsible for its deceleration, parachute descent, sampling activities and data relay. Once the probe's mission was accomplished, the main spacecraft would be free to pursue its orbital tour.

Saturn's largest satellite, Titan, dominated the planning of the primary mission's orbital tour. It is fully 50 times more massive than any other member of the retinue. With only one object sufficiently large for the gravity-assists required to contrive a progression of specific orbital geometries, the tour would necessarily involve many close passes of Titan. However, because Titan is of major interest, the science teams were delighted with this focus. At the start of the planning process, several notional 'Titan only' tours were hastily put together in order to explore the issues.[69] Towards the end of these tours, the moon's gravity was employed to arrange close encounters with some of the icy satellites. These studies established that it would be possible to achieve the mission's objectives involving the icy satellites if a total of six months of the four-year tour was devoted to this task.[70]

In order to design a tour, it was necessary to trade-off a multitude of conflicting requirements. Many of the science instruments imposed viewing requirements that translated into essential geometries, some of which are time-dependent. In Cassini's original design, there would have been two scan platforms carrying the particles and fields and the remote-sensing instruments, and these would have been able to pursue independent programmes simultaneously. The penalty for omitting the platforms to save money was operational complexity. A list of essential geometries needed by the atmospheres, rings, magnetosphere, and surface science working groups provided the basis for evaluating the trade-offs, and a few geometries were then decided to be so important that they were assigned 'must have' status.

One of the highest science priorities of the orbital tour was to set up occultations to measure the refraction of the radio signal as the spacecraft crossed the limb of the planet, or the flickering of the signal as the spacecraft passed behind the rings. At the time of Cassini's arrival, the ring plane will be 'open' about 24 degrees as viewed from the Earth, but by the end of the primary mission the plane will be almost edge on, so ring occultations will have to be secured early. If the inclination of Cassini's orbit within the Saturnian system is then progressively increased, the steepening angle will: permit the rings to be viewed face-on; create opportunities for using stellar occultations to profile the rings; improve viewing of Saturn's aurorae; increase the range of latitudes for atmospheric profiling during limb-crossings; and benefit the magnetospheric studies. Inclining the spacecraft's orbital plane will involve making high-latitude fly-bys of Titan, and this will enhance the radar mapping of the moon's surface. As the high-inclination phase is characterised by short orbital periods and apoapses near the orbit of Titan, the spacecraft will be confined to the inner part of the system during this time.

The icy moon specialists desire to inspect the smooth plains on Enceladus, the wispy streaks on Dione and Rhea, and the enigmatically dark leading hemisphere of Iapetus. Each of these observations imposes geometric constraints. For example, it is preferable to document a moon's dayside, and to secure both terminator and high-illumination coverage to facilitate topographic as well as surface reflectance studies. Enceladus, the highest priority, will need three fly-bys to map it comprehensively, and phasing these involves a sequence of 500- to 1,000-kilometre fly-bys, such as Enceladus-1, Iapetus, Enceladus-2, Dione, Hyperion, Rhea and Enceladus-3. As Enceladus orbits very close to Saturn, and Iapetus is beyond Titan, satisfying this list will involve some orbital gymnastics. Having such an inclined orbit, Iapetus is rather difficult to reach, and the timing of this encounter is therefore controlled by the rate at which the plane of the spacecraft's orbit is steepened. The fly-by geometry will be crucial. The Sun–Saturn–Iapetus angle must not be near 90 degrees, as the leading and trailing hemispheres would be fully illuminated, which would not be conducive to imaging the terrain at the transition. It will take several Titan fly-bys to 'stretch' the spacecraft's orbit in the direction needed for the Iapetus fly-by, and arranging this must not compromise the evolutions underway to address the requirements of other science teams. Another priority is to sample Saturn's magnetotail, so some highly eccentric orbits will have to be planned in which the apoapsis is oriented down-Sun. Also, the planetary atmospheres specialists wish to make whole-disk studies of Saturn under high illumination, so highly eccentric orbits will have to be arranged with the apoapsis up-Sun. In effect, to satisfy both of these tasks, it will be necessary to *rotate* the apoapsis through 180 degrees. A considerable number of gravity-assists from Titan will be required to achieve this rotation, over a period of almost a year, a major challenge during a primary mission of only four years' duration – less, in fact, if the time reserved for the Huygens mission is discounted.

There were also some general constraints. The spacecraft could not be called upon to make a manoeuvre associated with a fly-by if the Sun–Earth–Cassini angle would inhibit Doppler tracking by the Deep Space Network, so no targeted fly-bys could be scheduled for the 20-day intervals spanning Saturn's superior conjunction; that is, 8 July 2004, 23 July 2005, 7 August 2006 and 22 August 2008 during the primary mission. The minimum possible time interval between any pair of targeted fly-bys was 16 days, which is one Titan orbit. If *four* successive targeted fly-bys were contrived, there would have to be a hiatus of either two 32-day intervals or one 48-day interval in order to allow the ground team to take a break. On one of the Galileo spacecraft's fly-bys of Io, engineers had abandoned their Thanksgiving dinners to coax their vehicle out of a radiation-induced 'safing'. For Cassini, however, the tour designers were to prohibit manoeuvres, fly-bys or occultations in the 9-day interval running from the morning of the Saturday before the Christmas holiday to the end of the following Sunday! Planning Cassini's tour was therefore more than a matter of simply taking into account arcane factors of orbital dynamics.

The 'must have' requirements served as an effective filter for whittling down the range of possible options. Tours were grouped into classes based on the evolution of the spacecraft's orbital inclination and orientation, and their suitability was assessed by the Project Science Group. Two classes of tour – referred to as 'T9' and 'T18' as a result of their places in the arbitrarily ordered numerical series – were selected for detailed study by the incorporation of targeted icy satellite fly-bys and the various other mission objectives.[71] Even with computer support, creating an integrated tour was a time-consuming process lasting many weeks or in some cases months.

By early 1998 there were three candidates: T9-1, T18-4 and T18-5.[72] In general, T9-1 was the ultimate 'science tour', but it violated the constraints imposed by the ground system – the sheer number of targeted fly-bys would have 'burned out' the support teams. The T18 class provided a better balance between the scientific yield and mission operability requirements. The T18-4 tour satisfied the ground system requirements by increasing the time between Titan fly-bys, but did not fully address some of the 'must have' geometries. Although the T18-5 tour violated some of the fly-by frequency constraints, it satisfied the key science objectives. The time needed to thrash out the fine details meant that a commitment had to be made by the time that Cassini flew by the Earth in August 1999 and finally headed for the outer Solar System. It was decided to fly the T18-5 tour, and accept the heavier workload that it would impose upon the ground team.

The selected tour provided 44 close fly-bys of Titan, each of which would enable the radar to map a portion of the moon's surface through its obscuring atmosphere. It provided three particularly close fly-bys of Enceladus, and one each of Dione, Rhea, Hyperion and Iapetus. Of course, there will be many opportunities for remote studies of the many moons over the course of the mission. The tour also provided superior, long-term studies of the rings at varying illuminations, detailed global mapping of Saturn and, over time, direct sampling of all the key regions of the magnetosphere.

Science objectives
During its four-year primary mission, Cassini will study the physical, chemical and long-term temporal behaviour of the planet's atmosphere and magnetosphere, its icy moons, Titan and the ring system.

Saturn's atmosphere
In comparison to Jupiter's colourful and dynamic atmosphere, Saturn appears bland to terrestrial observers. During their fly-bys, the Voyagers resolved considerable fine structure using a variety of filters. Some of this detail was visible to the Hubble Space Telescope, but Cassini's ongoing remote-sensing will greatly improve our understanding of how atmospheric features develop, evolve and disappear. With luck, it will catch a prominent 'white spot'. As did the Galileo spacecraft at Jupiter, Cassini will investigate lightning, both by listening to the characteristic radio-frequency 'whistlers' and by tracking the sites of optical flashes on the nightside around into daylight in order to identify their sources. The distribution of active sites should yield insight into the vertical circulation system. Determining long-term variations in the temperature field across the planet's surface, measuring the composition and

physical properties of the cloud particles, and monitoring winds will also yield information on the deeper structure of the atmospheric circulation system.

Magnetosphere
The particles and fields instruments, and some others, will:

- accurately chart the structure of the magnetosphere;
- identify long-term variability;
- identify the composition, and sources and sinks of the charged particles, particularly with regard to the satellites;
- investigate particle–wave interactions;
- determine the magnetosphere's interaction with the solar wind;
- investigate the interaction between the magnetosphere and the ring system; and
- investigate the sources of the kilometric-radio emissions.

Icy moons
Although Titan will be Cassini's primary focus, over time it will also investigate the smaller icy moons in order to:

- map them to determine the processes of crustal formation and modification, in order to infer their individual geological histories;
- infer their internal structures;
- identify evidence of cryovolcanic activity;
- determine the composition of their surfaces, particularly the non-ice and organic-rich materials;
- investigate how they interact with the magnetosphere (one hemisphere is 'bathed' in charged particles and the other is partially shielded by being within the magnetospheric wake) with a view to identifying resulting surface processes;
- investigate the nature of Iapetus's 'dark' hemisphere; and
- determine whether Phoebe is a captured cometary nucleus.

Titan
Between them, Huygens's brief *in situ* report and Cassini's long-term remote-sensing of the moon should:

- accurately measure the composition of the atmosphere (including noble gases – argon in particular – which the Voyagers were not equipped to detect) and the isotopic ratios of the elements present;
- measure the vertical and horizontal distribution of trace gases (particularly the 'higher' hydrocarbon polymers);
- determine whether there are complex organic – 'prebiotic' – molecules;
- determine the energy sources driving the atmospheric chemistry;
- refine models of the upper atmospheric photochemistry, cloud physics and processes producing aerosols;

- measure the global temperature field and chart the wind flows and atmospheric circulation;
- determine whether lightning occurs;
- identify seasonal variations;
- place constraints on theories of the origin and evolution of the atmosphere;
- investigate how Titan's exosphere (which is mostly hydrogen) interacts with the planet's magnetosphere;
- investigate how the atmosphere interacts with the solar wind when the planet's magnetosphere contracts within the moon's orbit;
- determine the physical properties, composition and morphology of the surface (in particular, whether there are large bodies of fluid hydrocarbons); and
- refine our understanding of the moon's interior.

Ring system

Cassini's study of the ring system should:

- determine the size and composition of the ring material, and how it varies across the system;
- monitor long-term variability in the structure of the system, particularly the development and disappearance of individual ringlets, clumps, and 'braiding';
- identify embedded moonlets;
- determine the distribution of micrometeoroids and dust in the vicinity of the ring system;
- refine models of the 'spokes';
- investigate whether Enceladus is the source of the material in the 'E' ring; and
- investigate how material spirals down towards the planet, and how it interacts with the magnetosphere, ionosphere and atmosphere.

The opportunity to study the rings offered by the Voyager fly-bys was limited in time, viewing perspective, and wavelength coverage. Nevertheless, the result was what J.N. Cuzzi described as a "paradigm shift" in which the earlier view of the ring system as a relatively simple and static structure was supplanted by an appreciation of its tremendous complexity and dynamism. Cassini's sophisticated sensors should shed considerable light on this dynamism during its tour. J.A. Burns of Cornell has ventured that the Cassini results "may fundamentally alter our understanding of ring processes and even origins".

TARGET OF OPPORTUNITY

The initial science budget had not included investigating 'targets of opportunity' in the asteroid belt. However, a graduate student of Carl Murray's at the University of London, Tolis Christou, realised that Cassini would fly close to 2685 Masursky, an asteroid that had been named in memory of Harold Masursky, a renowned planetary geologist who died in 1990. As one scientist put it: on a mission of exploration was it realistic to fly past an object without bothering to switch on the cameras? Funding was secured to make limited observations.

The observations were made when the Sun–spacecraft–Earth angle was nearest to 90 degrees, between 7 and 5 hours prior to the time of closest approach, which was at 09:58 Universal Time on 23 January 2000. Although at the range of 1.6 million kilometres the asteroid was only a few pixels across, the exercise was a valuable test for three of the remote-sensing instruments. In addition to testing automated object-targeting and helping to measure the boresighted alignments of the Imaging Science System, the Visual and Infrared Mapping Spectrometer and the Ultraviolet Imaging Spectrograph, this fly-by provided a test of the spacecraft's ability to image a barely resolvable object, thereby effectively mimicking snapping Saturn's smaller moonlets.

In this case, the task was to determine Masursky's size, rotation period, albedo, and surface photometric properties. From Cassini's perspective, the asteroid was 15 to 20 kilometres across. One welcome result was the first high-resolution spectrum of an asteroid in the 2.5- to 5.0-micron range. From its location, Masursky had been expected to be similar to Ida and Gaspra (which had been investigated by the Galileo spacecraft) but a preliminary analysis suggested that this is not the case.[73]

ENCOUNTER WITH THE GIANT

In the final week of January 2000, project scientists congregated at JPL to further refine plans for the Jovian encounter and the Titan fly-bys. The first workshop was held to plan the fly-bys of Saturn's icy satellites, and planning began for a workshop that would be held at Oxford University in England to maximise synergy between the Cassini and Huygens teams.

In February, Cassini and Galileo undertook a Conjunction Experiment to observe the Jovian radio emissions stereoscopically, as Cassini had done with the Wind spacecraft during the Earth fly-by. Such observations could be made only when the geometry of the two platforms was favourable in relation to the planet. Another opportunity would present itself in May.

The Orbiter Science Operations Working Team met in March to define the plan for the 'Jupiter Subphase' of the interplanetary cruise, the Spacecraft Office started to refine the Saturn Orbit Insertion Critical Sequence, the Satellite Orbiter Science Team integrated the Iapetus observations into the orbital tour, and the Titan Orbiter Science Team worked out precisely how the radar would achieve a minimum of 25 per cent high-resolution coverage of Titan's surface. In early April, the Probe Relay Critical Sequence Team met to consider the results of the end-to-end test that had been made in February. No one seriously expected Cassini to be disabled in passing through the asteroid belt, but a collective sigh of relief was expressed when it emerged in mid April. Despite the potential hazard, NASA's record was now 7 for 7: because after Pioneer 10 had blazed the trail, its mate had followed it through, as had the two Voyagers and Ulysses and Galileo. "It's pretty routine," said Robert Mitchell. "There's a lot of material in the belt, but there's also an awful lot of space out there." A week later, Cassini crossed the orbit of Comet Wild 2. Meteor streams may be associated with the orbits of comets, but the Cosmic Dust Analyser did not report any hits. A full-scale model of the Huygens probe was the centrepiece of a two-day meeting of the European Geophysical Society's XXV General Assembly in Nice, France, in late April, at which more preliminary results from the Venus and Earth fly-bys were presented. *The Journal of Geophysical Research* agreed to publish the formal papers in a special issue entitled 'First Results from Cassini'. With Cassini nearing superior conjunction on the far side of the Sun, it was buffeted on 9 May by the shockwave from a 'coronal mass ejection'. Meanwhile, the planners agreed on Enceladus fly-by option E3 as the 'baseline' orbital tour, and scheduled a review in November 2006 to consider switching to the E3a option. The Satellite Orbiter Science Team continued to work on the allocation of time and resources for studying the icy satellites, in this case integrating the Dione fly-by. Consideration was given to the observations to be made during the Saturn Orbit Insertion period which, although a critical phase of the mission, would facilitate very-high-resolution imaging of the ring system and offer a never-to-be-repeated opportunity to sample the inner magnetosphere.

On 14 June, Cassini's main engine was fired for 5.8 seconds for a 0.6-metre-per-second trajectory correction manoeuvre. Voyager 2's imagery of Phoebe from a range of 2.2 million kilometres had shown that there are intriguing bright patches on its rather dark surface, but the resolution had been poor. Cassini is to inspect Phoebe from a much closer range on its way into the system, and this manoeuvre was to refine this encounter.

After the successful in-flight reprogramming of the Galileo spacecraft in order to overcome a variety of faults, it had been decided that the Cassini mission would adopt a strategy of launching the spacecraft with a core of basic software and develop and transmit upgrades when they were required. Referring to this 'just-in-time' process, Earl Maize, the manager of the Spacecraft Operations Office, said: "We get the right stuff done at the right time." In 1999, it had been decided to undertake a science campaign while passing through the Jovian system. In March and April 2000, software was uploaded to the Attitude and Articulation Control Subsystem to enable Cassini to orient itself using its system of reaction wheels, instead of firing its thrusters. In addition to saving propellant, this would facilitate much sharper imaging. At the end of July, the Command and Data Subsystem was upgraded to improve the processing of large amounts of scientific data and permit simultaneous use of the two solid-state data recorders. "The studies of Jupiter are a rehearsal of the process of planning and executing complex sequences of operations to share the spacecraft's data-handling capabilities among the various instruments," reflected Brian Paczkowski, the science planning manager. Having two spacecraft offered a unique opportunity to study the interaction between the solar wind and the Jovian magnetosphere. Initially, Cassini would monitor upwind and the Galileo spacecraft would simultaneously report from its vantage point inside the magneto-sphere. A few weeks later, as Cassini penetrated the magnetopause, Galileo would temporarily re-emerge into the solar environment at apojove. "Having two spacecraft there at once is, possibly, the only chance in our lifetime to simultaneously relate changes in the solar wind to conditions inside Jupiter's giant magnetosphere," pointed out Scott Bolton, who was on the teams for both Galileo's Plasma Wave Spectrometer and Cassini's Radio and Plasma Wave Spectrometer.

Meanwhile, in mid-August, the latest observations of Titan were reported at the General Assembly of the International Astronomical Union held in Manchester, England, confirming the fascinating nature of the intended landing site for the Huygens probe. At JPL, the Galileo scientists presented the latest observations of the Jovian moons to their Cassini colleagues in order to assist them in planning their own programme. On 25 August the Jupiter Readiness Review sought to resolve outstanding activities. In early September, with Jupiter looming, Cassini encountered submicron dust from Io's volcanic plumes which had been accelerated by Jupiter's magnetic field and sent into interplanetary space in the form of a collimated stream. On one day, the Cosmic Dust Analyser counted more than 250 hits, but this was fairly mild for such streams because the Galileo spacecraft's detector had been reporting 20,000 hits per day by this point in its approach to the planet.

Later in the month, in the never-ending series of meetings to review progress and to plan future activities, the Spacecraft Operations Office held its preliminary design review for the Probe Relay Sequence. As the Cassini and Huygens teams pored over the detailed information on how the relay was to occur, and correlated this with the end-to-end test of the system conducted earlier in the year, it became apparent that there was a serious flaw in the design which meant that not all of the data transmitted by the probe would be received by Cassini. In effect, when the technical specifications for the relay link had been agreed by the two teams, no allowance had been made for the fact that Cassini, inbound on its elliptical capture orbit, would be racing towards Titan so rapidly that the Doppler effect would shift the probe's frequency so far that the signal-to-noise ratio in the receiver's pre-tuned narrow bandwidth would degrade so much (some 10 decibels) as to render it unreadable. Such a frequency shift had been designed into the relay for the Galileo probe, but Cassini and Huygens had been designed by different teams and this subtlety had been omitted from the specifications – a fact that had gone unnoticed in the design review process.[74] It came to light only when the Deep Space Network's engineers, in simulating the transmission from the probe, realised that there was no compensation. Fortunately, the fault had been detected early enough for a 'workaround' to be devised so that the Huygens probe's science potential could be fully realised.

Meanwhile, on 1 October, Cassini activated its Imaging Science Subsystem and recorded its first view of Jupiter. Even at a range of 84 million kilometres, the clarity of the image was astonishing. "This spacecraft is steadier than any I have ever seen," enthused Porco, the leader of the team. "It's so steady the images are unexpectedly sharp and clear, even in the longest exposures taken in the most challenging spectral regions." The exceptional resolution was a tantalising taste of the system's scientific potential. Cassini showed Jupiter's cloud tops in exquisite detail. The narrow-angle camera's near-infrared imagery from 168 planetary rotations through to 9 December was sequenced into a movie.[75] Analysis of this unprecedented data set produced a few surprises. "This is the first movie ever made of the motions of clouds near Jupiter's poles," noted Porco, "and it seems to indicate that one notion concerning the nature of the circulation on Jupiter is incomplete at best, and possibly wrong." A

A view of Jupiter taken by Cassini on 7 December 2000. Note Europa's shadow on the disk (the moon itself is out of frame).

popular model posited that the alternating bands of east–west winds are the exposed edges of deeper, closely packed rotating cylinders extending north–south through the planet.[76] At the planet's surface, one would see only east and west winds, alternating with latitude symmetrically about the equator. The fly in the ointment was that the winds in the polar regions in the movie did not behave in this way. Perhaps Jupiter's wind pattern involves a mix of cylindrical structures near the equator and some other mechanism near the poles.

At first sight, the mottling in the polar regions appeared to be chaotic, but in fact this is not the case. Thousands of spots, each an active storm system larger than the largest terrestrial storm, were seen jostling one another as they streamed together in any given latitudinal band; only a few changed bands. "Until now, we didn't know the lifetime of these storms," pointed out Andrew Ingersoll of Caltech. Although some spots merged with one another, most persisted throughout the entire sequence. "The smaller and more numerous storms at high latitude share many of the properties of their larger cousins, like the Great Red Spot, at lower latitudes," noted Ingersoll. Why the storms last so long is a mystery of Jupiter's weather. Storms on Earth last approximately seven days before they break up and are replaced by other storms. The new data heightens the mystery, because it shows long-lived storms at the highest latitudes where the weather patterns are more disorganised than at lower

On 12 December 2000, Cassini snapped Io in transit across Jupiter's disk, casting its shadow on the planet to the east of the Great Red Spot.

latitudes. "Perhaps we should turn the question around, and ask why the storms on Earth are so short lived," Ingersoll mused. As often happened, in observing another planet, we gained insight into our own world. "We have the most unpredictable weather in the Solar System, and we don't know why."

Cassini's Composite Infrared Spectrometer measurements of the abundance and distribution of various gases in the Jovian atmosphere (including methane, acetylene, benzene and other hydrocarbons, water and carbon dioxide) would provide insight into the photochemical processes at work in the stratosphere. Galileo's radiometer was sensitive into the far-infrared, but it had not been able to chart the atmosphere with the spatial and spectral resolution provided by the CIRS.[77] The temperature fields of the stratosphere and the tropopause that forms its lower boundary would provide a measure of the zonal winds and the thermal anomalies resulting from 'atmospheric waves' with lifetimes ranging from hours to months.[78]

Although by this time the Galileo spacecraft had been orbiting Jupiter for five years, its communications capacity was limited by the fact that its high-gain antenna had not deployed properly and it had been unable to yield such intensive synoptic coverage. Cassini's fly-by was a welcome opportunity to ameliorate this aspect of its predecessor's mission. About 60 years ago, an atmospheric disturbance south of the Great Red Spot had given rise to three 'white ovals'. Two of these had merged in 1998, but Jupiter had been at superior conjunction at the time and Galileo had not been able to monitor their merger, but by a lucky fluke of timing Cassini was able to document the final merger.[79]

The Ultraviolet Imaging Spectrograph was providing an excellent combination of spatial, spectral, and temporal resolution data on Io's plasma torus. This first-ever imaging spectroscopy of the torus took the form of multiple overlapping exposures, each at a different emission wavelength. "We can see the entire donut of glowing gas in all its invisible colours," pointed out L.W. Esposito. In mid-November, the Visual and Infrared Mapping Spectrometer made its first observations of the atmosphere.[80] Between 18 and 23 November, Cassini was immersed in the shock wave from yet another coronal mass ejection. "Such major disturbances in the solar wind may well cause Jupiter's magnetosphere to flap around significantly," said Andrew Coates of the Mullard Space Science Laboratory, part of London University, leading the CAPS instrument's electron spectrometer team. In fact, as it dived back towards Jupiter from apojove, Galileo had no sooner entered the magnetosphere than the pressure of the solar wind so compressed the magnetosphere that the spacecraft was once again in the solar environment. As Cassini was to make a fairly distant Jovian fly-by, it was possible that it would remain outside the magnetosphere if the solar wind was very powerful. By the start of December, Cassini was within 30 million kilometres of Jupiter and was being accelerated by its tremendous gravitational attraction. Starting on 14 December, as Cassini and Galileo made particles and fields measurements, the Hubble Space Telescope, orbiting the Earth, monitored the Jovian auroral displays. The main stimulus for auroral activity is Io's presence deep in the magnetosphere, but the objective was to study the degree to which the magnetosphere's response to the gusty solar wind influenced the aurorae. "We know that the solar wind controls the terrestrial aurora," explained J.T. Clarke of the Department of Atmospheric, Oceanic and Space Sciences at the University of Michigan, leading the team using the Space Telescope Imaging Spectrograph, "but we are not sure how they influence the aurora on Jupiter." The insight thereby achieved would be able to be applied to the study of extra-solar planets, because most of the cases identified to date are 'Jupiter class' giants orbiting very close to their parent stars.

On 15 December, a fortnight from the fly-by, and while working autonomously, something prompted Cassini to deactivate its reaction wheel system and revert to using thrusters for attitude control. It continued its observational programme in this mode. The problem was not noticed until the routine telemetry downloading session on 17 December, at which time the engineers saw that one of the wheels had become sluggish. If the motor was commanded, the wheel took 5 to 10 times the nominal amount of force to act, therefore the system was ordered off again to enable the fault to be analysed. Although Cassini could have continued science operations employing its thrusters for attitude control, Robert Mitchell chose to save the propellant to avoid eroding the healthy margin that had been carefully built up for the primary mission, so on 19 December planned activities that would require the spacecraft to point in a specific direction were cancelled. "We're responding cautiously while we test the systems," Mitchell noted. The particles and fields measurements of the spacecraft's environment would continue unabated. Cassini adopted an orientation with its high-gain dish antenna pointed towards the Earth in order to maintain communications. On 18 December, Cassini passed Himalia, the largest of a group of outer moonlets believed to be captured asteroids,[81,82] at a range of 4.4 million kilometres. As Himalia had not been on Galileo's list, Cassini had been tasked to determine its size, rotation and composition. Although, at best, Himalia would span only 7 pixels and the resolution would be 25 kilometres per pixel, it would be a useful observation.[83,84] Himalia has the overwhelming majority of the group's mass, which suggested that the others (Leda, Lysithea and Elara) had been 'chipped' from it. Although several images were taken, showing that the side of Himalia that faced the spacecraft was about 160 kilometres across, the plan to take a series of images to determine its rotational period was frustrated by the decision to minimise manoeuvring. A recent study exploiting sophisticated software to identify moving objects had turned up a surprisingly large number of small moons orbiting far from Saturn,[85] so imaging Himalia was a rehearsal for future work, as even basic

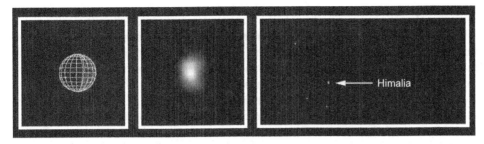

Cassini's route through the Jovian system took it within 4.4 million kilometres of Himalia, one of the small outer moons. Despite suffering an attitude-control problem, the spacecraft turned to record this unique view using a near-infrared filter. The inset shows the moon enlarged by a factor of 10 and a graphic of the illumination from the left. The resolution is about 27 kilometres per pixel. The dimension of the visible part is roughly 160 kilometres in the north–south direction, and in all likelihood it is non-spherical.

knowledge of the physical parameters of these moons will provide useful statistical information on the kind of bodies that are captured by giant planets.

Over the next few days, the engineers subjected the sluggish reaction wheel to a programme of tests. At first it continued to misbehave, but then, several days later, resumed normal operation. It was concluded that an irregular distribution of the lubricant in the motor might have caused the problem. "That's our leading theory, but we may never know for sure," Mitchell admitted. The wheels were commanded back on-line on 21 December, and monitored for a week while they maintained the Earth-pointing attitude. "Everything has been working smoothly," reported Mitchell after the trial, "so we'll resume all scientific observations." A few hours later, the spacecraft slewed around to aim its remote-sensing instruments towards Jupiter. On 29 December, the Galileo spacecraft continued its exploration of the Jovian system by making a fly-by of Ganymede while the moon was in Jupiter's shadow, providing an opportunity to measure how it cooled down in eclipse to determine the thermal characteristics of the non-ice components of its surface. While close to Jupiter, Cassini and Galileo made a coordinated study of how dust from Io's plumes is accelerated in the magnetosphere until it escapes the giant planet's gravitation.[86] Both vehicles carried dust detectors supplied by the Max Planck Institute for Astrophysics. On 29 December, a stream initially noticed by Galileo was seen 9 hours later by Cassini. The manner in which a stream swept over first one spacecraft and then the other was another way in which Cassini's passage was able to enhance Galileo's mission. When Cassini was launched in 1997, few had dared to hope that as it flew past Jupiter, Galileo, having survived triple its expected total radiation exposure, would still be essentially fully functional.

At 10:12 UTC on 30 December Cassini passed within 10 million kilometres of Jupiter. Although at a planetocentric distance of 135 radii, the slingshot nevertheless accelerated the spacecraft by 2,218 metres per second and placed it onto a trajectory that would result in an encounter with Saturn.

Early on 28 December, Cassini had encountered Jupiter's magnetosheath. This indicated that the magnetosphere had inflated once again in a lull in the solar wind. Although the particles and fields instruments reported numerous encounters with the bow shock between then and 3 January, the distant fly-by did not penetrate the magnetosphere. Coordinated observations with Galileo continued as Cassini flew on. "We're making the most comprehensive characterisation of the radio emissions from Jupiter ever," assured William Kurth of the University of Iowa and a member of the Radio and Plasma Wave Spectrometer team.[87,88,89,90] The turbulent bow shocks were also noted by Cassini's Plasma Spectrometer[91] and magnetometer.[92] The Ultraviolet Imaging Spectrograph[93,94] and the Hubble Space Telescope's Imaging Spectrograph monitored auroral activity at the time of closest approach.[95] "I've been observing Jupiter's aurora for 22 years and these images have provided an enormous amount of new and interesting data," said J.T. Clarke, leading the Hubble team. Although Io is primarily responsible for the Jovian aurorae, Cassini's Plasma Spectrometer showed that variations in the solar wind appeared to be correlated with fast fluctuations in brightness, known as 'polar cap flares'. "We collected more data in two weeks than we've amassed in several years," noted Hunter Waite, leading the Ion Neutral Mass Spectrometer team. "This will help us determine if our theories of how the aurorae behave are right." Beyond Jupiter, Cassini was able to observe the dark hemisphere, as the Hubble Space Telescope continued to monitor the dayside. "It's a rare chance to view Jupiter from two vantage points simultaneously. As Jupiter rotates and the solar wind changes, we can collect images and solar wind data without interruption. That is something we have never been able to do before," explained Waite.

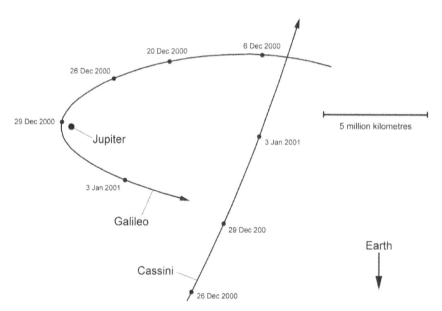

The trajectory of the Galileo spacecraft during Cassini's Jovian fly-by.

Although Cassini was able to observe Jupiter's four large moons only from afar, its Composite Infrared Spectrometer was sensitive to a wider wavelength range than Galileo's Near-Infrared Mapping Spectrometer, with improved spectral resolution, so despite the inherent trade-off of spatial resolution it was able to make useful observations of the composition and thermal structure of their surfaces. However, even at the closest point of approach, none of the satellites spanned more than a pixel to the Visual and Infrared Mapping Spectrometer.[96] In monitoring Io's volcanoes, Cassini saw plumes rising hundreds of kilometres above Pele and Tvashtar.[97] By chance, in November 1999 Galileo had seen Tvashtar spewing a fire fountain of lava some 2 kilometres into the sky from a curvilinear fissure. An eruption from a much longer fissure nearby was spotted by terrestrial observer Frank Marchis on 16 December 2000.[98] Remote imaging by Galileo shortly afterwards revealed that the plume from this new eruption had deposited a large ring of reddish pyroclastic similar to the halo that has adorned Pele since at least the time of the Voyager fly-bys. Although remote, Cassini's imagery of Io while the moon was in Jupiter's shadow was welcome because thermal emissions from 'hot spots' can be studied without the solar reflection that is present when the surface is illuminated.[99] In the case of Io, the faint auroral glows in the moon's tenuous volcanically-generated atmosphere that are induced by interactions with the charged particles flowing in the flux tubes could also be observed when in eclipse.[100,101,102] The Ultraviolet Imaging Spectrograph's imagery was sequenced into a movie that depicted Io's plasma torus gyrating in the extreme ultraviolet. "We're visualising the torus, and seeing it evolve and change in a level of detail that people have never seen before," said team leader L.W Esposito. Cassini also investigated the inner moons Metis and Adrastea, which the Galileo results had suggested were the source of the fine dust in the rings. Once across the planet's orbit, Cassini turned to enable the CIRS to observe the forward-scattered sunlight from the rings, and to employ filters to determine their chemical composition, in one case seeking the spectral signature of aluminosilicates. In effect, this was a rehearsal for the observations it would make of Saturn's rings.

The Magnetospheric Imaging Instrument detected a cloud of neutral atoms and ions pervading the interplanetary medium at least 25 million kilometres from the Jovian system. INCA images taken near closest approach were sequenced to yield a large-scale view of the compression and expansion of the magnetosphere in response to gusts in the solar wind.[103] By making coordinated observations the MIMI sensors were able not only to monitor fluctuations in the magnetosphere's shape, but also to map its chemical composition. "CHEMS was able to show that a significant portion of the particles in the cloud were sulphur and oxygen, with sulphur dioxide probably present as well," pointed out D.C. Hamilton, leader of this team. "Sulphur dioxide is the main gas emitted by volcanoes, indicating Io as the likely origin for much of the gas cloud." Once neutral atoms from the plumes are ionised, they are 'picked up' by the rapidly rotating magnetosphere. Although the ions are accelerated sufficiently to escape the planet's gravitational field, their electrical charges enable the magnetic field to retain them. However, free electrons also circulate in the magnetosphere, and if an electron neutralises an energetic ion it can escape. Once in interplanetary space,

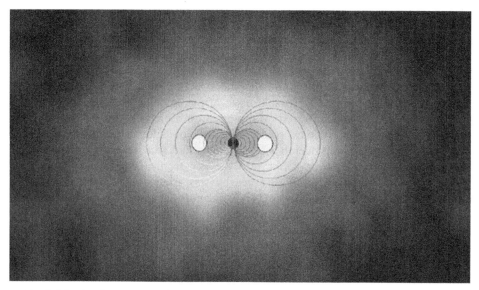

Cassini is the first spacecraft able to image the bubble of charged particles trapped within a planet's magnetic field. The Magnetospheric Imaging Instrument's Ion and Neutral Camera captured this view of the Jovian magnetosphere in early January 2001. For perspective, the planet's disk, Io's plasma torus, and a series of magnetic field lines are superimposed. (Courtesy of the Applied Physics Laboratory of Johns Hopkins University.)

many atoms are re-ionised, this time by solar ultraviolet, and CHEMS was able to detect them.[104,105] "This fly-by has been an excellent test of MIMI's capabilities," Hamilton reported, "and it has allowed us to make important refinements to the software."

As Cassini moved beyond Jupiter, it was able to study the magnetotail. It spent several days in the magnetosheath until noon on 9 January, when it slipped through the magnetopause, but this washed back and forth several times later in the day. The spacecraft did not finally enter the magnetosphere until the following day, by which time it was 14 million kilometres downwind. Galileo, half as far from the planet and trailing behind it, was also inside the magnetosphere when the solar wind gusted and the magnetosphere rapidly shrank, leaving both spacecraft in the solar environment. Cassini spent most of January and February skating along the magnetosphere's dusk flank. A 0.5-metre-per-second manoeuvre on 28 February served both to correct the small dispersions resulting from the fly-by and to perform preventative maintenance on the bi-propellant system's main engine, whose specifications require it to be fired for at least 5 seconds in every 400 days in order to clear any oxidation build-up. The post-encounter phase of the Jovian studies continued until early May 2001. Cassini spent much of this period with its high-gain antenna aimed at Jupiter. This phase of activity, which was conducted in concert with several terrestrial antennas, had two objectives. Firstly, Cassini's proximity to Jupiter meant that the planet's disk filled the antenna's narrow beam, and the thermal emission from the atmosphere offered a welcome opportunity to calibrate the radar's radiometer. Simultaneous multi-band observations by the Deep Space Network antennas provided a further check. Noting the radiometer's extreme sensitivity, the team that was planning the primary mission decided to accept 'targets of opportunity' such as Saturn's disk. Of more immediate interest to the Jovian magnetosphere specialists was that when in 'listen-only' mode Cassini could detect the synchrotron emission from the relativistic electrons trapped in the radiation belts concentrated above the equator, and the radar's data-processing system provided imagery. Such emissions have been studied at various frequencies by radio telescopes since the 1960s. Cassini was able to map the radiation belts emitting at a wavelength of 2.2 centimetres, which is denied to terrestrial telescopes because the thermal output from the planet swamps the synchrotron emission. Being closer, Cassini was able to differentiate between the synchrotron emission from the radiation belts in space and the emissions from the atmo-sphere.[106,107,108] This was done by nodding the spacecraft back and forth in order to scan across the 'target' several times, then rolling 90 degrees to repeat the procedure. The synchrotron radiation could be identified by its characteristic polarisation. The Very Large Array in New Mexico and the Goldstone–Apple Valley Radio Telescope in California made simultaneous observations at 20 and 90 centimetres.

"Cassini has been able to 'anchor' the high-energy end of the electron spectrum from Jupiter's radiation belts for the first time," said S.J. Bolton on 28 March 2001, when the preliminary analysis was presented to the European Geophysical Society in Nice, France. The various data sets were integrated to chart the energetic particle distribution within Jupiter's radiation belts. "We got some surprises," Bolton added. In particular, the highest-energy electrons were less populous than predicted.[109] As it turned out, the magnetosphere is asymmetric. "The dusk flank of the magnetosphere is a surprising contrast to the dawn flank," S.M. Krimigis explained at the American Geophysical Union in Boston in May 2001. This solved the mystery. The lopsided magnetosphere is leaky. There is an unexpected abundance of high-energy particles bleeding out of one side. These escaping electrons and ions might be riding magnetic field lines that are attached to the planet at one end and are waggling freely on the other, unlike the field lines closer to the planet which loop between its northern and southern magnetic poles. There was a dearth of the highest-energy electrons because they were leaking away.[110]

In retrospect, the most astonishing aspect of the Jovian encounter is that when Cassini was launched, budgetary constraints had imposed a flight plan in which the spacecraft would have made the fly-by with most of its instruments switched off!

Carolyn Porco was delighted with the Imaging Science Subsystem. "The camera has performed beyond our wildest imaginings – and that's saying something, because we've been imagining this for a decade now." And team member Carl Murray of the University of London was looking forward to Saturn, "I'm confident that ... the best is yet to come."

After a detailed study of the sluggishness suffered by one of the reaction wheels, it was decided that, on concluding its post-Jovian sequence in July, Cassini should minimise the use of the wheels and employ its thrusters for attitude control on the long cruise to Saturn.

When Cassini's high-gain antenna was employed as a radiometer, it could 'see' the radio emission from high-energy electrons in Jupiter's radiation belt. Since the magnetic field is inclined to the planet's rotational axis, the structure precesses.

BEYOND JUPITER

"After Jupiter, we're in completely uncharted territory," observed Andrew Coates of the CAPS instrument's electron spectrometer, as "particle measurements beyond Jupiter were poor on Pioneer and Voyager".

On 26 November Cassini began the second major campaign of its interplanetary cruise, this time seeking evidence of 'gravitational waves' passing through the Solar System. Albert Einstein had posited the existence of 'ripples' in the fabric of space-time that propagate across the Universe in response to the motions of ultra-massive objects in the farthest reaches of space. In the 1970s, indirect evidence had suggested that this was so, but no one had detected a wave. John Armstrong, an astronomer at JPL, headed an international team using spacecraft in interplanetary space to seek proof. The experiment involved the Deep Space Network monitoring the radio link to detect disturbances in the spacecraft's motion. For the duration of the test, the reaction wheel system was reactivated in order to precisely maintain the spacecraft's orientation. "We've tried this before with other spacecraft," pointed out Armstrong. Such an investigation is best done far from the Sun. An attempt using Galileo after it had passed through the asteroid belt had been inconclusive. Now far beyond Jupiter, Cassini would never be better situated. "This time we have new instrumentation on the spacecraft and on the ground that gives us 10 times the sensitivity. We're able to measure the relative velocity with exquisite accuracy." If the space-time between the spacecraft and the Earth is rhythmically compressed and stretched as a gravitational wave in a particular range of long wavelengths passes through the Solar System, this should be measurable. The apparatus to enable Cassini to use higher frequency radio transmissions was supplied by Luciano Iess of the University of Rome and Bruno Bertotti of the University of Pavia. By suppressing solar wind 'noise', it can permit more precise measurements of velocity changes for an unprecedented measurement of Doppler shifts. Sami Asmar, the supervisor of JPL's Radio Science Group, had overseen the upgrading of a large antenna at the Goldstone complex to communicate using this equipment. The experiment became the highlight of the uneventful 'Quiet Cruise Subphase'. The only significant distraction was a 1-hour 24-minute hiatus on 29 December when Cassini was occulted by the Moon. Indeed, when the test ended on 5 January 2002, some 90 per cent of the 40-day high-frequency data set had been secured. As research at the frontier of astrophysics, the confirmation of gravitational waves would probably lead to a Nobel Prize.

BLURRED VISION

Imagery of the star Spica (alpha Virginis) taken in March 2001 in order to refine the boresighting of the Imaging Science Subsystem's narrow-angle camera with the field of view of the Visual and Infrared Mapping Spectrometer revealed an excess of light in the camera's field. This was presumed to be a contaminant that had coated the lens during the Jovian fly-by, possibly thruster efflux during the manoeuvring while the reaction wheel system was inoperable. In July the camera returned its first shots of Saturn. The planet's disk spanned about 40 pixels, and the ring system 90 pixels. Titan was also visible, but less than 2 pixels wide. Although of historic significance for the mission, these images were for calibration purposes only. In boresighting the Composite Infrared Spectrometer in October, it was noted that there was a 'halo' of light around Spica's image, with some 70 per cent of the star's light falling outside a 5-pixel radius. When a heater was switched on, the halo faded, but the star's point-like image became smeared. To discover if this was 'flare' due to the Sun reflecting from another part of the spacecraft, images were taken in various orientations. As the results indicated that the problem directly affected the camera, an 'anomaly team' was formed. Possible causes were a contaminant on the optical system or a fault in the CCD detector. It was determined that the electronics had not suffered from excess radiation. The normal operating temperature for the camera was a chilly –90°C. A check of the records revealed that the 'haze' had developed after it had been heated to +30°C as a maintenance function after the Jovian fly-by. As the subsequent routine heating session had only raised it to −6°C, it was decided to use a second heater and warm it to +6°C. "We're beginning a series of decontamination cycles," said Robert Mitchell in December. "I'm very optimistic that we'll be able to correct the problem." The appropriate commands were added to the next routine increment for uplink on 14 January 2002. Several images of Spica were taken a few days later to provide a basis for comparison, and the heaters were then activated for a week, raising the temperature to +4°C. The star was imaged again and the results were somewhat better than those obtained after the first heating cycle, which was encouraging. "We're fully confident it is going to get better," announced Mitchell. Rather than run the temperature back up to the maximum +30°C, it was decided to augment the increment that would be uploaded in early March, and then maintain +4°C for about 60 days in the hope of completely eliminating the problem. Images of Spica taken in early May, after the camera had cooled to −40°C, showed that 6 per cent of the light was falling outside the 5-pixel radius. The specification for the camera's operating temperature allowed no more than 4 per cent. Another warming treatment was therefore started on 9 May, and ran for 60 days. In images taken at –90°C in mid-July the diffusion was 5 per cent and the star was "virtually the same" as prior to the onset of the problem. "We're very happy with what we're seeing," announced Mitchell, but added a caveat, "We still do not know for sure what caused the fogging, and probably never will." If it was due to efflux from the spacecraft's thrusters, the concern was that it might recur during manoeuvres in orbit around Saturn.

SAVING THE HUYGENS MISSION

In December 2000, the Huygens Communications Link Enquiry Board announced that while Cassini would receive the probe's tracking signal, the Doppler shift would degrade the telemetry subcarrier to barely 10 per cent of its intended strength. The problem was traced to an omission in the specification for the hardware.[111]

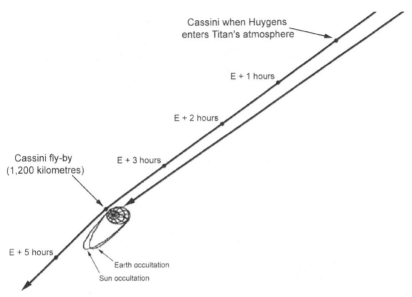

Cassini when Huygens
enters Titan's atmosphere

E + 1 hours

E + 2 hours

Cassini fly-by
(1,200 kilometres)

E + 3 hours

E + 5 hours

Earth occultation

Sun occultation

As originally planned, two days after releasing the Huygens probe on 6 November 2004, Cassini would have deflected the inbound leg of its 'capture orbit' to make a close fly-by of Titan on 27 November in order to receive the transmission from the probe as it entered the moon's atmosphere. However, due to a design oversight, the high speed of this straight-in approach would have Doppler-shifted the probe's transmission beyond the receiver's narrow frequency range.

"We have a technical term for what went wrong," pointed out J.C. Zarnecki, the leader of the Surface Science Package team. "It's called a cock-up!"

With the problem now understood, the Board set out to devise a solution that would allow the probe to be released as early as possible, while minimally changing the orbital tour in which so much planning had been invested. A number of actions were identified, each involving a degree of trade-off; these were not alternatives, they were options to be combined as deemed appropriate.

- *Determine the prevailing wind direction and speed on Titan*
  Since the probe's drift would affect the Doppler shift, this option envisaged delaying the probe's release until the second orbit to enable the direction and speed of Titan's prevailing wind to be determined. Delaying the release until the *third* orbit offered the advantage of refining the moon's ephemeris in order to reduce the pointing errors of the spacecraft's high-gain antenna, and thereby improve the gain towards the end of the descent by a factor of 5 or more. Voyager 1 had measured the winds in the upper atmosphere as flowing from west to east at speeds of 200 knots, and this was recently confirmed by NASA's Infrared Telescope Facility on Mauna Kea in Hawaii.[112] While there was a consensus that these conditions would probably persist, it was deemed unwise to risk the mission on a long-term weather forecast. It would be better to have Cassini verify the winds before releasing the probe.

- *Exploit clock bias*
The baseline mission called for a relative velocity of 5,534 metres per second during the relay. In-flight tests had revealed that the probe's transmitter clock was biased in a direction that would offset this by 1,200 metres per second. If the clock were to continue to drift, it might eventually compensate for the problem. However, the manufacturer could not confirm that the drift, which was temperature-driven, would continue at a predictable rate.

- *Increase data transitions*
The signal-to-noise ratio could be improved by up to 3 decibels by adding 'zero packets' in order to create a two-fold improvement in reception clarity, and because much of the data would be sent redundantly this would result in little or no data loss overall. This would be most effective at the beginning of the probe's entry, when Cassini would still be at its furthest point from its destination.

- *Improved assumptions about the probe's antenna patterns*
During its parachute descent, gusts in the wind would cause the probe to swing on its parachute like a pendulum. It would also spin on its parachute at a rate that would reduce from 15 revolutions per minute early in the relay to 1 revolution per minute towards its end, and the signal strength would vary with the angle to Cassini. The probe's antenna pattern was measured to determine the signal strength that could be assured for each 10- to 15-degree arc. The two redundant transmitters each had its own antenna. The signal via one transmitter was to be delayed by 5 seconds in order to preclude an overall loss of data in the event of a brief outage. One option was to endeavour to optimise this phasing to minimise interference between the antennas when the signal-to-noise ratio diminished towards the end of the descent.

- *Reduce the probe's descent time*
On the nominal timeline the probe's descent would take between 135 and 150 minutes. One option was to optimise this time for the performance of the radio link. Reducing the time the probe spent on its large parachute from 15 minutes to 5 minutes would sacrifice some of the early atmospheric data but would enable the probe to reach the surface more quickly and eliminate 10 to 15 minutes when the signal strength would be unusable.

- *Reduce the Orbiter Delay Time*
If the Orbiter Delay Time was reduced from 4 hours (as planned) to 3 hours, this would reduce the initial communication distance from 77,000 to 57,000 kilometres. This would increase the signal strength by 3 decibels at the start of the transmission, but complicate the final phase since Cassini would have to swing around to ensure that its high-gain antenna remained pointing at the probe. However, shortening the descent time would compensate for this complexity. If Cassini had to manoeuvre to point its antenna, it would be advisable to delay the probe's mission until the winds were confirmed, since

refined antenna pointing would yield a 5-fold increase in the received signal strength during the final part of the descent.

- *Reduce the probe's flight time*
  As planned, the probe was to be released 22 days distant from Titan. Reducing the flight time would enable Cassini to enter a shorter orbit, which would reduce the relative velocity during the relay. However, the additional manoeuvring to establish such an orbit would cost propellant that would otherwise be available to facilitate extending the orbital tour beyond the primary mission.

- *Undertake the probe's mission on a later orbit*
  With so many fly-bys of Titan on the tour, postponing the probe's release to a later orbit would preserve the early part of the tour, and therefore exploit the detailed planning already in hand. Orbits in the middle of the tour would permit a high-altitude fly-by that would reduce the Doppler issue. However, as long as the probe remained in place, it would partially obscure the field of view of several of Cassini's instruments. Postponing the probe's mission for so long would also cost up to 15 per cent of the available propellant averaged over the tour. The nominal plan included the option of delaying the probe's release until the second orbit. If this was pursued, it would enable Cassini to verify Titan's winds during its first fly-by, and because the plan was to use this fly-by to trim the apoapsis, the shorter second orbit would reduce the relative velocity during the relay.

- *Redesign the first two orbits*
  A shorter capture orbit would enable Cassini to release the probe on its third fly-by from an orbit having a period of about 32 days (twice that of Titan) and the slower fly-by would be acceptable. With careful planning, the fly-by geometry would enable Cassini to 'rejoin' the planned tour.

- *Raise Cassini's fly-by altitude*
  On the plan, Cassini was to receive the probe's transmission while inbound, then skim Titan's leading hemisphere at an altitude of 1,200 kilometres. During the relay, the relative velocity would be 5.5 kilometres per second. If the range was opened to 50,000 kilometres, the relative velocity would be cut to 3.4 kilometres per second and the Doppler shift would be acceptable. The increased range would significantly decrease the signal strength, however. If this option was pursued on the first orbit, then Cassini would be required to increase its deflection manoeuvre just after the probe's release, which would cost propellant. The great advantage of this option was that it would solve the entire problem, even if no other measures were pursued. However, it would require the early part of the tour to be revised to contrive an additional 'short' orbit to provide a Titan slingshot that would re-establish the planned tour.

A joint ESA/NASA Huygens Recovery Task Force was established to study the possible recovery options and decide how to proceed. It met at the European Space Technology Centre in Noordwijk in the Netherlands on 10 January 2001 to consider the various proposals for improving the performance of the relay link, such as better ground processing and error correction, and changes in the software for the probe and orbiter. A further meeting at JPL was arranged for the following week. Later in the month, a Probe Relay Test 'mini-sequence' was uplinked to calibrate the Huygens receivers on board Cassini. During the test, which was conducted between 31 January and 5 February, the Deep Space Network commanded Huygens through Cassini and received the probe's response via the relay link. Each session ran about 10 hours per day, while the spacecraft was above the Goldstone antenna's horizon. In essence, this was a repeat of the test of February 2000, but this time the effort was driven by the need to fully characterise the performance of the relay receivers, to provide the Task Force with precise engineering data and to enable the engineering test vehicle in Darmstadt to be adjusted to precisely mimic its spacefaring partner. The Huygens team expressed themselves as being "very delighted" with the results. A routine check on 22 March showed the probe to be healthy. It was awakened every 6 months or so. "As nearly as possible," explained Shaun Standley, an ESA systems engineer, "we put the probe though all the stages of the real descent system." Of course, with the probe cocooned inside its aeroshell some equipment could not be exercised, and none of the once-only devices could be fired. Nevertheless, each instrument was tested in real mode and its engineering data routed to Cassini for transmission to Earth. In April, members of the Spacecraft Operations Office attended a meeting convened at the Alcatel factory in Cannes, France, and the Huygens Science Working Team participated in the process for the first time. The primary task was to review the Task Force's progress in determining how the relay receiver's performance would be influenced by the signal-to-noise ratio, the received frequency, and the data bit transition probability. On this basis, the meeting then selected recovery scenarios for further study in the coming months.

When Cassini's prime Command and Data System detected that its backup had initiated a series of resets on 10 May, it 'safed' the vehicle by halting all sequences, powering off of its science instruments, adopting an attitude in which its high-gain antenna would act as a Sun shade, selecting the low-gain antenna for communication, and reducing the downlink data rate to 20 bits per second – precisely as required by its fault-protection software. The fault was found to be a missing 'telemetry mode' table on the backup computer. The timing was unfortunate, as an important test was scheduled to provide information for the Task Force, but rather than cancel the test it was decided to proceed by issuing real-time commands. The test was to evaluate the ability of the thrusters to maintain a specific orientation within the requisite deadband of 0.5 milliradians, as a preliminary to deciding whether to have Cassini slew around to keep its high-gain antenna pointing at the probe during the final phase of the descent. The test was successful. Meanwhile, having overcome the CDS issue, the engineers restored Cassini to normal operation.

The Task Force's meeting in Pasadena in mid-May was attended by members of the Orbiter Science Team to enable them to be briefed on the recovery options and their impact on the orbital mission. The Huygens team presented the results of their analysis of the various data-return scenarios, and how they would affect the probe's science. The low-altitude and high-altitude relay options were refined and issues for follow-on studies were defined. The Task Force met again in Noordwijk two weeks later. A summary report on progress towards solving the problem was presented to the full Project Science Group held in Oxford, England, in mid-June. Meanwhile the Atmospheres Working Group considered the implications of the possible trajectory changes identified by the Task Force, and an 'Apoapsis Splinter Group' was formed to investigate integration issues in apoapsis periods. The Titan Orbiter Science Team finalised the integration of the first ten Titan fly-bys, and its plan was presented to the Oxford meeting.

The large number of actions that might improve the situation were encouraging, since they provided flexibility. In essence, however, three different scenarios were considered: (1) slower low-altitude fly-bys, (2) low-altitude fly-bys with improved navigational performance and (3) high-altitude fly-bys integrated into the planned orbital tour. It soon became apparent that the propellant cost of contriving a slower low-altitude fly-by was prohibitive.[113] Furthermore, to have pursued such a solution would have required some (or all) of the subsequent orbital tour to be redesigned. To postpone the probe's mission to the second or third orbit of the original tour would enable Titan's ephemeris to be improved and increase the accuracy of the delivery and antenna-pointing, but it was decided that this would not produce a satisfactory mission.[114,115] Nevertheless, these studies established that it would be *possible* to deliver Huygens from an orbit with a 32-day period, and identified the manoeuvre locations and tracking requirements.

Four 'high-altitude delivery' options were assessed in the context of the planned orbital tour. Flying the early part of the tour as planned would enable the completed detailed planning to be pursued. Thereafter, a high-altitude fly-by could be contrived to deliver the probe. Doing so after the 10th Titan encounter would guarantee the early fly-bys of the icy moons, but the rest of the tour would have to be redesigned and it might not be practicable to arrange the anticipated later encounters with the icy moons.[116] Contriving a high-altitude fly-by after the 35th Titan fly-by would protect the Iapetus fly-by, but only, unfortunately, at the cost of the third and final inspection of Enceladus.[117] Another study showed that it would be possible to arrange a high-altitude fly-by on the second Titan fly-by and then initiate a series of 'clean up' manoeuvres in order to resume the tour at the sixth encounter, but this could only be done at the expense of valuable propellant.[118] An alternative would be to reconfigure the *early* part of the tour to arrange a high-altitude fly-by, and then rejoin the tour at the sixth encounter without such extensive manoeuvring.[119] An overriding factor was the two years that had been invested in developing a tour that would address *all* the science objectives, as this could not be lightly discarded. The Task Force published its decision on 29 June 2001. It chose a compromise that would not only overcome the communication issue but also rejoin the original tour within 8 months.

The primary and backup delivery timelines for the Huygens probe to Titan, which rejoin the original planned orbital tour with either the T_3 or T_6 fly-by, respectively.

The Saturn Orbit Insertion burn would be slightly longer than originally planned, in order to lower the apoapsis of the capture orbit and advance it by one month. The burn to raise the periapsis would be made on 23 August rather than 25 September, and be increased by 57 metres per second to 393 metres per second. As a result of these changes, the encounter with Titan on the inbound leg would be advanced by 32 days, from 27 November to 26 October. Cassini would then fly past the moon at an altitude of 1,200 kilometres with the probe still in place, and make observations to measure the winds to refine the antenna-pointing requirements for the final phase of the probe's relay. It would have another an opportunity to verify the winds during its second fly-by on 13 December.

After refining its trajectory on 17 December, Cassini was to release the probe on 24 December. Five days later, Cassini would perform the deflection manoeuvre in order to open the range of its fly-by on 14 January 2005 – the day the probe was to reach Titan – to 60,000 kilometres. Although the increased range would marginally diminish the received strength of the probe's signal, the offset fly-by would greatly constrain the Doppler shift because, although Cassini would still cross Titan's orbit at 6,800 metres per second, the radial component responsible for the Doppler shift would peak at 2,500 metres per second. It would decrease as the spacecraft drew alongside the moon, pass through zero, then increase in the opposite sense as the spacecraft withdrew towards the end of the probe's descent. Thus, the frequency drift ought never to leave the narrow band to which the receiver was tuned. Cassini was to rotate to maintain its high-gain antenna pointing at the probe's predicted position, taking into account both the spacecraft's fly-by and the probe's motion in Titan's atmosphere. Within the constraints of its firmware, which could not be altered,

An artist's impression of Cassini releasing the Huygens probe. Note that a certain licence has been taken concerning the range from Titan when this occurs.

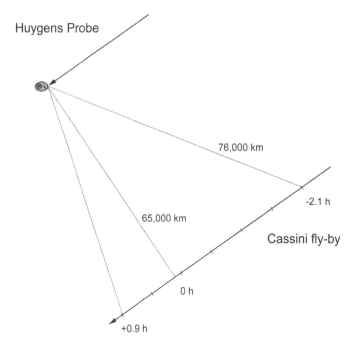

Huygens Probe

76,000 km

-2.1 h

65,000 km

Cassini fly-by

0 h

+0.9 h

In the revised fly-by to receive the data from the Huygens probe, Cassini was to make a distant pass in order to reduce the radial component of its relative velocity and thus minimise the Doppler effect, and slew around to keep its high-gain antenna aimed at the probe's predicted position during the descent in order to maximise the signal-to-noise ratio.

Huygens's software was to be rewritten to optimise its transmission characteristics. Taken together, it was hoped that these measures would significantly improve the signal-to-noise ratio, and result in little or no loss of data. On the original plan, the first Titan fly-by would have occurred on 27 November and the second on 14 January. Tightening the capture orbit enabled the insertion of a 'short' orbit for the probe's mission. After a series of manoeuvres, Cassini would be able to return to Titan on 15 February 2005 and resume its primary mission just in time for the first encounter with Enceladus.[120]

"This plan will allow us to meet *all* of the mission's scientific objectives," noted Robert Mitchell, "and has the additional advantage of giving us a close look at Titan before releasing Huygens."

"It's a fantastic solution," agreed J.C. Zarnecki.

"We were in hell, now we're in paradise," said Marcello Fulchignoni, a Huygens scientist and astronomer at the Paris Observatory.

The probe was required to dive into the atmosphere at a depressed angle of 65 degrees below local horizontal in order to establish a deceleration profile that would enable the onboard sensors to initiate the sequence of events for the programmed descent. The landing site was defined by the approach geometry. The target ellipse was nominally 200 by 1,200 kilometres, but would be able to be reduced if the winds were measured in advance. Although the longitude of the target would remain 190°W, the geometry of the revised fly-by caused it to be moved from about 10°N to 10°S and the timing would be such that Cassini would cross Titan's orbit behind, rather than ahead of the moon.

The cost of this elegant solution was the propellant for manoeuvres equivalent to an overall velocity change of about 95 metres per second, but owing to the excellent navigation a substantial reserve had been husbanded, and consuming some of this to exploit the planning for the primary mission was considered to be acceptable trade-off. In the event of a problem that prevented the probe's release, there was provision for postponing it by one revolution, to about 25 January, for the Titan encounter on 15 February. But if this was required, it would mean using more propellant – due in part to the probe still being in place during the manoeuvres – and it would impose a significant delay in rejoining the original tour. If all went as planned, however, the recovery plan would consume 25 to 30 per cent of Cassini's propellant reserve, and unless a further problem occurred that imposed additional manoeuvres, there was every possibility that the orbital tour would be able to be extended beyond 2008.

After Edward Weiler, NASA's Associate Administrator for Space Science, and David Southwood, ESA's Director of Science, accepted the revised plan, the Task Force met in Noordwijk once more to prepare its final report, which was issued on 27 July 2001.[121] The reinvigorated science teams set about adjusting their plans for the revised portion of the orbital tour. By the end of August, the navigators had an accurate trajectory for the first few orbits, including the changes required to execute the probe's mission.

In November, the Goldstone Deep Space Network antenna played the role of the probe in a full rehearsal of Huygens's descent to Titan's surface, transmitting to the relay receivers on Cassini a stream of data in precisely the same format as the probe would utilise, and continuously adjusting the frequency to simulate the Doppler effect of the fly-by, while simultaneously factoring out the spacecraft's actual motion during the test. As Jean-Pierre Lebreton observed, "We need to be *certain* that the modified mission will allow the receivers on board Cassini to operate within the narrow range of frequencies available." The tests took four days to complete. "The whole test was very smooth," reported Julie Webster, the deputy manager of the Cassini Spacecraft Operations Office at JPL, who was responsible for interfacing between the JPL and ESA relay teams. "We've taken major steps towards the validation of the Huygens Recovery Task Force design," agreed Earl Maize, her boss. "We tested a nominal mission scenario, and several deviations from it," pointed out Claudio Sollazzo, the Huygens Mission Operations Manager in Darmstadt. "We appear to have met all of our objectives." The data analysis would take months, but it was clear that the relay *would* work. Jean-Pierre Lebreton was very optimistic: "Even under the worst-case conditions tested – with significant deviations from the nominal parameters – we've shown that we will retrieve all the data that the probe will send to Cassini during its descent, and for at least 15 minutes on the surface."

RECENT STUDIES OF TITAN

The first solid hint of meteorological activity on Titan was reported in 1998 by a team led by C.A. Griffith, then based at Northern Arizona University in Flagstaff, in which whole-disk brightening in near-infrared data taken in 1995 was attributed to the presence of a methane cloud in the troposphere at an indeterminate position.[122] A re-examination of contemporary observations obtained by the Hubble Space Telescope was supportive.[123] Later observations by Griffith indicated such activity to be highly variable.[124,125] In 2002 M.E. Brown of Caltech and H.G. Roe, then at the University of California at Berkeley, independently announced that they had resolved tropospheric clouds in the south polar region.[126,127] It seemed that sunlight was heating the surface and causing convection that produced the methane clouds. At that time, Titan was enjoying its southern summer, with the solstice in October 2002, and its south polar region had been in continuous sunlight for several years. It was speculated that, allowing for some lag due to the thermal inertia of the surface, the convective activity would follow the point of maximum insolation and migrate back and forth between the poles with Titan's 29-year seasonal cycle. At 5,150 kilometres in diameter, Titan is too small to manage the sort of complex large-scale wind patterns that occur on Earth – the huge rotating structures that play a key role in transporting heat from the equator to higher latitudes would simply not fit. At higher altitudes, the haze changed on a seasonal basis.[128,129] A team led by Pascal Rannou of the University of Paris in France used the smog particulates as tracers to investigate the circulation of the atmosphere, and concluded that it is a meridional flow, rising above the warmest region of the surface and migrating at high altitude to sink in the winter hemisphere.[130] On 18 December 2003 Roe made the surprising discovery of a short-lived cloud in the mid-southern temperate zone.[131] They were soon a common occurrence, but it was not clear how they formed.

After mulling over the Voyager data, Carl Sagan and Stanley Dermott at Cornell suggested in 1982 that Titan might be totally covered with an ocean of methane.[132] If the sea was shallow, the tidal action on the sea bed would progressively slow the moon's rotation, which, by the conservation of angular momentum, would have the effect of circularising Titan's orbit. If there was indeed an ocean, the fact that the orbit was elliptical required it to be sufficiently deep to decouple the tides from the sea bed. Then in 1990 a team led by D.O. Muhleman reported radar observations by the Very Large Array showing the surface to be heterogeneous, which meant that if there was an ocean it was not global in extent.[133] When it was realised that the atmosphere was transparent in near-infrared wavelength bands, this heterogeneity was confirmed by a number of teams, some re-analysing previous data,[134] and others making new observations.[135,136,137] Once it was known that the albedo variation was synchronised with the rotation of the moon, indicating that the surface was visible, in 1994 a team led by P.H. Smith of the University of Arizona in Tucson used the Hubble Space Telescope to take near-infrared imagery that revealed the presence of a vast 'continent' in the equatorial region of the leading hemisphere.[138] The by-products of methane dissociation would form a haze of organic particulates that would collect on the surface, perhaps to a considerable depth, but an analysis by

Griffith of the spectroscopic signature (as far as this could be observed through the near-infrared windows) implied that 'icy bedrock' was exposed in places.[139]

With a diameter of 1,000 feet, the Arecibo dish in Puerto Rico is the biggest radio telescope in the world. It was already the world's most powerful instrument for radar studies when, in 1997, the year that Cassini was launched, it was upgraded to make it even more sensitive. Arecibo transmitted pulses at a wavelength of 12.6 centimetres towards Titan when Saturn was at opposition in 2001 and 2002 and, 2.25 hours later, received the reflection from the sub-Earth point. Observations were made over a 16-day period in 2001 and 9 days in 2002, providing a series of 'spots' as Titan slowly rotated, each integrated over an area 200–300 kilometres in diameter. A surface that was rough on the scale of the 'illumination' would scatter the energy and produce a diffuse echo, but a smooth surface would reflect it without scattering. Most of the echo power was in a diffusely scattered component, but in 75 per cent of the observations there was a small but specular component suggesting a mirror-like glint.[140] While this was the best evidence to date for the existence of lakes, it was not conclusive. However, as the team leader, Donald Campbell of Cornell University, explained, the glint "was being produced by something that could reflect only about 2 per cent of the signal that hit it. Such a poorly reflecting surface is characteristic of liquid hydrocarbon." While it was possible that these were reflections from very smooth ice, "part of the surface would have to be rough because the signal we'd get from an ultra-smooth icy surface would be much larger than was measured". As no specular reflections had been observed from Jupiter's icy moons, this argued in favour of the surface of Titan being very different. "If you take this data at face value," Campbell concluded, "I think there is a high possibility that Huygens will splash down." In a commentary on the Arecibo results, R.D. Lorenz of the University of Arizona at Tucson inferred that the glint was coming from sites 50–150 kilometres across.[141] He estimated that Titan could reasonably be expected to have 80 craters of 150 kilometres in diameter and thousands of smaller ones in which hydrocarbons could have formed lakes. Although there was no correlation between the glint and the infrared albedo, the radar was sampling at a much higher spatial resolution than the infrared imagery and would be able to see fine detail that would not be apparent in the infrared. It was also possible that the glint came from paleolakes that dried out long ago; provided that a lake bed remained smooth it would still produce a specular reflection.[142] If sites that gave a glint to Arecibo appeared rough to Cassini's 2.3-centimetre radar, this would rule out liquid, but the resolution of Cassini's imaging systems during its close fly-bys should enable liquid on the surface to be located by direct observation.

CRUISING ON

On 3 April 2002, in the first burn since the trajectory refinement after the Jovian fly-by, Cassini fired its main engine for 9.85 seconds. This was routine maintenance, as the engineers did not wish to allow more than 400 days to pass without firing the engine to clear its propellant lines. On this occasion it had been decided to test new

Cassini's first view of Saturn (with Titan), taken on 21 October 2002.

Manoeuvre Automation Software designed to simplify the process of drawing up the command sequence for a manoeuvre. As another test, the sequence began and ended with the high-gain antenna pointed at Earth, as would be the case when orbiting Saturn, rather than at the Sun, as it did during the interplanetary cruise to provide thermal protection. On 21 October Cassini's narrow-angle camera took filtered imagery for a natural-colour view of Saturn from a range of 285 million kilometres. "This is an emotional event for the mission," noted project scientist D.L. Matson. "We now have Saturn in our sights." However, the resolution was still inferior to that provided by the Hubble Space Telescope.

A fly-by of Phoebe on the way into the Saturnian system had always been on the flight plan in order to use that moon's gravity to slow the spacecraft, save propellant in the Saturn Orbit Insertion manoeuvre, and ultimately extend the mission at Saturn beyond 4 years. The baseline fly-by range was 52,000 kilometres, but this had been considerably reduced as a consequence of the 9-day delay in launching the mission, and a trajectory tweak inbound to Jupiter had refined that slingshot to further reduce it. "A small burn is required shortly before Phoebe, and another one afterwards, but if Phoebe hadn't been there we'd have had to make a fair-sized manoeuvre two or three weeks prior to Saturn Orbit Insertion as a system test, so the net cost of going closer to Phoebe is practically zero," pointed out Robert Mitchell. After meetings of the project science group in October–November, it was decided to make the fly-by at an altitude of 2,000 kilometres. Some had argued for a closer encounter in order to better chart the moon's gravity field and thereby gain insight into its internal structure, but this was dismissed owing to the possibility that dust in the vicinity of Phoebe could damage the spacecraft sufficiently to prevent it proceeding with the Saturn Orbit Insertion manoeuvre.

A new moon was discovered on 5 February 2003.[143] It was discovered by Scott S. Sheppard, David C. Jewitt and Jan Kleyna using a wide-field digital camera on the Subaru telescope in Hawaii. It was the 14th satellite to be discovered in an eccentric distant retrograde orbit, and the first such discovery for nearly 3 years. It increased to 31 the total of listed satellites. The International Astronomical Union designated it S/2003S1. On the assumption that its surface was dark, its diameter was estimated at about 8 kilometres.

On 17 June 2003, with Earth now within 5 degrees of the Sun as viewed from the spacecraft, Cassini began its radio science solar conjunction experiment. When the angle had reduced to 3 degrees, the Deep Space Network began to send sequences of 'no op' commands 10 to 20 times per day, as it had during each conjunction period after the Earth fly-by, so as to gather statistics on the reliability of reception of commands as functions of conjunction angle and heliocentric range. This data

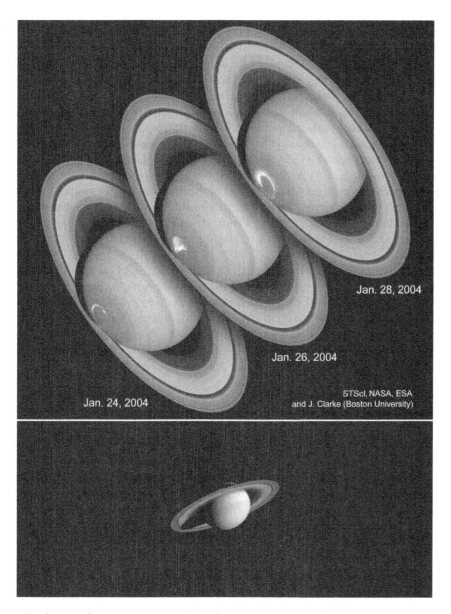

An image of Saturn taken by Cassini on 9 November 2003 (bottom), and an auroral display at the south pole in the ultraviolet viewed by the Hubble Space Telescope in January 2004. Note the different vantage points. The ultraviolet emission is due to solar wind electrons flowing in the magnetic field in the polar region and exciting atomic and molecular hydrogen. There was an increase in the electron flux on 25 January 2004.

would form the basis for deciding the duration of the hiatus in uplinking for the conjunction that would occur shortly after Saturn Orbit Insertion. On 1 May, Cassini fired its engine for 17.53 seconds to flush the propellant pipes for its 400-day routine maintenance. In this case, it also verified software intended for use during the orbital tour, and the ability of the Deep Space Network to upload and execute a burn during a single communications session. In September, after tests had indicated that there would be an adequate power margin, it was decided that the data-relay system of the Huygens probe should be powered up 4 hours prior to the calculated time of entry into Titan's atmosphere in order to warm up and stabilise the circuitry. On 10 September a minor manoeuvre tested the Manoeuvre Automation Software's ability to specify burns to use the thrusters of the reaction control system – as would be done often once the spacecraft was in orbit around Saturn. A 21.8-second burn of the bipropellant engine on 1 October demonstrated the yaw-steering that was to be utilised during Saturn Orbit Insertion, and the energy-based algorithm that would terminate that burn.

On 9 November 2003, one year after its 'first look', and now at a range of 111 million kilometres, Cassini took an image of Saturn showing the banded structure of its atmosphere. "For all of us who have worked for more than a decade preparing for this mission, seeing Saturn grow larger in Cassini cameras is a bit like the feelings children have as they come downstairs on Christmas morning to see what gifts are waiting for them under the tree," noted Anthony DelGenio of the Goddard Institute for Space Studies in New York City.

CASSINI APPROACH PHASE

As 2004 began, astronomers set up a program of imagery and spectroscopy using both the Hubble Space Telescope and ground-based telescopes to determine the state of Saturn's atmosphere as a prelude to Cassini's arrival. On 9 January the spacecraft initiated its own program. There had been occasional opportunities for science on the interplanetary cruise, but the approach phase was to involve science observations at a level of activity representative of that planned for the primary mission, in order to validate the operating modes. It started with the Dual-Technique Magnetometer, Plasma Spectrometer and Magnetospheric Imaging Instrument monitoring the state of the solar wind heading for Saturn's magnetosphere while the Radio and Plasma Wave Spectrometer monitored radio emissions from the planet's auroral activity and the Hubble Space Telescope took imagery of the ultraviolet emission. Large shocks in the gusty solar wind on 17 and 25 January changed both the aurora and the radio emission. In the case of Earth an aurora lasts for several hours, but on Saturn it can last for days and differ in character day by day, often moving, other times stationary. Viewed from space, an auroral display is a ring of light around the magnetic pole. In the case of Saturn the ring shrinks in diameter when an aurora becomes brighter and more powerful, then expands when it dims, whereas on Earth the ring stays the same size as an aurora brightens and temporarily fills the central area.[144,145,146] The light is emitted by atoms and molecules excited by electrons flowing in the magnetic field. In the case of Earth it is mostly from oxygen atoms and nitrogen molecules, but on Saturn it is from atomic and molecular hydrogen.

On 6 February Cassini began to take imagery of Saturn for optical navigation, and 3 days later returned the first of a sequence of weekly colour 'postcards' of the planet, which was then 66 million kilometres away. With the north pole tilted away from the Sun and the spacecraft travelling in the plane of the ecliptic, the view was of the southern polar region and the illuminated 'underside' of the ring system. As it closed in, Cassini took imagery for colour movies to monitor the temporal variation of the atmosphere and measure the wind speeds. Imagery also enabled the ephemerides of the known moons to be updated, and facilitated a search for hitherto undiscovered moonlets. There were no plans to try to image the recently discovered satellites out beyond Phoebe. The composition of the rings was to be investigated by imaging in the mid-infrared, and the distribution of the atomic hydrogen inside the magnetosphere was to be mapped by ultraviolet imaging. On 23 February Cassini took imagery of the 'F' ring. The data had to be magnified and contrast enhanced in order to display the 50-kilometre-wide ring, but several 'clumps' were readily apparent.

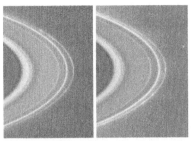

Two images taken 2 hours apart on 23 February 2004 contrast-adjusted to show the motion of clumps in the 'F' ring.

Prior to 1993 it was believed that interplanetary dust must be either 'primitive' material originating in interstellar space that was passing through the Solar System, or silicate from collisions between small bodies such as asteroids. However, the Ulysses spacecraft found streams of dust originating from the Jovian system,[147] and the Galileo mission later identified the source of this dust as volcanoes on the moon, Io.[148] In early 2004, Cassini sporadically encountered 'bursts' of dust as it closed in on Saturn. An analysis found that the particles escaping from the Saturnian system were between 0.01 and 0.1 micron in size. It was reasoned that while larger grains would be confined by the planet's gravity field, smaller grains would be confined by its magnetosphere; therefore only those grains in this size range could escape. The dust was emerging at a fairly constant rate, but Saturn's magnetic field and the solar wind were concentrating it into streams, through which, from time to time, Cassini flew, giving the impression of bursts.[149,150] Further analysis showed that the particles came from the outer part of the 'A' ring and were silicate, indicating that they are representative of the 'impurities' rather than the icy ring material.[151]

An inflight demonstration of the Huygens probe relay in late February exercised the commands that would be needed to receive the data from the probe when it entered the atmosphere of Titan. Prior to the test, all of Cassini's instruments were either switched off or muted – as they would be on the day. Continuous round-the-clock coverage was provided by the Goldstone, Madrid and Canberra stations of the Deep Space Network. The activity started with uplinking the sequences to the solid-state recorder. During the test, Cassini adopted the requisite relay attitude, recorded some 6 hours of simulated data, then turned its high-gain antenna to Earth for replay. The test ended with the delivery of the data to the Huygens Project Operations Centre at Darmstadt in Germany. The spacecraft then resumed its science activities.

In February, now 60 million kilometres distant, Cassini began to resolve discrete features in Saturn's atmosphere. On 19–20 March it was able to observe the merger of two storms in the southern hemisphere, each of which was about 1,000 kilometres in diameter. Both were drifting westwards, the more northerly one at twice the rate

The upper frames were taken over an interval of 26 days, starting on 22 February 2004, and show two Saturnian storms in an anti-cyclonic shear zone near 36°S, in which the flow to the north is westward relative to the flow to the south, causing the northern storm to catch, engage in a counterclockwise dance, and merge with its southern counterpart on 19–20 March. The lower frames were taken on 19, 20, 21 and 22 March. The merged feature was elongated in the north-south direction with bright clouds on either end, but within 2 days it had adopted a more circular shape and the bright clouds had become circumferential.

of the southerly one. As they met they spun around each other in a counterclockwise manner. The resulting storm was elongated in the north–south direction with bright clouds on each end, but within 2 days it had adopted a more circular shape and the bright clouds were in a halo forming its circumference. It was only the second time that a merger had been observed on Saturn. In addition, there was a distinctive 'dark' circular spot precisely at the south pole. This correlated with an observation by the Keck Observatory on 4 February showing the spot as a bright feature in the wavelength range 8 to 25 microns, indicating it to be warm.[152] In fact, it was the warmest place on the planet. The mystery was not that the polar region was warm – it had, after all, been in continuous sunlight for several years, heating the polar vortex, which was a persistent weather pattern akin to a jet stream. But both the distinct boundary of the vortex flowing around the pole and the hot spot at its centre were unexpected. If the increased southern temperatures were the result of the seasonal variation in sunlight, they ought to increase smoothly towards the pole, but this was not the case – the tropospheric temperature increased abruptly from 88K to 89K near

70°S, and to 91K at the pole, and the stratospheric temperature rose even more abruptly from 146K to 150K at 70°S, and to 151K. If the abrupt change at 70°S was the result of a concentration of sunlight-absorbing particulates trapping heat in the upper atmosphere, this would explain why the 'hot spot' appeared dark in visible light.

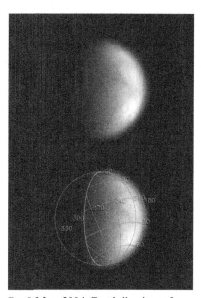

On 10 March 2004 Cassini made its first sighting of the shepherding moonlets of the 'F' ring: Prometheus (interior) and Pandora (exterior).

Meanwhile, on 10 March Cassini caught its first sight of the 'F' ring shepherds Pandora and Prometheus. An image taken on 27 March from a range of 47.7 million kilometres was notable because the rings spanned the frame of the narrow-angle camera.

On 1 April the readiness review of the critical event sequence for the Saturn Orbit Insertion manoeuvre recommended several new contingency measures. A week later there was an operational readiness test, as a dry run of all the nominal events that were to be undertaken on 2–3 June, immediately after Cassini had achieved orbit. In April, Cassini began to image Titan at wavelengths that would show methane clouds and the surface, and by mid month the narrow-angle camera was resolving details that previously could be seen only by the use of adaptive optics on large terrestrial telescopes. Despite the 38-second exposure, there was no discernible smearing – the spacecraft was a remarkably stable platform.

Meanwhile, JPL ran an end-to-end rehearsal of Saturn Orbit Insertion by using the Integrated Test Laboratory as a stand-in for the spacecraft. The first act on 9 April was to load the critical sequence into the solid-state recorders, then simulate the final trajectory correction prior to the critical sequence for Saturn Orbit Insertion, ending on 1 May with a simulation of the first orbit trim manoeuvre designed to 'clean up' small residuals of the insertion burn. A second operational readiness test, completed on 9 May, demonstrated that the team could recover from an anomalous insertion burn by defining and executing a substantial corrective manoeuvre within 3 days. Meanwhile, on 5 May, at a range of 29.3 million kilometres, Cassini imaged Titan at a resolution exceeding that attainable by the

By 5 May 2004 Cassini's view of Titan had a resolution exceeding that attainable by the best terrestrial telescopes. The southern pole was in full daylight.

best terrestrial telescopes. Later in the month, the spacecraft began to take movies to enable wind speeds to be measured from the motions of the clouds – data that was required by the Huygens mission. In parallel, the Titan Monitoring Project at the Keck Observatory took frequent high-resolution imagery to: (1) determine the typical size, frequency and life time of the clouds in the south polar region; (2) search for meteorological activity at other latitudes; and (3) compile the most detailed map possible of the surface to assist with planning the observations to be made by the spacecraft during its fly-bys once it was safely in orbit around Saturn.

Just why the atmospheres of Jupiter and Saturn have an alternating pattern of east–west winds varying in direction with latitude was a matter of conjecture. In contrast to Earth, whose weather is driven primarily by sunlight, the giant planets are still in a state of gravitational collapse and so have an additional energy source in the form of the heat that leaks from their interiors. The challenge was to understand the role of these interior energy sources in sustaining the strong jet-stream winds. In one theory, circulation was driven by solar heating of a shallow layer at the top of the atmosphere. In another theory the winds extended deep into the interior and were driven by the energy that leaks out. Neither theory accounted for the maximum wind speed occurring at the equator. One test was to measure the long-term sensitivity of the winds to variations in the amount of sunlight resulting from seasonal and other influences. Studies had shown Jupiter's winds to be insensitive to seasonal changes, but the case of Saturn was difficult because its cloud structure was 'muted' visually. When the Voyagers passed by they saw sufficient detail to enable wind speeds to be measured, and established the equatorial jet stream to be an astonishingly fast 1,700 kilometres per hour. The Hubble Space Telescope could track sufficient detail to facilitate further studies. Observations in 1996–2001 found that despite the change in season and the location of the shadow cast by the ring system, the jets far from the equator had not changed, which suggested that they were deeply rooted.[153] However, at just 990 kilometres per hour the equatorial jet stream was much weaker than in 1980–1981. One suggestion for the apparent change was that the structures observed by the Hubble Space Telescope were at higher altitudes, where the winds might well be slower. In mid-May 2004 Cassini's Composite Infrared Spectrometer made the first of a number of long integrations to measure the temperature fields across the planet's southern hemisphere.[154] It produced a three-dimensional chart of the winds in the stratosphere that confirmed that the equatorial winds diminished rapidly with increasing altitude above the level of the ammonia cloud tops, the gradient being as steep as 500 kilometres per hour for an altitude increase of 300 kilometres. This suggested that the Hubble Space Telescope had indeed seen structures at a higher altitude than those observed by the Voyagers, and that the apparent slowing was due to storms projecting their 'tops' to higher altitudes, where they were more readily seen from Earth. Cassini provided a clue to what might be maintaining the energetic jets. At an earlier stage it had found the region near 35°S to be particularly active – so much so, in fact, that it was dubbed 'storm alley'. One imaging sequence showed dark spots emerge from the upper-level outflow of a convective circulation structure, thereby yielding their energy to the jets – suggesting that this was how the energy escaping from the interior sustained the horizontal winds of the upper atmosphere.

A trajectory correction on 27 May 2004 refined the forthcoming Phoebe fly-by to an altitude of 2,000 kilometres. The burn lasted 362 seconds for a delta-V of 34.71 metres per second. As the regulator for the helium pressurant for the propellant tanks had developed a leak soon after launch, a latch was being used to isolate this system to preclude it overpressurising the tanks between firings. This burn could have been performed in 'blowdown' mode, which used the residual pressure in the tanks, but because the helium system would be required for Saturn Orbit Insertion and this had not been exercised since the Deep Space Manoeuvre in 1998, it was decided to fully pressurise the tanks in order to determine the leak rate. The latch was opened 100 seconds prior to the burn, and closed shortly afterwards. The results confirmed that the rate of leakage was unlikely to overpressurise the tanks during the much longer Saturn Orbit Insertion burn. This trajectory correction marked a milestone since, as chief navigator Jeremy Jones put it, "If the road to Saturn were a highway, Cassini would have passed the sign that says 'Saturn county line'."

In the final month of the approach, the spacecraft took high-resolution imagery to search for moonlets that might pose a hazard to the planned ring-plane crossings. Two hitherto unknown moonlets 3 to 4 kilometres in size were located 194,000 and 211,000 kilometres from the planet's centre – between the orbits of Mimas and Enceladus.[155] They were spotted by Sébastien Charnoz, a planetary dynamicist at the University of Paris in France, in imaging sequences taken on 1 June for the purpose. The International Astronomical Union labelled them S/2004S1 and S/2004S2, then later named them Methone and Pallene. Their discovery caught Charnoz by surprise, "I had looked for such objects for weeks while at my office, and it was only on holiday using my laptop that my program detected them; this tells me that I should take more holidays!"

Operational readiness tests using the Integrated Test Laboratory – in some cases lasting 2.5 weeks – continued into early June. These explored different scenarios for Saturn Orbit Insertion. In one case the burn was interrupted several minutes short of the planned duration, requiring a correction 2 days later of 150 metres per second to achieve an acceptable capture orbit.

PHOEBE FLY-BY

Phoebe is some 220 kilometres in diameter, and slightly irregular. It has an elliptical retrograde orbit of 18 months that is more or less in the plane of the ecliptic. A study during December 2003 and January 2004 in support of Cassini precisely measured the repeating 0.12-magnitude variation to determine the rotational period as 9.2735 hours (± 0.0015), which was an improvement over the previous measures by a factor of 10.[156] Although Cassini would have only one opportunity to inspect Phoebe, the moon's rotation would offer most of its surface to view during the fly-by. An albedo of 0.06 indicated that there was dark material on the surface. On the basis of the Voyager data, it had been classified as a captured C-type (carbonaceous) asteroid, but this analysis predated the surprising finding that the nucleus of Halley's comet was black and the discovery of the Kuiper Belt (or Trans-Neptunian) Objects which accreted far from the Sun and, as such, were 'pristine' material of the solar nebula, thus raising the possibility that Phoebe was either a captured comet or Kuiper Belt Object. Cassini's remote-sensing instruments were to study the surface composition. "This moon has a little ice and a lot of black material on its surface, but other than that we know very little," noted D.P. Cruikshank of the Ames Research Center and a member of the Visual and Infrared Mapping Spectrometer team. "This is a unique opportunity. We've never had a close-up look at an irregular low-reflectivity moon of any planet before, so we are prepared to be surprised."

The Phoebe fly-by marked the introduction of a 'live updating' facility, whereby a set of scientific observations were specified in terms of best-knowledge instrument pointing and timing, which was updated once the early imaging had been processed. The optical navigators worked around the clock in the days following the trajectory correction on 27 May to refine the fly-by geometry, and on 6 June issued this to the science teams, which gave the updates required to make their planned observations. An image taken on 7 June from a distance of 2.5 million kilometres with a resolution of 15 kilometres per pixel exceeded the best Voyager view of Phoebe. An image on 10 June from 658,000 kilometres with a resolution of 4 kilometres per pixel showed a dramatic sight looming.

Although an image taken from a range of 32,500 kilometres with the resolution of 190 metres per pixel showed Phoebe to have a heavily cratered surface with a great deal of albedo variation, it was just the 'taster', because the moon did not yet fill the narrow-angle camera's field of view. The morphologies of the overlapping craters of various sizes suggested an ancient surface. "It is very interesting that a lot of craters smaller than 1 kilometre in size are visible," observed, Gerhard Neukum, a member of the imaging team based at the Frei University in Berlin, Germany. "It means that, in addition to the large ones, many projectiles smaller than 100 metres have hit." It was open to question, however, whether these projectiles originated inside or outside the Saturnian system – that is, whether they were fragments of shattered Saturnian satellites or were of asteroidal or cometary origin. A part of the answer would relate to when Phoebe was captured by Saturn, and where it was previously. Although the Voyager imagery had shown bright areas on the generally dark surface, these proved to be as much as three times brighter than expected; in fact, sufficient to saturate the image at the exposure based on the mean albedo. The presence of a small crater with bright 'rays' implied that the dark material was a thin 'lag' layer that accumulated as the brighter water-ice was eroded by micrometeoroids, in which case the rays would darken as the lag regenerated.[157] There are small moonlets in distant irregular orbits around Saturn, some of which, as in the case of Phoebe, are retrograde and near the ecliptic. While these may have been captured independently, it was possible that they were related to Phoebe, and the presence of 50-kilometre-diameter craters on Phoebe prompted speculation that these moonlets were impact ejecta. The International Astronomical Union had named a bright feature in the Voyager imagery Leto Regio. When it was realised that this corresponded to the bright streaks in the interior wall of the largest crater, the crater was initially referred to as Leto, then officially named Jason.[158]

The fly-by on 11 June 2004 was at an altitude of 2,071 kilometres. The imagery at

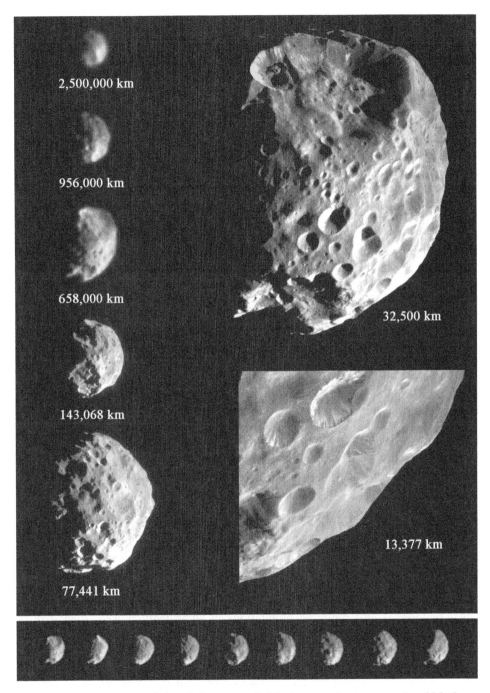

Views of Phoebe as Cassini made its approach. The non-synchronous way in which the moon rotates enabled the spacecraft to inspect most of its surface (bottom).

closest approach had a resolution of 12.3 metres per pixel, but was slightly smeared owing to the relative speed of 6.4 kilometres per second. The Visual and Infrared Mapping Spectrometer observed across its full spectral range from 0.5 to 5 microns to map surface composition at a resolution of 500 metres per pixel. The Ultraviolet Imaging Spectrograph measured surface reflectance in order to map the distribution of volatiles. The relative warmth of Phoebe's surface (in comparison to 'clean' icy satellites) enabled the Composite Infrared Spectrometer to attain a high signal-to-noise ratio in making compositional and thermal observations of both the dayside and nightside. The radar was expected to be able to penetrate to a depth of 20 centimetres, to characterise the physical properties of this material. The Visual and Infrared Mapping Spectrometer detected iron, bound water, trapped carbon dioxide, phyllosilicates, organics, nitriles and cyanide compounds consistent with a surface of cometary origin incorporating primitive materials of the outer Solar System.[159] In fact, Phoebe was one of the most compositionally diverse objects yet studied. The temperature map by the Composite Infrared Spectrometer showed the surface to be 110K, with day/night characteristics establishing the surficial material to be fluffy and porous.[160] The radar results were consistent with a dirty, rocky, icy surface. The presence of carbon dioxide was a clear indication that Phoebe was not a C-type asteroid. It must, in fact, have formed outside the orbit of Jupiter.[161] Bodies having a 'primitive' mixture of rock, ice and carbonaceous material would have been plentiful in the outer reaches of the solar nebula. However, Phoebe's actual point of origin is not clear, in part because so little is known of the Kuiper Belt Objects. Monitoring of the spacecraft's trajectory during the fly-by indicated the moon's density to be 1.63 g/cm^3, somewhat higher than expected, suggesting it has a significant amount of rock in its interior, but not as much as the best-studied Kuiper Belt Objects – for example, Pluto, its satellite Charon, and Triton (which may have been captured by Neptune) all have densities in the 1.9–2.1 g/cm^3 range.[162] As D.L. Matson of JPL said, one thing was certain: Phoebe was "a geologist's delight", and as an interloper in the Saturnian system it was a welcome bonus for the Cassini mission.

THE FINAL LEG

A small trajectory correction was made on 15 June to 'clean up' after the Phoebe encounter. Between 2 and 22 June, Cassini took imagery to make an albedo map of Titan spanning 360 degrees of longitude between 80°S and 35°N at a best resolution of 46 kilometres per pixel. There was no clear indication of impact structures, but the linear boundaries and preferred orientations of some features were suggestive of tectonism. Compiling the imagery as a movie of a complete 16-day rotation revealed that the clouds near the south pole formed preferentially in the late 'afternoon'.[163]

When C.D. Murray of the University of London, England, analysed a series of 19 narrow-angle camera images taken during an interval of 2.3 hours on 21 June to facilitate a search for new moons, he spotted a small body 1,000 kilometres from the 'F' ring, and within 300 kilometres of the orbit of the outer shepherd, Pandora. The International Astronomical Union designated this S/2004S3. It was not immediately

This albedo map of Titan from latitude 30°N to 80°S was produced by combining imagery taken by Cassini between 2 and 22 June 2004, and gave an improvement of almost a factor of 3 over terrestrial observations. As the rotation of the moon is synchronous with its orbital motion, longitude is measured westward of the centre of the Saturn-facing hemisphere.

apparent whether this was a moon 5 kilometres in size, or merely a transient clump of ring material.[164,165] In studying four frames taken 40 seconds apart, 5 hours later, for a photometric study of the 'F' ring, Joseph Spitale at the University of Arizona at Tucson found another object interior to the ring; this was designated S/2004S4.

After the critical sequence for Saturn Orbit Insertion had been uplinked and stored in Cassini's solid-state recorders, it was read back to verify that it had been received correctly. As a further verification, the sequence was allowed to execute part of the set up, turning the spacecraft to the appropriate attitude, transitioning attitude control from the reaction wheels to the reaction control thruster system, and configuring the instruments. It was activated for real on 22 June, and Cassini adopted a 'quiescent' state that prohibited instrument state changes. At this point, the 70-metre antennas of the Deep Space Network began continuous coverage – which was to be maintained through Saturn Orbit Insertion to the start of solar conjunction a week later. Whereas a 'normal' command sequence is designed to take note of an anomalous situation and 'safe' the spacecraft, a 'critical' sequence is required to 'push on' as best it can with preprogrammed recovery options. Unless something were to go badly amiss, Cassini would not be actively commanded during this time, but merely monitored.

The location of the 'bow shock' where the solar wind accumulates in front of the magnetosphere varies with the strength of the solar wind which, being gusty, causes the magnetosphere to inflate and deflate. Cassini first encountered this feature some 3 million kilometres from Saturn, which was about 50 per cent further out than for the fly-by missions. The shock washed back and forth over Cassini a total of seven times over an interval of several days before the vehicle finally entered the relatively quiescent magnetospheric environment on 28 June. The Plasma Spectrometer found the outer region of the magnetosphere to be populated by solar wind plasma, but the charged particles closer in were evidently derived from the icy satellites and/or the rings.

SATURN ORBIT INSERTION

An early proposal to fit Cassini with a medium-gain steerable antenna to enable it to maintain communications during manoeuvres that required the high-gain antenna on the spacecraft's axis to be pointed away from Earth had been precluded by mass and cost constraints. The plan had therefore been to make the Saturn Orbit Insertion manoeuvre 'in the blind'. However, following the loss of the Mars Polar Lander and CONTOUR missions while manoeuvring in this mode (with the result that there was no telemetry from which to infer how they were lost) NASA directed that a means be devised to monitor the Saturn Orbit Insertion burn in order that, in the event of the mission being lost, it would be possible to determine whether this derived from an onboard or external cause. One option was to hold the high-gain antenna pointing at Earth throughout the manoeuvre, but the off-path angle of the thrust vector, which would vary as the burn progressed, would impose a tremendous penalty in terms of propellant – so much so, in fact, that it would use up almost all the margin that had been built up in the hope of extending the mission beyond the 4-year primary tour of the Saturnian system. Furthermore, since such an off-path burn would last longer, it would curtail the time available for science activities while north of the ring plane. In any case, there would still be gaps in the telemetry when Cassini was occulted by the rings, particularly the 'B' ring, which is the densest part of the system. Also, the fact that the reporting cycle was 62 seconds meant that the onset of a problem that led to a catastrophic failure would likely be missed. Because even a particle the size of a grain of sand could be disabling, Cassini was to face its 4-metre-diameter high-gain antenna forward to serve as a shield during the ring-plane crossings. Another option was to turn after the ascending crossing to briefly point the antenna at Earth to report the status of the spacecraft prior to adopting the optimal attitude for the imminent burn; if Cassini were to be lost during Saturn Orbit Insertion, this might establish whether damage suffered crossing the ring plane was a contributing factor. The constraint on the science activities would be less severe. In November 2002 it was decided to adhere to the original plan, but have one of the low-gain antennas transmit a strong carrier signal prior to, during, and for some time after the Saturn Orbit Insertion burn. However, because Saturn and Earth would be on opposite sides of the Sun, this signal would not only be weak but also noisy, and hence be unable to carry even low data-rate telemetry. The best that could be done was for the Deep Space Network to monitor the Doppler on the signal, to provide real-time indications of the start, progress and termination of the manoeuvre. Immediately after the burn, the spacecraft was to turn to issue a status report. The procedure for using a low-gain antenna in this way was demonstrated on the trajectory refinement on 27 May.

As project manager Robert Mitchell told reporters on the day before Saturn Orbit Insertion, "Everything has to go just right. The burn must run for all 96 minutes, the turns must be at the right time, the computer must keep the critical sequence going even in the event of something unexpected happening – it has been programmed to continue even in the event of an emergency. With a one-way light-time of 1 hour 24 minutes, we had to teach the spacecraft to take care of itself. We don't want it to call home if a problem arises, we want it to keep going. This is precisely what we've told it: 'Don't stop, keep going until you've put in all 96 minutes of burn'." At the start of the burn the spacecraft's velocity relative to Saturn would be 24.26 kilometres per second and, although the burn would slow it by 626 metres per second, at shutdown its speed would be 30.53 kilometres per second. Asked by a veteran space journalist why the spacecraft's velocity would be higher at the end of the 'braking' burn than at the start, Mitchell explained, "Orbit insertion is sort of like applying your brakes while driving your car down a hill. Although you have got your foot on the brakes, you still pick up speed as the steep gravity well draws you in." The objective of the burn was to slow the spacecraft by just enough to enable it to be captured instead of making a slingshot.

Since the Saturnian system was tilted at 26.75 degrees to its orbit around the Sun, which was itself inclined at 2.5 degrees to the plane of the ecliptic in which Cassini was travelling, and Cassini's trajectory would take it within a planetocentric distance

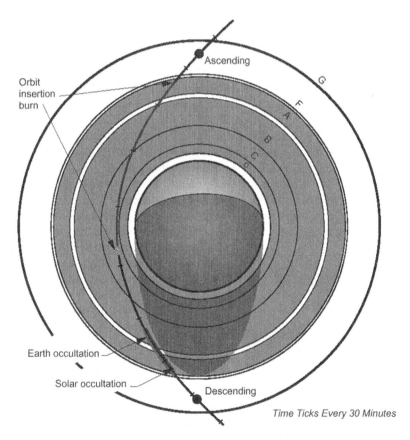

A diagram showing Cassini's trajectory while north of the ring plane, annotated to show the ignition and shutdown points of the orbital insertion manoeuvre.

of 1.3 radii, it, like Pioneer 11 and Voyager 2, was to cross the ring plane by passing through the 30,000-kilometre-wide gap between the 'F' and 'G' rings at 2.627 radii. This had been examined for hazards by terrestrial telescopes, and by Cassini during the approach phase, and if any threatening debris had been detected this would have resulted in a switch to a back-up plan in which the crossing point would be relocated outside the orbit of Mimas – resulting in a less scientifically satisfying orbital tour. The decision to proceed as planned had been taken on 19 May, a week ahead of the trajectory refinement that set up the crossing points.

At 23:47 Universal Time on 30 June, Cassini turned its high-gain antenna away from Earth to face forward to act as a shield. For JPL the loss of the signal became evident at 18:11.[166] One hour later, the spacecraft crossed the ring plane northbound at a point some 19,000 kilometres beyond the 'F' ring. The Radio and Plasma Wave Spectrometer detected the 'puffs' of plasma created as the spacecraft was struck by micron-sized grains at a relative speed of 22 kilometres per second. The impact rate started to increase dramatically 2 minutes prior to the crossing, peaked at more than 1,000 hits per second at crossing, then returned to the previous level 2 minutes later.

At 19:30 Cassini adopted the attitude for the Saturn Orbit Insertion manoeuvre. Its bipropellant propulsion system had demonstrated its performance on the Deep Space Manoeuvre, which was only slightly shorter than the nominal duration of the Saturn Orbit Insertion burn. The most efficient burn in terms of propellant would be centred on the moment of closest approach to Saturn, but in late 2002 it had been decided to advance the start by 30 minutes in order to provide a more favourable geometry for observing the ring system over a period of 75 minutes while north of the ring plane. This was significantly longer than for a 'centred' burn, and although the cost was 28 metres per second, advancing the manoeuvre would also provide time in which to rectify any issues that might arise. For example, if engine 'A' were to malfunction, the computer would shut it down and start engine 'B' to continue the burn, with the switch-over taking 10 minutes. However, as project scientist D.L. Matson pointed out, "Getting the spacecraft into orbit is the priority; getting the science is a bonus."

A wall display in the control facility at JPL showed a chart of the Doppler on the transmission from Cassini's low-gain antenna. At 19:36 the horizontal line deflected downwards, indicating both that the burn had started on time and was progressing as planned. The spacecraft pivoted at 0.008 degree per second through a total 'steering angle' of 46 degrees in order to maintain the thrust vector aligned with the Saturn-relative velocity, to increase the efficiency of the burn and use less propellant than if the direction had remained fixed. Particles and fields measurements were to be made prior to, during, and after the manoeuvre. The Dual-Technique Magnetometer was to measure the strength and direction of the magnetic field, as

A display showing how the Doppler on the transmission from Cassini followed the predicted trace during the orbital insertion manoeuvre.

irregularities in the field close to the planet would provide clues to the structure of its deep interior.[167] The Radio and Plasma Wave Spectrometer sampled the ionosphere and listened for radio bursts produced by lightning in the atmosphere. At 21:02 Cassini reached its closest point of approach, 20,000 kilometres from the cloud tops and 80,230 kilometres or 1.3 radii from the centre of the planet. Precisely on time at 21:12, the trace on the Doppler plot ended its downward slide, indicating that the burn had been nominal. Cassini then turned to point its high-gain antenna at Earth to make a 1-minute status report.

After a journey through interplanetary space of 3.5 billion kilometres lasting 6.7 years, Cassini was in a highly elliptical 'capture orbit' inclined at 17 degrees to the equatorial plane of the Saturnian system. At 21:31, while 16,000 kilometres 'above' the ring system, it started its remote-sensing observations. This sequence required it to adopt two attitudes for oblique viewing in order to maximise its coverage during the limited time available to investigate the structure, composition and temperature of the rings from this vantage point; this was the only time in the primary mission that Cassini would be in such close proximity to the rings, and the observations were a priority scientific objective. While south of the ring plane, Cassini had viewed the illuminated face, but now, on the shadowed side, the system was seen in silhouette. Meanwhile, at 21:57 Cassini passed behind Saturn's limb as viewed from Earth, and 3 minutes later entered the planet's shadow. At 22:32, as it passed back into sunlight and ended its Earth occultation, Cassini aimed its high-gain antenna forward for the descending ring-plane crossing at 22:58, then turned for remote-sensing of Saturn's southern hemisphere. At 24:00 Cassini again aimed its high-gain antenna at Earth in order to downlink the accumulated engineering and scientific data – a task that took almost 20 hours.

"I have been working on this mission for 14 years, and I shouldn't be surprised, but it is remarkable how startling it is to view these images," pointed out C.C. Porco on first seeing some of the high-resolution imagery taken north of the ring plane. "It is mind-boggling, just mind-boggling." An image of a part of the 'A' ring illustrated wave patterns produced by the gravitational perturbations of nearby moons causing the particles within the rings to jostle one another in their individual orbits. The ring system had changed little since the Voyager era, but the relative brightnesses of the major discrete ringlets of the 'D' ring had changed significantly, and one had migrated 200 kilometres closer to the planet. In fact, the 'D' ring would be revealed to be much more complex and dynamic than believed, and observations from various phase angles would establish the presence of more than one population of particles in this

This view of the unilluminated side of the 'A' ring taken immediately following the orbital insertion manoeuvre showed patterns induced by the gravitational effects of nearby moonlets.

region.[168] Data taken by the Magnetospheric Imaging Instrument during the Saturn Orbit Insertion period revealed the presence of a hitherto unsuspected radiation belt between the 'D' ring and top of the planet's atmosphere. Although the spacecraft did not fly through this region, the instrument was able to make the discovery because it sensed charged particles by their electromagnetic emissions rather than by directly sampling them.[169] The Ultraviolet Imaging Spectrograph documented compositional differences in the 'A', 'B' and 'C' rings.[170] The 'A' ring was 'dirty' near the inside and icier outside. The Encke Division was also dirty. There was no ice in Cassini's Division, the 4,700-kilometre-wide gap that separates the 'A' and 'B' rings. The 'B' ring was mainly ice. The 'C' ring was dirty towards the inside. There were also thin ringlets of dirt embedded throughout the system. Similar indications were noted by the Visual and Infrared Mapping Spectrometer.[171] It found Cassini's Division to contain a non-ice material with spectral characteristics similar to the dark material that was present on Phoebe. The 'F' ring seemed to be more dirt than ice. The Composite Infrared Spectrometer gave the most detailed temperatures to date for the rings, showing the temperature on the shadowed side to vary. The opaque parts, most notably the outer part of the 'A' ring and the entire 'B' ring, were cooler, while the translucent parts, such as Cassini's Division and the 'C' ring, were warmer. The 'A' ring was 90K, the outer 'B' ring was 70K, the inner 'B' ring was 90K, and the 'C' ring was 110K. This could be explained in terms of sunlight penetrating the sparsely filled rings to warm the material on the 'far side'.[172] The Plasma Spectrometer found a hitherto unknown low-energy plasma trapped by magnetic field lines that threaded through Cassini's Division. In one image, C.D. Murray of the University of London in England found a hitherto unknown ringlet, which the International Astronomical Union designated R/2004S1.[173] Estimated at 300 kilometres wide, this ringlet was 138,000 kilometres from Saturn's centre, just beyond the 'A' ring and coincident with the orbit of Atlas, implying a relationship between the moonlet and the ring.

The gravitational interactions between the small moonlets that orbit in and just beyond the outer part of the ring system transfer angular momentum in a manner that causes the moons to move outwards and the rings to sag towards Saturn, which implied that the rings had a life time of several hundred million years. However, if there was a collisional cascade at work in which large moons struck by asteroids and comets were broken into smaller moons, and so on, the result would be a supply of particles to sustain the system beyond the time in which it would otherwise fade. Imagery taken north of the ring plane

This view of the 325-kilometre-wide Encke Division taken shortly after Cassini had passed back south of the ring plane shows ringlets within the gap and the 'scalloped' nature of its inner edge.

showed the inner edge of Encke's Division to be scalloped. This was later found to be caused by the fact that the moonlet Pan – responsible for the 325-kilometre-wide gap – was in an eccentric orbit. The pattern enabled the mass of the moonlet to be calculated, which in turn led to a density of 0.5 g/cm^3. Observations of the perturbations of Atlas gave a similar value. As solid water-ice is 0.93 g/cm^3, this meant these objects were porous, which was consistent with their being reaccretions of ice. The ongoing reaccretion of ring material would greatly extend the life time of the system.[174] In total, the material is equivalent to an icy moon with a diameter of 200 kilometres – less than half that of Mimas. The debate about whether the rings were material that never accreted as a moon, or the debris of a moon that was shattered, was therefore seen to have been futile, since the system is in an essentially steady state, with different incarnations of moonlets and rings present at different times. As L.W. Esposito of the University of Colorado at Boulder put it, "The individual rings and moons we see are ephemeral, but the phenomenon persists for billions of years."

FIRST IMPRESSIONS OF TITAN

On 2 July 2004, some 30 hours into the capture orbit, designated revolution '0', Cassini passed Titan at a range of 339,120 kilometres. Although non-targeted, this fly-by, referred to as T_0, provided a welcome 'first look' prior to the close encounter scheduled for October. At closest approach, the frame of the narrow angle camera was tighter than the moon's disk, and because the spacecraft was south of the plane of the Saturnian system it offered a view of the illuminated south polar region. The atmospheric haze proved to be so efficient at scattering light that it degraded the resolution from the ideal 2 kilometres per pixel to a more modest 10 kilometres per pixel. At an impromptu press conference the next morning, D.L. Matson of JPL delightedly announced, "Although the images appear bland and hard to interpret, we

Views taken over an interval of several hours on 2 July 2004 showing the evolution of clouds in Titan's south polar region.

have seen Titan's surface with unprecedented clarity." E.P. Turtle of the University of Arizona at Tucson noted, "There are linear features, circular features, curvilinear features – this suggests geological activity, but we don't know how to interpret them yet." Cassini documented the motions of methane clouds over a period of 8 hours. Images taken by different filters indicated that the clouds tops were at an altitude of 30 kilometres, in the upper troposphere. All but one of the clouds were located near the 80°S latitude line; the exception being in the temperate zone at latitude 38°S.

Although the Visual and Infrared Mapping Spectrometer took a spectrum for each pixel, the atmosphere had to be 'subtracted' before the composition of the surface could be studied, and this analysis was impaired by the fact that so few characteristic absorption features fell within the narrow spectral windows at which the atmosphere was transparent. Reporting the preliminary study, K.H. Baines, a team member at JPL, said, "We see dark areas of relatively pure ice, and brighter areas with a much higher amount of non-ice materials – such as simple hydro-carbons. This is *different* from what we expected." In fact, it was precisely the opposite of what had been expected, since it suggested that the dark areas were not hydrocarbon lakes. But further analysis cast doubt on this interpretation, prompting team leader R.H. Brown of the University of Arizona at Tucson to retract the claim that the dark areas were water-ice, and to suggest instead that the bright and dark areas had similar bulk compositions and were simply mantled by different materials. One surprise was that Titan glowed both day and night, powered by the emissions from atmospheric methane and carbon monoxide. A solar-driven fluorescence by methane in the upper atmosphere had been expected, but the night-time glow was a surprise.

During its fly-by on 2 July 2004, Cassini took this low-resolution view of the surface of Titan in the 1–2-micron wavelength range.

The most important observations from the T_0 encounter provided temperature and composition data required to validate the model of Titan's atmosphere used to plan the entry of the Huygens probe. On 14 July JPL hosted the Huygens Probe Mission Risk Review, with the objective of assessing the risks and risk-mitigation measures for the probe's mission.

A GLIMPSE OF IAPETUS

On 3 July 2004 Cassini turned its attention to Iapetus. The range was 3 million kilometres, but the imagery was welcome since it provided the first glimpse of detail on Cassini Regio, the dark area on the leading hemisphere that was featureless in the Voyager imagery, showing a number of large impact basins the presence of which implied that the crust dated to an early period of heavy bombardment. The ratio of the diameter of the largest basin to the diameter of the satellite was about 0.3, which was comparable to Tethys (Odysseus; 0.4) and Mimas (Herschel; 0.3).

A long-range view of Iapetus taken on 3 July 2004, showing, for the first time, detail on Cassini Regio.

CAPTURE ORBIT APOAPSIS

There was a hiatus from 5 to 12 July 2004 for solar conjunction with the Sun within 3 degrees of the line of sight, during which Cassini maintained its high-gain antenna facing Earth and the Deep Space Network uplinked sequences of 'no op' commands for periods of 5 minutes 10 to 20 times per day in order to accumulate statistics on the reliability of the uplink in such conditions. Meanwhile, the spacecraft recorded magnetospheric and plasma science observations. On 13 July Cassini emerged back into the solar wind. While the Plasma Spectrometer monitored the state of the solar wind, the Ultraviolet Imaging Spectrograph observed the auroral activity at Saturn's south pole.

By the end of July, the Sun was 20 degrees off the line of sight and the tracking by the Deep Space Network was sufficiently improved to enable the Periapsis Raise Manoeuvre to be specified on 19 August. The bipropellant propulsion system burn on 23 August lasted 51 minutes 8 seconds and increased the spacecraft's speed by 392.9 metres per second. The speed beforehand was 325 metres per second relative to the planet – a mere 1 per cent of its speed at the conclusion of the orbital insertion manoeuvre but, as with anything thrown 'up' in a gravity field, it had slowed as it climbed. The manoeuvre was designed to set up the trajectory for the fly-by of Titan on 26 October and lift the periapsis on 28 October by 300,000 kilometers in order to clear the ring system. After the 88-minute Deep Space Manoeuvre in 1998 and the 96-minute Saturn Orbit Insertion burn, this was the third most significant firing of the engine. As this was to be the last fully pressurised burn of the primary mission (subsequent firings would made in 'blowdown' mode, using the residual pressure in the propellant tanks) the latch on the helium flow was left open for 33 minutes after the burn in order to build up the pressure in the propellant tanks, then closed to isolate the system from the leaking outflow regulator. On 27 August, at apoapsis at a range of 151 planetary radii, some 9 million kilometres, Cassini headed back in on revolution 'a' of the tour designed to overcome the Huygens relay problem. On 8–9 September the Titan Atmosphere Model Workshop met at the Goddard Space Flight Center to integrate the atmospheric observations of the T_0 fly-by with ground-based supporting work. On 19 October the mission critical event readiness review assessed the results of the Titan Atmosphere Model Workshop.

SATURN'S LIGHTNING STORMS

At the time of the Voyager fly-bys, the sporadic radio emissions from Saturn were inferred to be due to electrostatic discharges in storms in the equatorial region, which was then in the shadow of the ring system. Cassini's Radio and Plasma Wave Spectrometer first detected the discharges in July 2003, at a distance of 161 million kilometres. On its Earth fly-by, the instrument had detected lightning at a range of 89,200 kilometres, measured from Earth's surface. The mystery was why Saturn's emission was almost 1 million times as intense as that from terrestrial lightning. It tended to occur in episodes lasting several hours, in some cases involving hundreds of bursts. Although the episodes often recurred every 10 to 11 hours, which matched the length of Saturn's day, for some reason the activity would cease for long periods and then resume. It was not until the spacecraft had closed in sufficiently for the narrow-angle camera to resolve atmospheric detail that it became possible to seek

An image taken in September 2004 of the 'dragon storm' in Saturn's mid-southern latitudes, whose lightning produced radio emissions that were detected by Cassini.

evidence to establish that the radio bursts were correlated with an atmospheric structure. In early August, M.D. Desch of the Goddard Space Flight Center told C.C. Porco that the Radio and Plasma Wave Spectrometer had noted an intense episode in mid-July, and asked whether the imaging team had seen anything unusual. As Porco explained later, "There was a bigger than average 'white storm' in the southern hemisphere at that time, in a latitude range that we'd come to call 'storm alley'. We caught it with the wide-angle camera, but didn't have enough visible observations of the storm to be confident of a correlation." When the storm reappeared in September, the radio team was alerted and, in fact, the Radio and Plasma Wave Spectrometer was in the process of monitoring an episode that was even more intense than that in July. This convective storm was so large and bright that the imaging team named it the 'dragon storm'. The fact that over an 18-day interval the largest storm was brightest when the radio emission was strongest – with the storm always in the same position during a burst – indicated that it was the same phenomenon. However, instead of the burst being detected when the storm rose over the horizon from Cassini's point of view and peaking when it was on the meridian facing the spacecraft, the emission started before the storm rose and halted as it crossed the terminator into daylight.

EARLY RING STUDIES

In late September 2004, Cassini made a movie in search of ring spokes, but found none. Spokes comprise dust particles less than about 1 micron in size that collect electrostatic charges in the plasma environment of the rings, and become subject to electric and magnetic forces. In certain conditions, the particles become negatively charged and are levitated *en masse* 100 kilometres away from the surface of the rings for brief periods, forming dark features against the illuminated side of the rings and bright features against the shadowed side. However, the light-scattering geometry changes as a result of Saturn's axial tilt and orbital motion, with the formation of spokes being more likely when the Sun is near the ring plane. The spokes had last been seen on 18 October 1998 by the Hubble Space Telescope, when the angle of illumination was 15 degrees.[175] But at the time of Cassini's arrival the rings were more 'open' to the Sun and conditions were less favourable.[176]

The manner in which light from Xi Ceti flickered during an occultation enabled the density of the ring material to be measured, and revealed the presence of two density waves in the 'A' ring

On 6–7 October, with Cassini inbound and 6 million kilometres from Saturn, the high-speed photometer element of the Ultraviolet Imaging Spectrograph sampled at 1,000 counts per second to observe an occultation of the 4th magnitude star Xi Ceti by the ring system. Occultation measurements are particularly well-suited to the low to intermediate optical depth regions of the 'C' and 'F' rings and gaps such as the Cassini Division, but less so to the regions of higher optical depth. There were great variations in the distribution of the material. The particles in individual ringlets were 'bunched', with the distance between the presence or absence of material at the edge of some ringlets – especially in the 'C' ring – being as little as 50 metres. The sharp edges dramatically illustrated the dynamics that constrain the ring processes against the tendency of material to spread out to occupy nearby empty space.[177] This was the first of 80 such observations to be made during the 4-year primary mission.

'WHITE DOTS' ON IAPETUS

On 17 October 2004 Cassini passed within 1.1 million kilometres of Iapetus, and saw a chain of 'white dots' on western Cassini Regio. In fact, these were hinted at in the Voyager imagery.[178] What was surprising, was that the line seemed to continue across Cassini Regio as a dark linear feature. The 'white dots' appeared to be peaks some 10 to 20 kilometres in height.[179]

An image taken on 22 October was exposed for 180 seconds in order to view the Saturn-facing hemisphere lit by 'Saturnshine', and the stability of the spacecraft

The imagery of Iapetus taken between 15 and 20 October 2004 (with grids showing the viewing angles) provided the best view yet of this enigmatic moon, and the chain of 'bright dots' on the western periphery of Cassini Regio.

ensured that there was no blurring. This view showed the dark 'halo' on the bright terrain east of Cassini Regio in unprecedented detail.

DISCOVERING POLYDEUCES

Six narrow-angle camera frames taken over a 3-hour period on 21 October 2004 led to the discovery of S/2004S5.[180,181] A search of earlier Cassini images traced it back six months.[182] Subsequently named Polydeuces, it is about 5 kilometres in size and is a Trojan of Dione. Trojans occupy the gravitationally stable Lagrange points 60 degrees ahead of, or in trail of, a body. Three Trojan moons had already been found in the Saturnian system:

An image of Iapetus lit by 'Saturnshine' on 22 October 2004, showing the 'halo' to the east of Cassini Regio.

Telesto and Calypso with Tethys, and Helene with Dione. Helene's position oscillates, but it remains within 15 degrees of the leading point. Polydeuces can evidently stray as much as 30 degrees from the trailing point.[183,184] Cassini would pass Polydeuces at a range of only 150 kilometres in February 2005, but this was not realised in time to arrange an inspection, since encounter sequences had to be specified well in advance.

FIRST CLOSE FLY-BY OF TITAN

At an altitude of 1,174 kilometres above the surface of Titan, the T_a fly-by on 26 October 2004 was 300 times closer than the T_0 encounter in July. In fact, it was so close that Cassini would skim the topmost layers of haze. The highest priority was to perform the first direct sampling of the atmosphere to measure the thermal structure and the densities of molecular nitrogen and methane – the two principal atmospheric constituents. This data was required to plan the Huygens mission. The Composite Infrared Spectrometer was to measure the stratospheric temperature versus pressure (and hence density) to contribute to the validation of the altitudes at which Huygens would be programmed to deploy its parachutes. The extent to which Cassini had to work to maintain its attitude would provide 'drag' data, to review the planned 950-kilometre minimum fly-by altitude. Electron and ion observations were to be made by the Plasma Spectrometer, both while skimming the atmosphere and while passing through the 'wake' that the slowly orbiting moon left in the planet's rapidly rotating magnetosphere. The Magnetospheric Imaging Instrument was to characterise the ion composition and charge state of the outer atmosphere. The Radio and Plasma Wave Spectrometer was to monitor Titan's ionosphere, to establish how it interacted with Saturn's magnetosphere. The Dual-Technique Magnetometer was to study how the moon interacted with the magnetosphere to determine whether the body had its own

A full-disk mosaic of Titan centred on 15°S, 156°W, taken as Cassini approached on 26 October 2004. The surface detail was clearest where the line of sight was directly down through the atmosphere, because the contrast was progressively reduced as the viewing angle passed through haze towards the limb. The bright area to the right of centre was the 'continent'. The intended Huygens landing site was left of centre. There were bright clouds near the south pole.

magnetic field. During the approach, Cassini was to image Titan every 15 minutes to make a movie of clouds. Next would be a 4-colour full-disk mosaic at 2 kilometres per pixel. Closer imagery at resolutions between 200 to 600 metres per pixel would provide a regional view of the western part of the 'continent' (recently nicknamed Xanadu) and the dark area to its west, where the Huygens probe would be targeted. The Ultraviolet Imaging Spectrograph was to determine the composition and distribution of haze aerosols. The Visual and Infrared Mapping Spectrometer was to study aerosols, cloud structures, the recently discovered methane fluorescence, and seek insight into the composition of the surface. The radar would take microwave radiometry to seek organics, and scatterometry to measure surface roughness in the hope of proving the presence of liquid on the surface.

The speed of the fly-by was 6 kilometres per second. In the 15 minutes prior to and following closest approach, the Ion and Neutral Mass Spectrometer analysed the outermost haze. Terrestrial observations had identified only 19 types of molecule but laboratory experiments implied that there should be more. The bulk composition and thermal structure of the atmosphere was much as it had been in 1980–1981, but the instrument detected many more 'heavy' organics than expected – including propane, benzene and diacetylene – indicating either that the *in situ* rate of creation of these species was faster than expected or that mixing was more vigorous than the models predicted.[185] The Magnetometer established that if Titan had an intrinsic magnetic field, this must be extremely weak.[186] The Plasma Spectrometer determined that the magnetospheric electrons that irradiate the upper atmosphere supplement the weak sunlight in promoting the chemical processes. The Ultraviolet Imaging Spectrograph revealed that the upper atmosphere glows due to magnetospheric electrons exciting nitrogen atoms, molecules and ions. (Note that this is a different glow to that in the infrared by methane lower in the atmosphere.) A fault resulted in the loss of most of the Composite Infrared Spectrometer's data, but the information that was received identified gaseous hydrocarbons in the 'hood' over the northern (winter) pole that were absent in the southern hemisphere – undoubtedly the result of a seasonal effect. The Ion and Neutral Mass Spectrometer measured the $^{14}N/^{15}N$ and $^{12}C/^{13}C$ isotopic ratios. The nitrogen ratio suggested that as much as 75 per cent of the nitrogen that was originally present in the atmosphere had been lost to space. Against expectation, however, the Plasma Spectrometer saw no sign of nitrogen escaping to form a torus matching Titan's orbit. The nitrogen ions that it did detect were located in the inner magnetosphere, and were associated with the 'E' ring.[187] As water-ice is an efficient carrier of primordial noble gases, their absence suggested that the nitrogen in Titan's atmosphere derived from the volatisation of ammonia-ice by the heat of accretion as the moon formed, with the ammonia later being dissociated by solar ultraviolet. This indicated that the largest Jovian moon, Ganymede, which was comparable in size to Titan, accreted in a part of the solar nebula that was too warm for ammonia-ice, and hence is airless, whereas the nebula from which Titan formed, being further from the Sun, was cool enough to contain ammonia-ice and methane-ice and yet too warm for a nitrogen clathrate containing trapped primordial noble gases.

A press conference was held at JPL on 27 October to give the 'first impressions' of the imagery. One surprise was that the view of the surface was better in the near-infrared, in part due to the low contrast at visible wavelengths. There was a complex interplay between the bright and dark albedos. Intriguingly, there were 'streaks' that were suggestive of the movement of (dark) material across the surface, and from the fact that these consistently pointed in the direction of the prevailing zonal circulation it was inferred that this was windblown material. It had initially been presumed that the organic fall-out of particulates from the haze would be 'sticky', but recently V.I. Dimitrov and A. Bar-Nun had suggested that on their long, slow descent through the atmosphere the particulates would 'harden' and, being non-sticky, would pile up on

On 26 October 2004 Cassini took this mosaic of Titan's surface at a wavelength of 2 microns showing a possible cryovolcanic site dubbed 'the snail'. Also shown are views of this feature at various wavelengths across the near-infrared (bottom) and an interpretation.

the surface like very fine dust.[188],[189] It had also been thought that since the energy of insolation was too weak to drive winds in Titan's lower atmosphere the surface must be stagnant, but then it was realised that gravitational tides induced by Saturn would generate 'atmospheric tides'.[190] In fact, Saturn's tidal influence on Titan is 400 times greater than that of our Moon on Earth. "Tides apparently dominate the near-surface winds because they are so strong throughout the atmosphere, top to bottom. Solar-driven winds are strong only high up," pointed out R.D. Lorenz.[191] In Titan's dense atmosphere and weak gravity, even a surface wind averaging a speed of 1 kilometre per hour could readily 'bounce' grains along the ground in a process called saltation.

While circular albedo features were apparent, none was unambiguously an impact crater. In view of a statistical case for the occurrence of impacts, this was surprising. Nevertheless, imagery by the Visual and Near-Infrared Mapping Spectrometer of the dark area west of Xanadu taken just prior to closest approach with a resolution of 2 kilometres per pixel showed an infrared-bright circular feature 30 kilometres across at 8°N and 142°W. At its centre was a dark feature interpreted as a caldera on the summit of a dome that rose several hundred metres above its surroundings, with two elongated wings extending to the west that seemed to be overlapping sheets from separate flows. The dome could be a diapir that broached the surface. In view of its appearance, the feature was dubbed 'the snail'. "We all thought that volcanoes must exist on Titan," said B.J. Buratti, a member of the team at JPL, "and this is the most convincing evidence to date." The significance of this, as Christophe Sotin of the University of Nantes in France pointed out, was that cryovolcanism could replenish the methane content of the atmosphere.[192]

The initial results from the radar were announced on 28 October. The imaging

The first part of Cassini's synthetic-aperture radar track taken on 26 October 2004 to be made public was of an area of 478 by 250 kilometres centred at 54°W, 50°N. The fact that this area had not yet been observed optically made the interpretation of the radar albedo speculative.

track ran east to west at mid-northern latitudes, forming a thin 'noodle' some 4,500 kilometres long that covered an area equivalent to a mere 1 per cent of the moon's surface. Since a synthetic-aperture radar provides its own illumination, the geometry of a surface influences the degree to which there is a reflection; it is not like taking a picture in sunlight. Viewing angle is a factor because a surface that is tilted towards the spacecraft will appear brighter than one that is tilted away. Also, a surface that is rough at a scale comparable to the radar wavelength will scatter the beam and thus appear brighter owing to the number of favourably oriented reflectors. In addition, different materials have different reflectance: rock is more reflective than ice, which is more reflective than organics. It therefore requires a 'practiced eye' to interpret the albedo variation of a 'radar image'.

The first view published covered 478 by 250 kilometres, which was only a small fraction of the overall track. The interconnected dark spots on the generally bright albedo were consistent with a very smooth or highly absorbant solid, or conceivably a fluid. Another view showed a complex geological surface with a variety of terrain types, including sinuous bright features that cut across the darker albedos. However, because these areas had not yet been viewed by the optical instruments there was no infrared albedo information to factor into an interpretation of the radar imagery. The surprise was the absence of impact craters. A study by R.D. Lorenz had predicted that there ought to be 200 craters exceeding 20 kilometres in diameter per million square kilometres of surface.[193] There would be very few craters less than about 10 kilometres in diameter, as the objects that would excavate such small features would burn up in the dense atmosphere without reaching the ground. Statistically speaking, it was reasonable to expect 40 craters with diameters larger than 20 kilometres to be on the imaging track, and at a spatial resolution of 300 metres per pixel these should have been visible. As impacts must have occurred, the absence of craters indicated that the moon had an 'active' surface. And, as J.I. Lunine told reporters, "there are perhaps hints of features that might be craters that've been eroded or buried; perhaps by organics". On the other hand, as Lorenz freely admitted, "We see impact craters all

A section of the synthetic-aperture radar track taken on 26 October 2004 centred at 82°W, 50°N, at a resolution of 300 metres per pixel. It suggested a complex surface geology.

A section of the synthetic-aperture radar track taken on 26 October 2004 centred at 30°W, 45°N covering an area about 150 kilometres square, showing what might be a cryovolcanic flow.

across the Solar System, and our one big fear was that Titan might look like Jupiter's moon Callisto, which is an old dead world covered in impacts." As in the infrared, there were features in the radar imagery that looked as if they might be of cryovolcanic origin, including one, 180 kilometres in diameter, that bore a similarity to a Venusian 'pancake' dome.[194,195] A very convincing example was a large bright branched and lobate feature on dark terrain.[196,197,198] "It might be something that flowed, or something that was carved by erosion," said Lorenz. "It's too early to say really, but it looks very much like something oozed across the surface." The Ion and Neutral Mass Spectrometer found argon-40, an isotope derived primarily from the decay of potassium-40, which has a half-life of 1.28 billion years. The outgassing of this volatile species from the core where the rock (and hence the potassium) should reside, was evidence that Titan had undergone significant cryovolcanic activity during its history.

Although Cassini took movies to study clouds, monitoring by H.G. Roe's team indicated that the mid-latitude clouds cleared up 2 days prior to Cassini's arrival and reappeared 2 days after the fly-by! "This demonstrates", said Roe, "how [Earth-based] observations are important for putting the Cassini observations into context." That is, the Cassini team would not have been able to infer that T_a had viewed Titan on an unusually clear day.

The radar altimetry was surprising. The area that was scanned was extremely flat, with the range of elevation being confined to 150 metres over a 400-kilometre track,

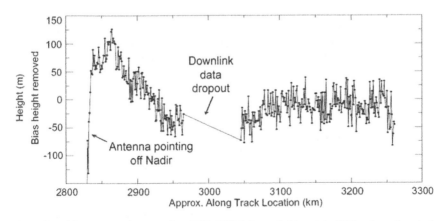

The radar altimetry track centred at 5°W, 25°N showed this part of Titan's surface to be remarkably level.

and reducing to just 50 metres across the latter part of the track. But since altimetry, radiometry and synthetic-aperture imaging could not be done simultaneously, there was no indication of what this area 'looked' like in terms of albedo. One speculation was that if organic particulates were blown into low-lying areas to form level plains, there might be areas that looked just like dry lake beds but had formed without the involvement of liquid. Might these plains have produced the specular glints detected by the Arecibo radar?

A VIEW OF TETHYS

On 28 October 2004 Cassini flew by Tethys at 250,000 kilometres. The trajectory provided a view of the trailing hemisphere, which was almost saturated with craters. At 1.5 kilometres per pixel the resolution was better than that of the Voyagers. An image taken on 15 December from 564,000 kilometres showed a new perspective on Ithaca Chasma.

Two views of Tethys: (left) the trailing hemisphere (north to the top) on 28 October 2004 showing the cratered surface and part of Ithaca Chasma, and (right) the Saturn-facing hemisphere on 15 December 2004 with a large impact basin with rings 300 and 130 kilometres in diameter adjacent to another section of Ithaca Chasma.

MORE RING STUDIES

Imagery of the ring system taken on 28 October 2004 revealed the presence of S/2004S6 near the inner edge of the 'F' ring, but it was not clear whether this was a moonlet or merely a clump of ring material.[199] A narrow ring, 300 kilometres wide, designated R/2004S2, was found between the 'Atlas ring' (R/2004S1) and the orbit of Prometheus. An image on 29 October showed a faint 'streamer' of material drawn out from the 'F' ring towards Prometheus. In fact, as subsequent study would reveal, Prometheus, the larger of the two shepherds, not only generated 'knots', 'kinks' and 'clumps' in the ring, but also created 'channels' or 'gores' in the material interior to the core of the ring. These structures were induced by the gravitational effects of the moon as it approached and receded from the ring during its 14.7-hour elliptical orbit. The perturbed particles took on elliptical orbits that led them to oscillate to and fro across the ring, with the result that one orbital period after the encounter there was a dark channel in the interior part of the ring, and, since these took time to dissipate, they formed a repeating pattern along the ring.[200]

An image (left) taken on 29 October 2004 shows a fine strand of material linking the 'F' ring to the inner shepherd, Prometheus; an image (right) of Prometheus taken on 7 June 2005; and a mosaic (bottom) taken on 13 April 2005 showing how this moon produces a dark channel in the ring every 14.7 hours as it pursues its eccentric orbit, rhythmically approaching and receding from the ring.

HUYGENS TESTS

On 2 November 2004 a 10-day end-to-end test of Huygens using the Integrated Test Laboratory at JPL began to exercise the command sequences planned for the interval 15 December 2004 to 16 January 2005 to perform the probe's independent mission. On 15 November the Titan Atmosphere Model Working Group convened at JPL to discuss the results of the T_a fly-by. Its analysis was delivered to the Huygens team on 23 November to steer the final planning review. Huygens had flown the interplanetary cruise in a state of hibernation, being awakened for a 3-hour test of its engineering and instrumentation systems every 6 months or so. The last check prior to arriving in the Saturnian system was in March 2004. A test was made on 15 July to confirm that it had not been damaged in transiting the ring system; another test was performed on 14 September and, for the final time, on 23 November. In addition, on 18 September the output of each of the five batteries was verified. This cleared the way to proceed with the option to have the probe warm up the radio relay system several hours before it entered Titan's atmosphere. On 21 November Cassini reached apoapsis at 78 planetary radii to conclude revolution 'a' and start revolution 'b'.

SECOND FLY-BY OF TITAN

The T_a fly-by had revealed there to be a layer of detached haze 150–200 kilometres higher than the expected maximum.[201] "The change in the detached haze over the 25 years since Voyager implied either the photochemical processes that make the haze, or the atmospheric circulation that distributes it around the planet, may change with the seasons," noted Robert West of JPL. "It will be a challenge for models to be able to predict how and where these detached hazes occur." This prompted concern that the planned minimum altitude of 950 kilometres was too low. The problem was that if the spacecraft were to encounter significant drag and had to use its thrusters to maintain the requisite attitude, and were to be overwhelmed in doing so, it would be obliged to 'safe' itself, which would terminate all science activities. Going low for better data at the point of closest approach therefore had to be traded against the risk of losing the post-encounter science. In late November it was decided to raise the altitude of the T_5 fly-by to 1,025 kilometres, as a compromise between the 1,174 kilometres of T_a and the nominal minimum in order to reassess the risk. As a side-effect of this revision, the altitude of the T_4 fly-by was lowered by 1,000 kilometres to 1,500 kilometres.

The redesign of the trajectory to relay the Huygens data cost about 95 metres per second of delta-V, which used up about 40 per cent of the margin predicted (prior to that) for the entire mission. Nevertheless, the remaining reserves would probably be sufficient to accommodate likely contingencies and still facilitate an extended mission. But if Huygens were to fail to release at its first deployment opportunity, it would cost about 25 metres per second to set up the backup date. If it were to prove impossible to release the probe, it would cost another 55 metres per second to carry it around for the remainder of the mission. One benefit of redesigning the early orbits of the tour was a fly-by of Iapetus on 1 January 2005 at a range of 60,000 kilometres. However, the way in which Huygens was to reach Titan meant it had to be released prior to apoapsis on revolution 'b', and follow a ballistic arc through apoapsis to encounter Titan 21 days later, inbound on revolution 'c'. As initially planned, the probe would have undertaken its independent mission rather earlier, and Iapetus would not have been in the vicinity when it was in transit. To address concern about how Iapetus might perturb the free-falling probe, it was decided on 4 October to revise the Probe Targeting Manoeuvre on 16 December, release the probe on 25 December, and perform the Orbiter Deflection Manoeuvre on 27 December in order to advance the fly-by of Iapetus to 31 December, thereby opening the range of the closest approach by the probe to 121,128 kilometres. This would degrade Cassini's imaging of Iapetus, but the opportunity would still be a considerable bonus because, on the original tour, the closest the spacecraft would have approached this satellite in advance of its targeted encounter in September 2007 would have been 415,000 kilometres. A further bonus of revising the tour for the Huygens relay was a close encounter with Enceladus on 17 February 2005. The downstream consequence of the revision designed to avoid Iapetus was that the altitude of the T_b fly-by was reduced by 1,000 kilometres, which required that the science sequences be redefined to suit the revised geometry.

For two days on the way in to the T_b fly-by on 13 December 2004, Cassini shot a movie of cloud motions. The mid-latitude clouds that were absent for the T_a fly-by had returned. The fact that one of the mid-latitude clouds appeared to have been drawn out by the prevailing wind into a streak strengthened the possibility that it was associated with a surface feature.[202] The fact that it was moving at only several metres per second suggested that it was at low altitude. However, H.G. Roe noted that if the cloud had formed in the wake of a mountain, it might 'stand' in the same manner as cloud banks stand offshore on Earth, in which case its motion would not provide a measure of the local wind speed.

One key task was to determine whether the albedo variations correlated with surface topography and, if so, how. The T_a and T_b fly-bys were at similar 'phase angles' (i.e. the angle between the Sun, Titan and Cassini) with the Sun 'behind' the spacecraft. While this was ideal for albedo studies, there were no shadows to reveal topographic relief. In fact, even under oblique illumination there might not be shadows because the haze was so efficient at scattering sunlight as to render the illumination at the surface more or less uniform. However, as K.H. Baines of the Visual and Infrared Mapping Spectrometer noted, "By the time you get out to 2 microns the sky is clear. We believe we'll see shadows at 2.0 and 2.7 microns, so there is a good chance that we could find some topography." On the other hand, it would be difficult to interpret near-infrared shadows or stereoscopic imaging at wavelengths

An image of mid-latitude clouds at 38°S on Titan in December 2004, one of which extends for about for 250 kilometres.

where the theoretical resolution was degraded by the haze. Altimetry was the only unambiguous measure of topography, but this had to be correlated with surface albedo to interpret the results. However, for economic reasons Cassini had been built without a scan platform for its remote-sensing instruments, and with the high-gain antenna on the main axis and the pallet of instruments affixed to the side, it was impossible to make simultaneous remote-sensing and radar observations of a target. Although both had been assigned time during the T_a fly-by, there were to be no radar observations on T_b, and no observations would be possible on T_c because the high-gain antenna would be required to receive the probe's transmission.[203] To correlate optical albedo with radar imagery and altimetry it would be necessary to combine data taken during many fly-bys, but this would not only take time, there would be no guarantee of being able to make follow-up observations of an interesting feature seen by one instrument by the use of another instrument. To make a start, it was decided to take altimetry across near-infrared light-dark boundaries on T_3 to definitively establish their relative elevations, and take optical imagery on T_4 of the areas that were imaged by the radar on T_a to see what some of those features actually looked like.

As Cassini encountered Titan on the T_b fly-by at essentially the same position as on T_a, the best imagery was again of the Xanadu area. In some places the boundary of this infrared-bright feature was quite sharp, and in others it was fuzzy; in some places it was fairly straight, and in others it was irregular. Because the bright and dark areas were very similar when viewed in the half-dozen near-infrared 'windows' available to the Visual and Infrared Mapping Spectrometer, there was still no good differentiation in terms of their compositions. Nevertheless, it was presumed that the dark areas were organic haze fall-out, perhaps windblown, that had accumulated in low-lying regions. There was still no unambiguous evidence for impact craters, but there was a possible multiple-ring form. When Cassini made its closest approach at an altitude of 1,192 kilometres, it imaged the Huygens target for stereoscopic mapping in combination with the imagery taken on the T_a fly-by. Outbound, the Visual and Infrared Mapping Spectrometer observed the north polar region, which had been in continuous darkness for a number of years. The Ultraviolet Imaging Spectrograph observed occultations of lambda Scorpii (Shaula) on Titan's southern limb and alpha Virginis (Spica) on the northern limb during egress to profile its atmosphere in the altitude range 450 to 1,600 kilometres. The Spica observation was made using thrusters to control instrument pointing, and was marred towards the end of the sequence by drift off axis. The Shaula observation was controlled using the reaction wheel system, which accurately pointed throughout. Six species were identified: methane (CH_4), acetylene (C_2H_2), ethylene (C_2H_4), ethane (C_2H_6), diacetylene (C_4H_2) and hydrogen cyanide (HCN). The abundances of complex hydrocarbons and hydrogen cyanide peaked sharply at about 800 kilometres, and decreased in the 750- to 600-kilometre range; the only detectable organic below that level was methane.[204] The asymptotic kinetic temperature at the top of the atmosphere was 151K. The temperature profile of methane placed the mesopause at 615 kilometres, where the temperature was 114K. The results confirmed the atmospheric model not only in the entry trajectory of the Huygens probe, but also in the design of its heat shield and parachute systems, and were fed into the calculation to refine the minimum altitude for future Cassini fly-bys.

A LOOK AT DIONE

Cassini's non-targeted fly-by of Dione on 14 December 2004 at a range of 72,500 kilometres provided a view of the trailing hemisphere containing the bright 'wispy' features. The 'quick look' data was the gossip of the 5–9 December meeting of the American Geophysical Union in San Francisco, California. Surprisingly, instead of strips of ice that had been extruded from fissures, the bright streaks were found to be grooves whose cliff-like walls of exposed ice were brighter than the 'contaminated' surface.[205] The grooves cut, and therefore post-dated, the large craters, but, at least in the area imaged, they were not interrupted by large craters. Since the grooves occurred in terrain that was generally darker than average for this body, the question became the nature of the low-albedo material.

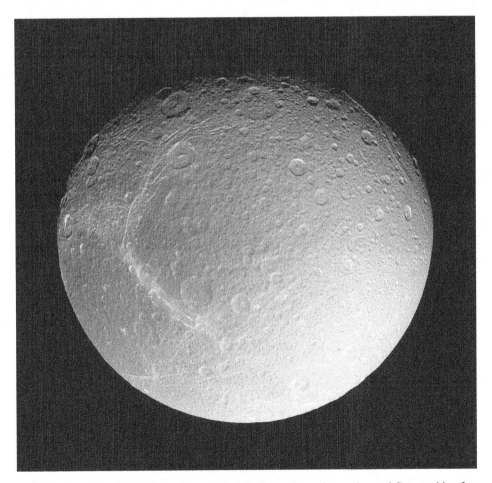

An image taken on 14 December 2004 of the 'wispy' terrain on the anti-Saturn side of Dione.

A high-resolution mosaic on 14 December 2004 showing that the 'wispy' terrain on Dione was not the predicted ice extrusions, but tectonism that had exposed cliffs of 'clean' ice.

HUYGENS RELEASE

On 16 December 2004 the Huygens management team reviewed the status reports for both Cassini and Huygens and the final report of the Mission Risk Review, then took the formal decision to proceed with the probe's release as per the plan. Later in the day, the Probe Targeting Manoeuvre was executed. It was a main engine burn of 84.9 seconds for a delta-V of 11.937 metres per second, and it placed Cassini on a collision course with Titan. The critical sequences for the Probe Relay were uplinked on 19–20 December, and the requisite value was loaded into the Master Timer Unit on Huygens. This triple-redundant 'alarm clock' – which was to power up the probe's key systems several hours prior to atmospheric entry – was started on 21 December. The next day, Cassini refined its trajectory by a burn of 20 millimetres per second. On 23 December the Probe Release Approval Meeting gave permission to proceed. The probe was to be released on 24 December (02:00 GMT on 25 December). As Huygens had no means of controlling its attitude, Cassini was to release it at precisely the angle for safe entry into Titan's atmosphere. The release mechanism was to spin the probe up to a rate of 7.5 revolutions per minute for stability in free-fall. Pyrotechnics would then sever the physical and electrical links to the probe, and a set of three springs would push it away at 0.35 metre per second – a delta-V that had been allowed for in calculating the Probe Targeting Manoeuvre. As soon as the probe was away, Cassini turned to point its high-gain antenna at Earth to report the event. Although the probe was magnetically 'clean' in its powered down state, it had a small residual magnetic signature. The fact that the Dual-Technique Magnetometer could detect this field enabled the instrument to measure the rate at which the probe was made to spin, and then verify its departure. "Detecting the spin was immensely reassuring," said David Southwood, director of science programs for the European Space Agency. "Not only did it show Huygens was rotating correctly but, because the spin is directly related to the departure velocity, that it was ejected at the right speed." A 5-by-5-frame mosaic of wide-angle images was taken 12 hours later, and to everyone's delight the probe was visible as a dot on the central frame. The navigation team used this and a narrow-angle image taken on 26 December to confirm that the probe was on course to penetrate the atmosphere of Titan at the requisite angle. On 27 December Cassini executed a main engine burn of 153.4 seconds for a delta-V of 23.7 metres per second. This Orbiter Deflection Manoeuvre was designed to adopt the trajectory required for the T_c fly-by on 14 January 2005 at a range of 65,000 kilometres, during which it was to receive the transmission from the probe. Revolution 'c' began with apoapsis on 31 December at 59.7 planetary radii. Following instructions, 7 days prior to the probe's arrival Cassini terminated all non-essential activities, including science observations.

An image taken by Cassini's wide-angle camera on 25 December 2004 showing the departing Huygens probe at a distance of 18 kilometres.

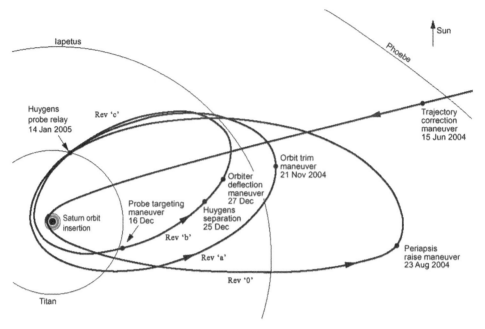

A diagram of the manoeuvres made by the Cassini spacecraft following Saturn orbit insertion in order to facilitate the Huygens mission.

IAPETUS REVEALED

A meeting at JPL on 20 December 2004 reviewed the science plan for the fly-by of Iapetus. The sequence was uplinked on 30 December for execution the next day, and as Cassini passed the moon at a range of 123,400 kilometres it took imagery at a resolution of 1 kilometre per pixel. The dark linear feature detected in October was not only seen to span the width of Cassini Regio but, because it was visible on the limb, it was also identifiable as a ridge.[206] It was basically continuous, and simply rose from the adjacent dark terrain without 'foothills', although it was formed in some parts by individual mountains. Where an impact had demolished a part of the ridge the bright rim of the crater suggested that it was made of ice.[207] It was subsequently determined that the ridge ran precisely along the equator, and extended beyond Cassini Regio. An analysis of the Voyager imagery had shown Iapetus to be an imprecise sphere. While it could be described as an ellipsoid whose half-axes were 767 × 742 × 713 kilometres, its shape was irregular, with substantial depressions and bulges, which was unusual for a body of such a mass.[208,209,210] If the ridge was a compression structure created as the moon's interior shrank, then this would be expected to form along the circumference of the widest diameter (as was the case), after which tidal effects would have tipped the spin axis perpendicular to its orbital plane and synchronised its rotation. The field of view included the northern polar region, where dark 'feathery' streaks on the bright terrain adjacent to Cassini Regio suggested that the dark material had been delivered to the surface ballistically. Indeed, to space artist Don E. Davis the streaks appeared "as if they had been airbrushed". The craters in this area shared a pattern of albedo that suggested that the interior walls facing Cassini Regio had been exposed to 'fall out' that had come from the centre of the leading hemisphere, leaving their 'shielded' walls bright. But, of course, opinion differed as to whether this material was of endogenic or exogenic origin.

An Arecibo study at a wavelength of 12.6 centimetres of the bright hemisphere in January 2002 and of the dark hemisphere in January 2003 gave the surprising result that the radar could not tell the two hemispheres apart.[211] Furthermore, even the bright side was unusual. As team member G.J. Black of the University of Virginia explained, "It is known that the bright side is mostly water-ice, but we find it does

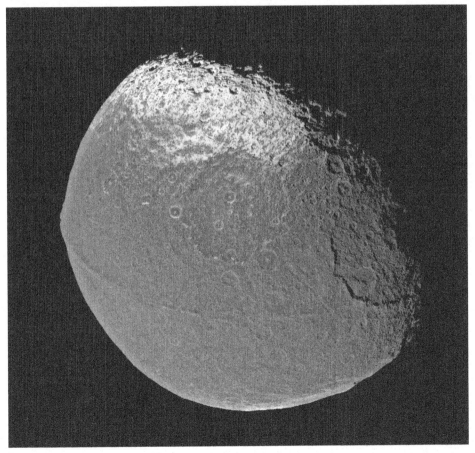

A mosaic of images taken on 31 December 2004 showing the leading hemisphere of Iapetus, and considerable detail on Cassini Regio.

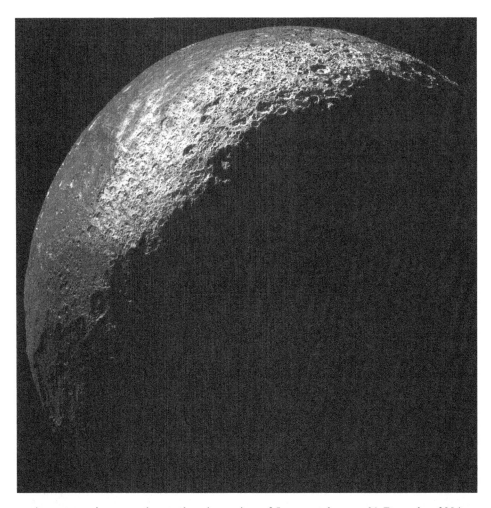

A crescent view over the north polar region of Iapetus taken on 31 December 2004, showing the 'feathery' albedo features.

not reflect the radar like any other icy satellite that we've studied with the radar; it appears much less reflective." If there was ammonia present, the ice would be bright optically, but would not reflect microwaves as strongly as 'pure' water-ice. The fact that Cassini Regio was indistinguishable "could mean that on the dark side there is merely a thin coat of some darkening material on the ammonia-laden water-ice, just like an inch of dirt over clean snow. Such a coat would not have much effect on the radar reflection, which sees the underlying ice." If this dark material is such a thin veneer, then to preclude excavation of the bright substrate it must either have been applied very recently, or its deposition must be ongoing and at a sufficient rate to mask new exposures of ice. Although the summits of the peaks on western Cassini Regio were bright, they were exceptional. A massive landslide from the 15-kilometre-

A close up of Iapetus taken on 31 December 2004 showing a landslide covering half of a 120-kilometre-diameter flat-floored crater located adjacent to the 15-kilometre-tall scarp that forms the rim of a 600-kilometre impact basin.

tall scarp that formed the rim of a large basin in eastern Cassini Regio had spilled across the floor of a crater contained within the basin, and the exposed face of the scarp was dark.

The resolution of the Visual and Infrared Mapping Spectrometer was such that its image of the moon spanned only four pixels, but the fact that one pixel was wholly on the bright area and another was wholly on Cassini Regio facilitated a basic comparison of their compositions. The bright terrain appeared to be water-ice contaminated by a small amount of organic (tholin) material, and Cassini Regio was best modelled as a mixture of tholin, a hydrogen cyanide polymer, a small amount of water-ice and ferric oxide.[212] This matched fairly well a model based on telescopic data.[213] The fact that Cassini Regio was also laced with carbon dioxide (probably a photochemical product that was trapped in either water-ice, a mineral or a complex organic solid) implied that the carbon-bearing material was 'swept up' by the moon after its rotation had become synchronised. The Composite Infrared Spectrometer indicated that the dark material was loose and fluffy.[214,215] The Ultraviolet Imaging Spectrograph determined that a distinctive far-ultraviolet absorption feature for water-ice was strong for both Phoebe and the bright side of Iapetus, but extremely weak for Cassini Regio.[216] Nevertheless, this did not reject Phoebe as the source of the dark material on Iapetus, because the volatiles could have been liberated at emplacement, thereby concentrating the non-ice material. If Phoebe is a captured body from the frigid outer Solar System, then on becoming trapped in the warmer Saturnian environment it might have undergone a period of outgassing and supplied the material which has disfigured Iapetus. Intriguingly, although Cassini Regio is centred precisely on the leading hemisphere, it does not reach the poles, and in the equatorial zone it extends beyond both limbs onto the trailing hemisphere. Also, the spectrum of the central area is somewhat 'redder' than the equatorial extensions, whose spectrum is 'flatter', suggesting a progressive transition.[217] If the dark material is indeed of exogenic origin, then it was deposited after the ridge formed, and after the moon adopted its current spin axis and synchronised its rotation with its orbital period – a timescale that might shed light on when Phoebe was captured.

Outstanding questions include:

- Why is Iapetus in an inclined orbit?
- Why is it so large for an 'outlier' in the system?
- Why is it so different from similarly sized Rhea?

On 10 September 2007 Cassini will fly within 1,300 kilometres of Iapetus to image it at a resolution 100 times better than that achieved so far, and hopefully this will provide insights into these mysteries.

HUYGENS REACHES TITAN

On the original schedule, the Master Timing Unit would have powered up Huygens about 45 minutes before the predicted moment of contact with Titan's atmosphere, but the schedule had been revised to do this with 4 hours remaining, to enable the radio system to warm up and further reduce the Doppler problem. With 90 minutes to go, Cassini started the command sequence by which it was to turn to aim its high-gain antenna towards the probe, record the transmission, and finally turn back to Earth to report its success. The probe penetrated the outermost fringe of the moon's atmosphere at an altitude of 1,270 kilometres and a velocity of 21,000 kilometres per hour on a trajectory dipping at an angle of 65 degrees below the local horizontal. The temperature of the forward heat shield rapidly rose to 1,800°C. In terms of real time, entry occurred on 14 January 2005 at 09:06 GMT. If the probe had attempted to send a signal at that moment, this would have taken 1 hour 7 minutes to reach Earth and arrived at 10:13 GMT,[218] but because the probe was enveloped by plasma the transmitter was not yet active. The deceleration reached its peak after 3 minutes, by which time the speed had been slowed to 1,800 kilometres per hour. The major optical observatories with a view of Titan watched for any sign of a fireball trail. Although nothing was detected, Fred Chaffee, director of the Keck Observatory, pointed out, "It was worth getting up in the middle of the night for this historic moment." Antonin Bouchez, a member of the team observing in the near-infrared in search of thermal radiation from the probe's passage, pointed out, "the observations provided the best images that characterise the satellite at the moment of the probe's entry". About 5 minutes after first contact – by which time Huygens was at an altitude of 180 kilometres and had slowed to 1,500 kilometres per hour, corresponding to Mach 1.5 in the local environment – a mortar deployed the 2.6-metre-diameter pilot parachute. This pulled off the back cover of the heat shield and drew out the 8.3-metre-diameter main parachute, which rapidly slowed the rate of descent to 320 kilometres per hour. At an altitude of 160 kilometres the heavy forward heat shield was released to fall away and the scientific instruments started to report data.[219]

In addition to receiving and storing the data transmitted by the probe, Cassini was to facilitate the Doppler Wind Experiment by measuring the Doppler on the carrier signal due to the relative motions of the spacecraft and the probe. Although the signal from the probe was not designed to be detected on Earth, while the mission was being developed and flown the capabilities of radio telescopes had improved so dramatically that it was now possible to 'listen in' to the transmission. In order to reconstruct the trajectory of the parachute descent, the European Space Agency had teamed up with the Joint Institute for VLBI in Europe (JIVE) to monitor the probe's transmission. The signal would be too weak to extract the data, but Very-Long-Baseline Interferometry processing of the carrier signal received by radio telescopes across the globe would enable the probe's position to be monitored with an accuracy of about 1 kilometre. In addition, JPL had equipped several radio telescopes with apparatus to enable them to measure the radial velocity relative to Earth. In combination, these observations would enable the probe's parachute descent to be plotted in three dimensions, and its landing site located. Cassini would be best positioned to measure the zonal component of the wind, while the telescopes coordinated by JPL determined the meridional component.

At about 10:23 GMT, some 10 minutes after entry, the Robert C. Byrd Green Bank Telescope in West Virginia, operated by the National Radio Astronomy Observatory, detected the carrier signal, which served to confirm that Huygens had survived the deceleration. The Doppler was as predicted for the nominal descent, indicating that the main parachute had successfully deployed. Since Titan was soon to set below the local horizon, Green Bank would not be able to follow the entire descent, but the similarly equipped Parkes Observatory in Australia would be able to monitor the final portion of the descent, as well as any transmission from the surface. Of course, the fact that the probe was transmitting did not in itself mean that its instruments were functioning. Furthermore, since Cassini was not in communication with Earth, it was not possible to confirm that it was listening to the probe and recording the data. The scientists would have to wait until Cassini swung around to point its high-gain antenna to Earth and replayed the transmission.[220] Even so, given the audacity of the mission, the omens were good.

Meanwhile, having descended to an altitude of 120 kilometres, the probe replaced its main parachute with a smaller 3-metre-diameter canopy designed to increase the rate of descent through the tropopause, the coldest part of the atmosphere. For most of the descent, the winds were west-to-east (with the rotation of the moon), and the maximum speed of 120 metres per second occurred at an altitude of 125 kilometres. In a study of imagery from previous fly-bys, the motions of tropospheric clouds near the south pole had confirmed that the atmosphere circulated at a rate exceeding that of Titan's 16-day axial rotation – a phenomenon known as super-rotation.[221] This was predicted a decade earlier by a computer model made by Anthony DelGenio of the Goddard Institute for Space Studies in New York. As his colleague John Barbara put it, "To discriminate clouds from the surface features, I used imagery of the same region taken at different times and subtracted them from each other. When I did this, time-variable clouds stood out as regions of changing brightness." Severe wind gusts and vertical shear in the altitude range 100 to 80 kilometres gave Huygens a wild ride, tilting it by as much as 60 degrees and threatening to collapse its parachute. The atmosphere was almost stagnant in the altitude range 80–70 kilometres, then the winds picked up again, reaching a steady 40 metres per second at 60 kilometres, after which they gradually tailed off.[222]

The only instrument to be activated for the deceleration phase was the Atmospheric Structure Instrument, which was to indirectly infer the pressure and temperature of the detached haze layers; when the forward heat shield was released, it switched to direct sampling.[223] The temperature and density of the upper atmosphere were greater than predicted. There was an ionospheric layer between 140 kilometres and 40 kilometres, with the maximum electrical conductivity at about 60 kilometres. A sensor might have detected a radio burst from lightning, but there was no acoustic evidence of thunder.[224] The Gas Chromatograph Mass Spectrometer took four samples, at 140, 85, 55 and 20 kilometres.[225,226] In addition, the mass spectrometer subsystem analysed the samples taken by the Aerosol Collector and Pyrolyser in the stratosphere and upper troposphere.[227] It had been expected that the probe would emerge from the base of the optically thick haze at an altitude of about 70 kilometres and provide the Descent Imager and Spectral Radiometer with a view of the surface, but this instrument did not get its first glimpse until 45 kilometres, and the scene remained murky until 30 kilometres.[228] In fact, the number density of haze particles increased from an altitude of 150 kilometres all the way down to the surface. Once the imaging 'triplets' had been received on Earth, they were to be mosaicked into a series of panoramas taken at successively lower altitudes and improving spatial resolutions. At 25 kilometres, the upward-viewing spectrometer indicated that methane was becoming a major constituent of the troposphere. Direct sampling showed a constant ratio of methane to nitrogen in the stratosphere and upper troposphere, but in the lower troposphere the ratio began to increase. At 8 kilometres the methane reached saturation, and remained so down to the surface. The fact that the densities of the haze aerosols decreased as their sizes increased implied that they were being coated by liquid. Evidently the motes in the lower troposphere were 'seeding' the rain drops. The $^{12}C/^{13}C$ isotope ratio indicated a continuous (or perhaps episodic) replenishment of the methane in the atmosphere. A biological source was ruled out. On Earth, living organisms have a preference for ^{12}C, and methane made biologically is enriched with this isotope. This serves as a biomarker in the chemical study of rock. There was no enrichment of ^{12}C in Titan's atmosphere. The $^{14}N/^{15}N$ isotope ratio was consistent with the measurement at the fringe of the atmosphere by Cassini's Ion and Neutral Mass Spectrometer, and supported the hypothesis that the lighter isotope was leaking to space. Argon-40 was confirmed. Argon-36 was also detected, but only at a trace level. Argon-38, xenon and krypton were all absent. The haze aerosols proved to contain ammonia and hydrogen cyanide, which was significant because ammonia was not present in gaseous form. This indicated that the aerosols must include the results of chemical reactions that gave rise to complex organic molecules – they were not simply condensates.

At 10 kilometres, the downward-looking infrared spectrometer was to start to take reflectance spectra to determine the composition of the surface in the vicinity of the landing point. At about 7 kilometres, the wind speed decreased to a mild 1–2 metres per second and the direction became variable, possibly indicating that the probe had descended into a convective region. At an altitude of 700 metres the probe switched on its 20-watt lamp in order to 'fill in' the wavelengths of sunlight that were filtered out when passing through the atmosphere, to improve the spectroscopy. To the imager, the result resembled an automobile shining its headlamps into fog, suggesting there was a drizzle of fine droplets of liquid methane.

Just before Cassini released Huygens, Mark Leese of the Surface Science Package team at the Open University in England had pointed out, "It is interesting that all the possible landing scenarios we envisaged still exist: a crunch down onto ice, a softer squelch into solid organics, and a splashdown in a lake of hydrocarbons." The rate of descent at surface contact was not expected to exceed 5 metres per second, which proved to be the case. The fact that radio telescopes continued to receive the probe's signal after the Doppler indicated it had come to rest showed that it had survived the impact.

Cassini re-established contact with Earth at 15:04 GMT and sent a brief update to indicate that it had received the probe's radio transmission. This prompted hearty applause in the European Space Agency's Huygens Project Operations Centre at Darmstadt in Germany. At 15:07 GMT Cassini started the first playback, a process that took some 3 hours. This indicated that the transmission from the probe had begun 15 seconds earlier than predicted, which meant that (in effect) the arrival and braking phase had been nominal. Cassini had recorded 2 hours 27 minutes of descent data. The probe's five batteries had been specified for a *minimum* duration of 153 minutes, this being the predicted maximum descent time plus 3 minutes on the surface, although it had been hoped to sustain 30 minutes of surface activity. In fact, the 64-metre radio telescope at Parkes was still receiving the signal when Titan set there at 16:00 GMT. There was initial concern that the prolonged transmission might be the result of one of more of the instruments failing to draw power, but (as the telemetry showed) the interior of the probe had been warmer than expected during the parachute descent, and this would have improved the performance of the batteries. After the probe reached the surface, Cassini recorded 1 hour 12 minutes of data until it exited the probe's transmission lobe – which occurred shortly before the spacecraft passed below the probe's horizon.

Huygens had two transmitters, known as the A/B channels. In terms of hardware, the only difference was that the A-channel included the ultrastable oscillator for the Doppler Wind Experiment. The original plan was to send all data over both channels, to provide redundancy, but the bandwidth required by the Descent Imager

Preliminary findings of the Aerosol Collector and Pyrolyser and Gas Chromatograph Mass Spectrometer on the Huygens probe.

and Spectral Radiometer was such that rather than repeat all of the imagery and spectral data it had been decided to send spectral data through both channels and divide up the imagery in order that, if all went well, twice as many pictures would be available. It was soon realised that the A-channel of the probe's transmission was 'missing'. Had it not been sent by Huygens? Or had Cassini not received it? The engineering telemetry would show. The loss of half of the imagery was frustrating as it would leave holes in the panoramas, but was by no means disastrous. The Doppler Wind Experiment was the real victim, because it relied on the A-channel. As Michael Bird of the University of Bonn, Germany, leading the experiment, admitted, "I have never felt such exhilarating highs and dispiriting lows than those I experienced when Green Bank reported the signal indicating 'all is well', only to discover there was no signal on the A-channel!" However, the data from the radio telescopes would provide a basis for pursuing the intended research.

The first image from the downward-facing Descent Imager and Spectral Radiometer of the Huygens probe to be made public, showing Titan's surface from an altitude of 16 kilometres.

The European Space Agency began its 'first impressions' news conference at 19:45 GMT, after the first full replay of the probe's transmission. The first image to be released was taken at an altitude of 16.2 kilometres, and had a resolution of about 40 metres per pixel. As hoped, Huygens had come down over a boundary between the light and dark areas. The image seemed to show a network of dark drainage channels on a bright area leading to a shoreline, with several offshore islands on the dark area. The wind seemed to have carried the probe out over the dark area. When the image that everyone was yearning to see (the post-landing view) was presented, it showed a solid surface littered with rocks! It was not a body of liquid. Nevertheless, the overwhelming impression was of the dry bed of a lake or shallow sea. Since the camera was very close to the ground, the sense of perspective was deceptive: the rocks were not the boulders they appeared to be; they were actually no more than pebbles and cobbles ranging up to 15 centimetres in size, with the nearest no more than 1 metre away, and they were assuredly not silicate rock but water-ice. The fact that they were rounded and on top of darker fine-grained material suggested that they were erosion products washed down the drainage channels onto the plain. The imager continued to function after

The surface of Titan as the Huygens probe saw it at ground-level.

landing, and repeated views of the same scene raised the prospect of detecting motion. There were enticing hints of liquid trickling between the pebbles, but the way in which the data had been compressed to minimise the bandwidth requirement meant that the 'lost' information could not be regenerated, making the imagery distinctly fuzzy. In retrospect, it would have been advantageous if one in every 10 pictures had been sent at full resolution. Nevertheless, there was evidence of erosion around the pebbles that was suggestive of fluvial activity. When was the last methane rain storm? Had this liquid drained into a porous surface layer? With the sunlight being both scattered by the haze and absorbed by the methane, the uniform lighting was 1,000 times dimmer than a sunny day on Earth, and it cast an orange glow over the landscape. The infrared reflectance spectrum measured for the surface was unlike any other in the Solar System: the 'red slope' in the optical range was consistent with tholins, and there were absorption features indicating water-ice, but the 'blue slope' in the near-infrared was due to an unidentified constituent. The processes were evidently different to those on Earth, but the landscape was, as T.V. Johnson of JPL put it, "strangely familiar". It was apparent that by characterising the interface between the light and dark areas, Huygens would yield the 'ground truth' needed to interpret the ambiguous near-infrared remote sensing by Cassini and from Earth. The sense of perspective would certainly have been much poorer if the probe had landed on one terrain without providing a glimpse of the other.

At a press conference on 15 January it was revealed that the loss of the A-channel was due to a software error. Although the fault was on Cassini, the European Space Agency was responsible for the mistake, which was, as David Southwood's deputy, Jacques Louet, put it, "as simple as throwing a switch". In effect, Cassini had not been told to switch on the receiver for the A-channel. "We made a stupid mistake," Louet admitted. "It's a classic example of the most simple of things escaping review because they're so simple." The error had not been discovered during the 'rehearsal' in March 2004 because, with the probe mounted on Cassini, the receivers were not

John Zarnecki (left), Toby Owen, Marty Tomasko, Jean-Pierre Lebreton and David Southwood at a press conference to discuss the preliminary findings of the Huygens probe.

The externally mounted ACC-E sensor of the Surface Science Package measured the force as the Huygens probe made contact and settled. The internally located ACC-I sensor measured the 'g' of the deceleration.

operated during the test. In view of the nature of the error, it was fortunate that *both* channels were not disabled. Imagine if the terrestrial radio telescopes had confirmed that the probe made its transmission, and it transpired that Cassini had not been listening!

The data from the Surface Science Package indicated that the probe impacted at 4.5 metres per second, with a 15-*g* deceleration over an interval of 40 milliseconds. The preliminary indication from the penetrometer that projected from the base of the package was of a thin brittle crust over a material with the consistency of wet sand. To R.D. Lorenz, who designed the instrument, the signal suggested a material with the characteristics of crème brûlée, a sweet food made of custard with a surface of hard-burned sugar. J.I. Lunine ventured that the sharp response interpreted as a crust was the result of the penetrometer hitting an icy pebble and nudging this aside prior to contacting the actual surface. This was supported by the full analysis.[229] The penetrometry and accelerometry measurements indicated that the surface was neither hard like solid ice, nor readily compressible like a blanket of fluffy aerosols. It was a 'soft' surface, for which plausible candidates were: (1) a solid granular material with little or no cohesion; (2) a 'mud' of ice grains from impact or fluvial erosion wetted by liquid methane; and (3) a 'tar' of fine-grained ice and photochemical products. Huygens had slipped along the surface for several seconds prior to coming to a halt tilted at an angle of 10 degrees with its base dug in about 10 centimetres, and the imager viewing to the south. It then slowly settled by a few millimetres, increasing its tilt by a fraction of a degree. All sensors in the Surface Science Package operated, but the refractometer, permittivity and density sensors gave little useful data because they were present to deal with a splashdown. The speed of sound at the surface was determined to be 180 metres per second. Acoustic soundings during the final 100 metres of the descent indicated that the surface around the point of impact was fairly smooth, being an undulating topography with a vertical range of about 1 metre.

The fact that Huygens embedded itself in the soft surface and continued to operate for over an hour provided a welcome bonus. The surface temperature measured by the probe was 93.65K ($\pm$0.25), in excellent agreement with the value of 92K inferred from remote sensing during the Voyager 1 fly-by. The lamp for the spectrometer was still illuminated when the relay was concluded, and would have provided a source of heat. In addition, the inlet of the Gas Chromatograph Mass Spectrometer on the base of the probe was held at $+90°C$, making Huygens by far the warmest object on Titan.[230] There was a significant increase in the methane fraction several minutes after landing, suggesting that heat from the probe was boiling volatiles out of the surface material. The lens of the downward-looking imager was obscured at touchdown because it became embedded. The repeating views from the oblique and side-looking imagers provided enticing hints that the probe was causing material to splutter across the ground. The implication was that occasional heavy methane rain drained from the elevated terrain onto the low-lying ground, where it rapidly soaked into the porous material, which then served as a reservoir for slow evaporation. The volatiles that the Gas Chromatograph Mass Spectrometer detected after landing were rich in organics that had not been noted during the descent. Although the signal was dominated by methane, there was ethane and a possibility of benzene and cyanogen (both results of methane and nitrogen chemistry) and carbon dioxide, all of which were indicative of a complex chemistry in the surface material.

It was not immediately realised, but the prolonged transmission following landing provided insight into the wider context. Part of the radio signal was reflected off the ground, and the manner in which this interfered with the line-of-sight beam served to

The manner in which the radio signal from the Huygens probe was reflected by the ground enabled the character of the surface to be inferred.

A mosaic of images showing a system of channels cut by surface fluid draining from the (bright) elevated terrain onto the (dark) low-lying plain.

characterise the surface for some 2 kilometres westward of the landing site, showing it to be fairly level and littered – just as at the landing site.[231] Analysis of the panoramic imagery taken during the descent revealed that while some of the dark channels on the bright terrain had a dendritic form that suggested they were cut by flowing surface fluid, the short stubby ones that joined at perpendicular angles were more probably produced by liquid flowing beneath the surface by the process of sapping. Overall, therefore, the site appeared to be the counterpart of a terrestrial desert that is usually, but not always, dry; with arroyo drainage onto a playa plain. Were the glints observed by the Arecibo radar from plains such as this?

A reconstruction of the probe's descent based on observations by terrestrial radio telescopes and a comparison of the probe's own imagery with that taken by Cassini during fly-bys prior to and after the Huygens mission, enabled the landing site to be located within a radius of several kilometres centred on 192.4°W, 10.2°S, very near to the point predicted by the approach trajectory and atmospheric model.[232,233]

A LOOK AT ENCELADUS

On 15 January 2005, as Cassini approached periapsis, it imaged Enceladus from a range of 367,000 kilometres at a resolution of 2 kilometres per pixel and showed the leading hemisphere, which had not been viewed by Voyager, to be an extraordinary landscape of swirling curvilinear wrinkles. This indicated that, on this moon, it was not the smooth terrain that was atypical, it was the heavily cratered 30 per cent of the surface. The contact dividing these terrains was remarkably sharp – indeed, it sliced across craters. Although there *were* craters on this 'wrinkled terrain', they were so sparse as to suggest that it was extremely young. Imagery of Mimas the next day from 107,600 kilometres caught the large crater Herschel on the terminator.

CASSINI RESUMES ITS TOUR

A few days before Cassini reached apoapsis at 59.3 planetary radii on 1 February it performed a main engine burn of 120.1 seconds for a delta-V of 18.68 metres per second to put it on course for revolution 3 of the *nominal* tour, with some revisions to the timings: on the T18-5 plan, the T_3 fly-by of Titan was to have been at an altitude

A prediction of the areal coverage of the down-facing imager on the Huygens probe during its descent superimposed on an image of the intended landing site, together with (top) a contextual view taken on 26 October 2004.

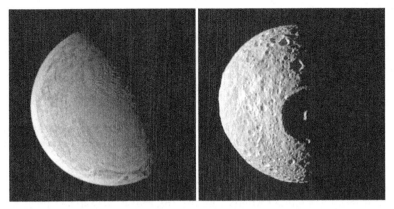

An image (left) of the leading hemisphere of Enceladus (north to the top) taken on 15 January 2005 showing a distinctive 'wrinkled' terrain devoid of impact craters. An image (right) of the leading hemisphere of Mimas on 16 January 2005 showing the crater Herschel on the terminator.

of 1,194 kilometres, but when the tour was revised to overcome the Huygens relay issue this was lowered to 955 kilometres, then the change to prevent Iapetus from perturbing the probe had the side-effect of raising it to 1,577 kilometres. As the revised geometry ruled out the Earth occultation in which the attenuation of the spacecraft's radio signal was to have been used to profile the atmosphere in the near-equatorial zone – the only such opportunity on the primary tour – it was decided to reassign this observing time to the radar. The T_3 fly-by on 15 February 2005 was to provide coordinated (although not simultaneous) optical and radar imaging observations. For 2 hours around the time of closest approach, the radar was to image several areas, including a part of Xanadu, this being the first location to be

A radar image taken on 15 February 2005 showing an 80-kilometre-diameter circular feature at 16°W, 11°N, that was clearly an impact crater surrounded by a blanket of ejecta. It was later named Sinlap.

imaged by radar that was unambiguously bright in the near-infrared. The radar was also to take altimetry to determine whether the extreme flatness measured on the T_a fly-by was typical. The optical imaging provided full-disk mosaics, a 5-by-5-frame mosaic of the western part of Xanadu and the dark area where the Huygens probe landed, plus two areas in support of the radar. The radar imaging track ran essentially parallel to that from the T_a fly-by, but offset to the south. It included an 80-kilometre circular feature that was clearly an impact crater surrounded by a blanket of ejecta.[234] However, without direct altimetry it was not possible to know whether it had been reduced to a palimpsest.[235] In addition, there was a 440-kilometre-diameter feature that

A radar image situated northeast of Xanadu taken on 15 February 2005, showing the 'cat scratches' suggestive of dunes, and a part of the 440-kilometre-diameter circular feature nicknamed 'Circus Maximus'.

could be a multiple-ringed basin that, in view of its size, was nicknamed Circus Maximus; it was later named Menrva by the International Astronomical Union. A radar-bright channel system on the adjacent terrain ran across its southwestern 'rim'. To the west were groups of dark linear features spaced 2–3 kilometres apart that ran for hundreds of kilometres. These 'cat scratches' gave the impression of being dunes of windblown material.

By this point, Cassini had provided one distant look at the south polar region and three close inspections of the equatorial region, with two radar imaging tracks which together covered less than 3 per cent of the surface. The first step in photogeology is to objectively classify surface units by their appearance. The radar imagery to-date indicated a variety of units, including: (1) a homogeneous unit that made up 50–60 per cent of the T_a swath and 20 per cent of the T_3 swath; (2) the radar-bright Xanadu unit; (3) features unambiguously caused by impact; (4) sinuous features; (5) a radar-bright mottled unit with a gradational boundary; (6) a radar-bright lineated unit; and (7) a radar-bright unit that had lobate, sheet and digitate characteristics.[236,237,238,239,240] The application of interpretations to surface units is subjective and therefore open to dispute. Nevertheless, it was clear that (as in the case of Earth) the surface of Titan had been shaped by tectonism, erosion, winds and (seemingly) volcanism – although in Titan's case, cryovolcanism.

Also on the T_3 fly-by, the Composite Infrared Spectrometer made full-hemisphere thermal maps to study the circulation of the upper atmosphere. The stratopause was at an altitude of 310 kilometres, where the temperature was 186K. The stratosphere was coldest in the northern hemisphere, which was in continuous darkness, with the zonal winds reaching 160 metres per second. At mid-to-high northern latitudes the wind had inhibited mixing and produced an isolated vortex around the pole. Being cold, the stratosphere in this area was sinking, drawing down and concentrating the organics created in the upper haze.[241] As regards the mid-latitude clouds seen in the southern hemisphere, where solar heating was greatest and the rising air flow of the pole-to-pole circulation system was located, C.A. Griffith, now at the University of Arizona at Tucson, suggested that the clouds occurred along a given line of latitude because they were the result of the global circulation system.[242] This case was based on the fact that 40°S coincided with the abrupt cut-off in the layer of diffuse particles that surrounded the south polar region, and which might well cause a change in circulation at this latitude. A general circulation model by Pascal Rannou at the

The 'cat scratches' in detail.

University of Paris in France took into account the haze cap at the summer pole and indicated that the thick haze would inhibit surface heating at high latitudes and cause the updrafts and convergence driven by heating of the surface to occur at temperate latitudes.[243] In effect, therefore, the mid-latitude clouds on Titan would correspond to the distinct bands of cloud that form near Earth's equator. On Titan, such updrafts would cause the formation of clouds, the tops of which would be drawn out for several thousand kilometres by the strong zonal wind. "It is no coincidence that Titan's south polar cap of smog extends from the pole to 40°S; exactly where the methane cloud bands appear," argued Griffith. However, H.G. Roe proposed that the mid-latitude clouds were the result of cryovolcanism.[244] Of 24 clouds observed over a 2-year period using telescopes fitted with adaptive optics, all were at low altitude (10 to 35 kilometres) and most were clustered near 350°W. A likely mechanism for geographic control was the episodic release of methane by the surface. One way to produce a cloud was to impart heat and induce convection, with the methane carried by the rising airflow being chilled and forming droplets. The other way was to increase the humidity by releasing methane into the atmosphere, which suggested the presence of a major site of cryovolcanic activity. As the clouds produced by this site were drawn out by the west-to-east wind, that latitudinal band would become humidified, making it easier for sites situated downwind to produce clouds. Roe ventured that this was "the first strong evidence for currently active methane release from the surface". As later observations by Cassini would reveal, there is a range of tall mountains in this vicinity, suggesting the clouds formed as the rising airflow condensed methane.

SATURN'S DEEP CLOUDS

Starting with periapsis on 17 February 2005, Cassini's Visual and Infrared Mapping Spectrometer began a campaign in which it imaged Saturn at 5 microns to utilise the heat continuously leaking from the deep interior to illuminate the cloud structures in silhouette. Previously the deep clouds had been sought by imaging in sunlight, but the view had been obscured by the upper level hazes and clouds. At 5 microns it was possible to map both the day and night sides, and there was a mass of structure.[245] As team member K.H. Baines put it, "Unlike the hazy broad global bands of clouds regularly observed in the upper atmosphere, many of the deeper clouds appear to be isolated, localised features." There was a large variety of sizes and shapes. They also behaved differently, and were made of ammonium hydrosulphide or water rather than ammonia. Observations of these clouds offered a means of both measuring the wind speeds at this deeper level and making a three-dimensional chart of the circulation. In keeping with the finding that wind speeds above the ammonia clouds decreased with increasing altitude, the winds at this 'low altitude' (estimated as 30 kilometres beneath the ammonia clouds) were some 275 kilometres per hour faster. The substantial wind shear at depth in the equatorial zone was a point of similarity with Jupiter. This mode of observation proved so fruitful that it was decided to make further such observations during subsequent periapsis passes.

ENCELADUS DISCOVERIES

The revision of the orbital tour in order to recover from the Huygens relay problem gave a bonus in the form of a non-targeted encounter with Enceladus on 17 February 2005 at a range of 1,264 kilometres. The imagery of the southern part of the trailing hemisphere at a resolution of about 100 metres per pixel revealed the smooth terrain to bear faults, fractures, folds and troughs with vertical relief of about 1 kilometre. Tracking by the Deep Space Network suggested that the rock/ice ratio of the interior was higher than expected, increasing the estimated density from the 1.0 to 1.3 g/cm^3 of the Voyager era to 1.6 g/cm^3. The Visual and Infrared Mapping Spectrometer showed the surface to be pure water-ice. As R.N. Clark of the US Geological Survey in Denver, Colorado, put it, "The spectra look like laboratory fabricated water-ice." Ammonium compounds and carbon dioxide had been expected, but if present were only in trace amounts. The Ultraviolet Imaging Spectrograph determined the texture of the surface by exploiting the fact that water-ice appears dark at wavelengths shortward of 160 nanometres but brightens in the 160–190-nanometre range; the wavelength of this transition being dependent on the grain size. The results indicated coarsely grained ice near the fresh fractures and ridges, suggestive of fresh 'snow'. The magnetospheric and plasma science instruments analysed the space around Enceladus to investigate the putative association between the moon and the 'E' ring, which extends from the orbit of Mimas (at 3 planetary radii) out beyond the orbit of Dione (at 8 planetary radii) and has its greatest density in the vicinity of Enceladus's orbit.

Revolution 4 began with apoapsis at 44.3 planetary radii on 27 February, but there were no encounters with Titan this time.

When Cassini flew within 500 kilometres of Enceladus on 9 March to make the first targeted fly-by of the primary tour, it made a discovery – the moon possessed a tenuous envelope. During both close fly-bys, the Dual-Technique Magnetometer had observed Saturn's magnetic field being 'bent' around the moon by electric currents generated by the interaction of the magnetosphere with neutral atoms of gas. In addition, when neutrals were ionised by plasma they were 'picked up' by the magnetic field and produced oscillations at frequencies that enabled them to be identified as O^+, OH^+ and H_2O^+ ions, indicating the presence of water vapour. The gravity of Enceladus is so weak that its escape velocity is a mere 212 metres per second. Because gas would readily leak away, there would be a net motion away from the moon – making the envelope an exosphere rather than an atmosphere. The fact that the envelope persisted for the interval between the two fly-bys suggested that it was being replenished. The source could be outgassing from the interior through fractures in the icy crust, geysers or cryovolcanism. When Voyager revealed Enceladus to have undergone extensive resurfacing, the evident youth of the smooth terrain prompted speculation that it might still be active – and the presence of a gaseous envelope was the first evidence that this might be so. The Magnetospheric Imaging Instrument also detected an interaction between Saturn's magnetosphere and Enceladus that was consistent with a gaseous envelope. The High-Rate Detector of the Dust Analyser had recorded thousands of 'hits' on each fly-by, but as Thanasis Economou, a team member from the University of Chicago, noted, "It will take a few more fly-bys to

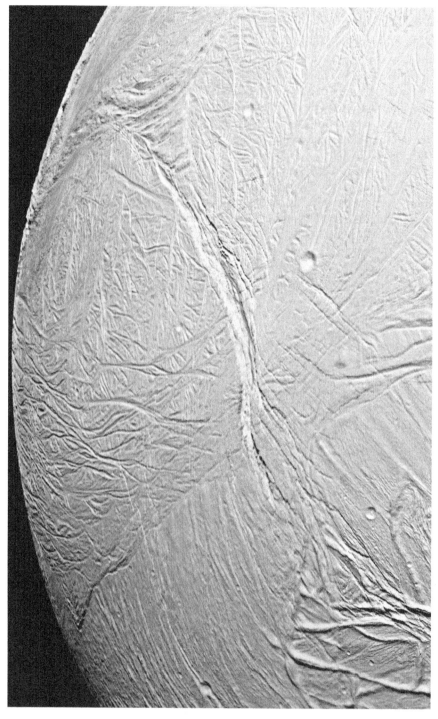

A mosaic of Enceladus taken on 17 February 2005 of the intensely fractured trailing hemisphere viewing obliquely from about 50°S towards the equator.

distinguish whether this flux is originating from the 'E' ring or a source on Enceladus."
On 21 April it was decided to lower the July fly-by from the planned altitude of 1,000
kilometres to 175 kilometres, to improve the sensitivity of the magnetospheric and
plasma science instruments and directly sample the gaseous envelope.

Cruising

Inbound on 7 March 2005, Cassini snapped the wispy streaks on Dione's trailing
hemisphere from a range of 1.6 million kilometres. At periapsis, north of the ring
plane several hours after the Enceladus fly-by on 9 March, it passed within 82,800
kilometres of Tethys, allowing high-resolution imaging. Revolution 5 began with
apoapsis at 44.4 planetary radii on 19 March. The T_4 fly-by on 31 March was at an
altitude of 2,404 kilometres, and provided the best optical view to date of the areas
lying north of Titan's equator that had been imaged by the radar during the T_a and
T_3 fly-bys, in particular the region to the east of Xanadu. On 29 March, after
reviewing plans for the fly-bys of Tethys and Hyperion in September, it was decided
to reduce the altitude at Tethys from 32,000 to 1,500 kilometres and at Hyperion

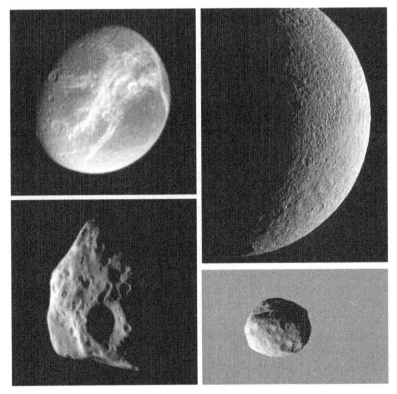

The 'wispy' terrain on the trailing hemisphere of Dione imaged on 7 March 2005 (upper
left); the anti-Saturn hemisphere on Tethys on 11 March 2005 (upper right); Epimetheus
on 30 March 2005 (lower-left); and Janus passing in front of Saturn on 25 September
2006 (lower right).

from 1,000 to 500 kilometres. Although this would affect revolutions 14 through 17 and cost about 8 metres per second, it would greatly improve the Tethys data. On 29 March Cassini flew by Epimetheus at 62,000 kilometres. An image taken the following day from 75,000 kilometres gave the best view yet of this irregularly shaped moonlet, showing a large number of 'softened' craters. The Visual and Infrared Mapping Spectrometer indicated its surface to be predominantly water-ice. Epimetheus and its larger partner, Janus, regularly exchange orbits. Their low densities suggested that they were loosely consolidated accretions rather than chips off larger bodies.[246] Revolution 6 began with apoapsis at 38.0 planetary radii on 6 April. At periapsis on 14 April Cassini made a long-range inspection of Pan, and the next day turned its attention to Mimas and Calypso.

A 'HOT SPOT' ON TITAN

The T$_5$ fly-by on 16 April 2005 at an altitude of 1,026 kilometres was the first 'low pass' of the tour, although for safety it was made a little higher than the originally planned 950 kilometres. In addition to offering improved data on any intrinsic magnetic field and the moon's interaction with its environment, this fly-by was to enable the Ion and Neutral Mass Spectrometer to gather more data to study the composition of the thermosphere and ionosphere and the thermal structure at various local times and latitudes in ambient magnetospheric conditions. It found a wide variety of complex hydrocarbons and carbon–nitrogen compounds (i.e. nitriles) with masses ranging up to as many as *seven* carbon atoms; methane, the simplest hydrocarbon, has one. The amount of ethane and octane was surprising, as it had been thought that in such a cold environment large molecules would sink soon after formation. The altitude selected for this fly-by marked a compromise designed to gather data to assist in determining the lowest safe altitude, and the data was fed into the Titan Atmosphere Model. The geometry provided an opportunity to re-examine a number of features on the Saturn-facing hemisphere, this time with coordinated observations in the visual and near-infrared. In particular, on the T$_b$ and T$_3$ fly-bys Cassini had seen a bright arcuate feature 550 kilometres wide at 85°W, 30°S. Owing to its shape and orientation, it was nicknamed 'the smile'. The International Astronomical Union later named it Hotei Arcus. On the T$_4$ and T$_5$ fly-bys the Visual and Infrared Mapping Spectrometer saw a bright feature in this vicinity in the 5-micron band. Combining the observations showed that the infrared feature was bounded to the south by the arc. Since this wavelength tends to represent thermal emission and this was the brightest point on the moon, it prompted speculation that this might be a 'hot spot' marking a recent impact or cryovolcanic eruption. Another possibility was a 'stationary' cloud of methane released by such a terrain feature, although its spectrum was different to other clouds the instrument had seen. But when microwave radiometry established there to be no temperature difference between the putative 'hot spot' and its surroundings, this ruled out both active cryovolcanism and a mountainous feature – in the latter case, due to the fact that in the troposphere temperature decreases with increasing altitude, and the summit of a mountain would be colder than the surrounding terrain.[247] When a search

An image of Titan taken on 16 April 2005 in the wavelength range 1.7–5.0 microns revealed a '5-micron hot spot' (lower left). A full-disk image taken on 10 December 2004 showed a bright arcuate feature located southeast of Xanadu (upper left) that was nicknamed 'the smile'. A projection of the near-infrared (lower middle) and visual (lower right) views revealed that the 'hot spot' was contained within this arc. On the 16 April 2005 fly-by Cassini sampled the uppermost atmosphere and found a rich mix of complex hydrocarbons (upper right). Note also the two streaky clouds in the full-disk image at 38°S.

of the Keck Observatory archives showed that this feature had been present for at least 3 years, this argued against a stationary cloud. "If this is a cloud," said J.W. Barnes of the University of Arizona, "it would have to be a persistent ground fog." The 5-micron feature would therefore seem to be a surface marking of some as yet unidentified type; perhaps a contamination of the ice.[248]

CONTINUING RING STUDIES

In April 2005 Cassini began to use a succession of Titan encounters to steepen the inclination of its orbit. Revolution 7 began with apoapsis on 23 April at a distance of 40.6 planetary radii. On 28 April Cassini made a main engine burn of 131.6 seconds for a delta-V of 20.5 metres per second in order to revise the altitude of the fly-by of Enceladus in July down to 175 kilometres. There was a non-targeted fly-by of Titan on 4 May at a range of 864,305 kilometres. Six frames over a 16-minute interval on 1 May revealed S/2005S1 orbiting in the Keeler Gap, which is 250 kilometres inside the outer edge of the 'A' ring.[249] The moonlet's existence had been inferred from its gravitational effects – notably the July 2004 sighting of 'spikes' and 'wisps' in the outer edge of the gap. Once it had been spotted, it was traced back to imagery taken on 13 April and its orbit was confirmed to be very near the exact centre of the gap. It was about 7 kilometres in size. Named

An image taken on 2 May 2005 showing the 7-kilometre-sized moonlet designated S/2005S1, whose presence within the Keeler Gap has shaped the edges of the gap.

Daphnis, this was only the second moonlet to be discovered within the rings; the first being Pan in the Encke Division.

Cassini's tour was designed to provide occultations in order to probe the structure of the rings. In addition to stellar occultations that occurred when the spacecraft's instruments viewed star light filtering through the rings, the primary mission included 20 radio occultations, at which times the spacecraft flew behind the rings as seen from Earth. The first radio occultation occurred on 3 May 2005.[250,251] The Deep Space Network monitored a strong carrier signal issued by the spacecraft. To exploit the fact that variously sized particles would attenuate different frequencies to differing extents, S-Band, X-Band and Ka-Band transmissions were made simultaneously. The results showed that particles less than about 5 centimetres in size were abundant in the outer part of the 'A' ring and in the 'C' ring, but scarce in the inner part of the 'A' ring and across the 'B' ring. This correlated with the compositional variations noted while Cassini was north of the ring plane immediately after the Saturn Orbit Insertion manoeuvre. The results also showed that the inner and outer parts of the 'B' ring had ringlets that were hundreds of kilometres wide and varied greatly in the amount of material that they contained, and also that the 5,000-kilometre-wide core of the 'B' ring contained several bands of material that were almost 4 times as densely packed as in the 'A' ring and 20 times as in the 'C' ring. The 'B' ring contrasted with the smoother 'A' ring and with the 'C' ring, the outer part of which comprised many dense narrow sharp-edged ringlets. There were more than 40 density waves in the 'A' ring, many of which were in the outer region, near the moons that orbit just beyond. "Some of these waves have been seen in other Cassini observations," pointed out Essam Marouf of San Jose State University, "but not in this large number, and not with this exceptional clarity."

Revolution 8 began with apoapsis at 40.6 planetary radii on 12 May 2005, but there were no encounters with Titan this time. On 24 May and 11 June Cassini's Visual and Infrared Mapping Spectrometer was able to monitor occultations of omicron Ceti, the long-period variable star called Mira, by the outer part of the ring system. The data showed an unexpected asymmetry – more star light passed through at points during the egress phase than when the star was equidistant from Saturn on ingress. This indicated that the ring material was arranged in narrow strips, radiating outward at a skewed angle. This was a case of the 'wakes' that form when the gravitational attraction of the particles in the rings competed with the tendency of the tidal forces of Saturn to draw them apart. "When the wakes are viewed almost end-on, the 'A' ring appears at its most transparent," explained P.D. Nicholson of Cornell University, "and when they are viewed from the side the ring becomes almost opaque." This was the most direct evidence so far of gravitational wakes.[252] Owing to their narrow width of about 100 metres, it was impractical to image them optically. Although it had been hoped that Cassini would confirm this phenomenon, its position 1.6 million kilometres from the planet, with only 3.5 degrees between the ring plane and the line of sight to the star, made the effect pronounced. "We weren't expecting it to be so obvious," Nicholson reflected.

By this time, having observed several stellar occultations, the Ultraviolet Imaging Spectrograph team knew that the 'A' ring contained clumps.[253] The spacing between the clumps was greater than the widths of the clumps themselves, which ranged in size from 'automobiles' to 'haulage trailers' – too small to be resolved by Cassini's cameras. "This is the first time we've been able to measure the size, orientation and spacing of these particle clumps," noted team leader L.W. Esposito. The individual clusters were largest near the middle of the 'A' ring, and smaller towards the edges. "If we could get close enough, the clumps would appear as short flattened strands of spiral arms with very few particles between them," explained Joshua Colwell of the University of Colorado at Boulder. Loosely bound by the gravity of the material that they contained, these structures would be periodically torn apart by Saturn's gravitational tides. This lent support to the theory that the ring material is not simply degrading towards ever finer material, but is also being recycled by being reaccreted into small bodies.

In fact, an analysis of imagery taken shortly after Cassini passed south of the ring plane on 1 July 2004 had shown traces of gravitational disturbances in the 'A' ring indicating the presence of four small embedded moonlets, each of which was calculated to be of the order of 100 metres in size. The difference between the sizes of these objects and Pan (30 kilometres) and Daphnis (7 kilometres) was that they were too small to sweep a clear channel in the rings. They were able to be spotted only

An analysis of imagery taken by Cassini on 1 July 2004 after Saturn Orbit Insertion revealed the presence of intermediate-size moonlets in the body of the 'A' ring.

because their gravitational wakes were superimposed on a particularly smooth part of the 'A' ring. The fact that they were found on images which, between them, covered a tiny fraction of the system implied that more remained to be found.[254] Furthermore, as Matthew Tiscareno of Cornell University said, "These intermediate-sized particles tell us that Pan and Daphnis are probably just the largest members of the ring population, rather than interlopers from elsewhere."

'LAKE ONTARIO' ON TITAN

Revolution 9 began with apoapsis at 40.6 planetary radii on 30 May 2005, and on 6 June provided a non-targeted fly-by of Titan at a range of 425,552 kilometres. This was only the second opportunity to view the south pole at a high angle of inspection, the first having been T_0. Imagery taken to make a movie of the clouds near 80°S revealed an oval patch that was considerably darker than anything else in its vicinity and had smooth boundaries suggestive of a shoreline. A.S. McEwen of the Lunar and Planetary Laboratory of the University of Arizona in Tucson opined, "I'd say this is definitely the best candidate we've seen so far for a hydrocarbon lake." It was 230 kilometres in length and 70 kilometres in width, making it comparable to

A view of the south polar region of Titan taken on 6 June 2005 showing clouds and a dark surface feature suggestive of a hydrocarbon lake. It was subsequently named 'Lake Ontario'.

Lake Ontario in terms of area. The idea that it was methane rain run-off was supported by fact that at latitude 75°S it was in the currently cloudiest region. "An alternative explanation," warned E.P. Turtle of the University of Arizona, "is that this is a lake that has dried up, leaving behind dark deposits." Regardless, the International Astronomical Union named it Ontario Lacus.

Long range views of Hyperion taken between in October 2004 and February 2005.

EARLY VIEWS OF HYPERION

At various times during its early orbits, Cassini imaged Hyperion from distances ranging between 1.3 and 1.6 million kilometres at spatial resolutions exceeding that of the Voyagers, but these were merely tasters. Imagery taken between 9 and 11 June 2005 as the range decreased from 815,000 to 168,000 kilometres made a movie that strikingly demonstrated the moon's irregular shape. There were craters

Views of Hyperion taken between 9 and 11 June 2005.

apparent to the limit of resolution of about 1 kilometre per pixel, many of which were sharply defined, with intriguing dark material on their floors.

Hyperion posed a challenge to the imaging team, whose automated software for mosaicking imagery worked best with a sphere, and therefore had difficulty with this irregular body. A variety of bulk density estimates had been made over the years, but Cassini provided an opportunity to determine this with a fair degree of accuracy. The fly-by provided a measure of its mass and (its shape notwithstanding) its volume, from which a density of 0.6 g/cm^3 was calculated. Solid water-ice is 0.93 g/cm^3, and if Hyperion was an icy body then it was so porous that 40 per cent of its volume was vacant. Its irregular shape would seem to derive from its being a loose accretion of icy fragments, rather than a solid body that had been so battered as to have had large chunks smashed off it, or indeed, was itself a solid irregularly shaped chip off a larger object. That Hyperion is just below the size limit at which internal pressure will crush loosely packed ice and close the pore spaces to form solid ice is evident from the fact that it is somewhat larger than the irregular moonlets Epimetheus and Janus, and smaller than the spherical moons Mimas and Enceladus. Precisely why there should be a reaccreted object at this location in the Saturnian system is a mystery.

MORE ENCELADUS DISCOVERIES

Revolution 10 began with apoapsis at 40.7 planetary radii on 17 June 2005. There were non-targeted fly-bys of Titan on 22 June at a range of 920,703 kilometres and on 26 June at 874,627 kilometres. Revolution 11 began on 5 July. While there were no encounters with Titan this time, Cassini returned to Enceladus on 14 July, with its point of closest approach over the southern hemisphere providing the first view of the south pole, recently discovered to be marked by arcuate features dubbed 'tiger stripes'.[255] During the fly-by, the Dust Analyser was to determine the composition of the particles in the vicinity of the moon; the Magnetospheric Imaging Instrument was to investigate how these particles interacted with the electrons and ions in Saturn's magnetosphere; the Ion and Neutral Mass Spectrometer was to sample the neutral atoms and positive ions in the immediate vicinity of the moon; and the Ultraviolet Imaging Spectrograph was to monitor an occultation of gamma Orionis (Bellatrix) on the approach in order to measure the density of the gaseous envelope along that line of sight. A similar observation in February had given no result, most likely because the star observed at that time, lambda Scorpii, was rather faint in the ultraviolet. The Visual and Infrared Mapping Spectrometer was to characterise any non-ice materials on the surface, and the Composite Infrared Spectrometer was to produce thermal maps of both the sunlit and darkened hemispheres. The geometry was specified prior to the discovery of the 'tiger stripes', but they were well observed. The imagery revealed that the south polar region had a complex history. Remarkably, there were icy boulders up to 100 metres in size littering an area that was notable for lacking the finely grained frost that was otherwise ubiquitous. In addition, this area contained the largest exposures of coarsely grained ice fractures to be found anywhere on the moon. Parallel ridges and valleys appeared to define a

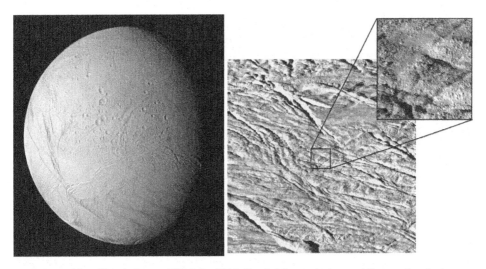

Approaching Enceladus on 14 July 2005 Cassini imaged the anti-Saturn hemisphere centred at 42°S, showing the long 'tiger stripes' at the south pole (left). An image from an altitude of 208 kilometres taken just before closest approach, and smeared due to the relative motion, showed the south polar terrain to be highly fractured and strewn with blocks of ice between 10 and 100 metres in size.

boundary around the south polar terrain which, judging from the absence of impact craters, was extremely young.[256,257]

The existence of the gaseous envelope was confirmed. On this occasion, Cassini penetrated the electrically conducting region and the Dual-Technique Magnetometer found the envelope to be *asymmetric*.[258] "It's a strange atmosphere," said team leader M.K. Dougherty of the University of London. "It seems to be concentrated at the south pole, and the best way to match our observations is with almost a cometary jet coming off the south pole." The Dust Analyser saw a peak in the number of fine particles coming off the surface when Cassini was at an altitude of 460 kilometres, about 1 minute prior to closest approach – which was 172 kilometres above the coordinates 23°S, 325°W. The Ion and Neutral Mass Spectrometer measured a large peak in the water vapour density about 35 seconds prior to closest approach, which corresponded to reaching the trajectory's most southerly latitude at an altitude of 270 kilometres. It was inferred from these results that Cassini had passed through the fringe of a cloud of vapour concentrated over the south pole. The results from the Plasma Spectrometer and the Ultraviolet Imaging Spectrograph were consistent with this interpretation. The variation of vapour density with altitude indicated that there was a localised source on the surface. The persistence of the cloud over an extended period in this weak gravity field indicated continuous replenishment. In fact, the process that issued water vapour and the process that ejected particulates were aspects of the same phenomenon, which was indeed similar to the venting of a comet. The Dust Analyser established that the 'E' ring comprised water-ice particles in the size range 0.3 to 2 micrometres.[259,260] Terrestrial studies had shown that whereas the

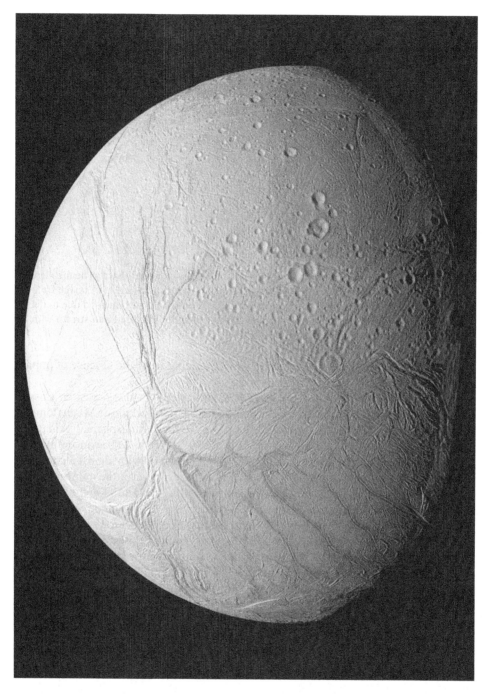

As Cassini closed in on Enceladus on 14 July 2005 it took images to be assembled into a high-resolution mosaic, featuring the 'tiger stripes' at the south pole.

material in the 'F' and 'G' rings had a power-law distribution that was characteristic of debris from impacts, the material in the 'E' ring was of a uniform size.[261,262] It had long been thought that Enceladus was responsible for the 'E' ring, and this was now seen to be the case. But where was the source of this material? The Composite Infrared Spectrometer provided the vital clue. Enceladus has such a high albedo that it reflects most of the incident sunlight, making its surface the coldest of the Saturnian satellites. The temperature at the subsolar point was expected never to exceed 80K. Even though the south pole was in continuous sunlight, the oblique angle of illumination suggested that it should be significantly colder. This fly-by gave the instrument its first view of the south pole, and it took data at wavelengths between 9 and 16.6 microns. The team were amazed to discover the south pole to be the warmest place on the moon. Although the average across the polar area was just 85K, in the 'tiger stripes' it was 145K.[263] As J.R. Spencer, a member of the team at the Southwest Research Institute in Boulder, Colorado, pointed out, "This is as astonishing as if we had flown by Earth and found Antarctica to be warmer than the Sahara." Evidently, the south polar region was being warmed by heat escaping from the interior. "And so the plot thickens!" observed C.C. Porco. "The fact that Enceladus is so alive, and yet Mimas, the moon next door and roughly the same size, is so dead, is really testing our understanding of the internal workings of planetary satellites. But we're not complaining; we like it this way!"

On the fly-by of Enceladus on 14 July 2005 the Composite Infrared Spectrometer revealed the south polar region to be anomalously warm.

Cruising

On 27 June 2005 Cassini imaged Tethys at a range of 500,000 kilometres, catching the giant crater Odysseus on the terminator. On 14 July it documented Rhea from a distance of 250,000 kilometres, and provided a new perspective on Epimetheus from 88,000 kilometres. Revolution 12 began with apoapsis at 40.8 planetary radii on 24 July, in the hiatus from 20 to 27 July for solar conjunction during which Cassini's activities were restricted to magnetospheric and plasma science observations and the data was stored for later replay. On 1 August it imaged the south polar region of Dione from 270,000 kilometres, Rhea at 255,000 kilometres, and caught Pan orbiting in the Encke Division. A non-targeted fly-by of Mimas on 2 August at 62,700 kilometres gave the best view of this moon, and showed there to be a landslide in Herschel crater. Non-targeted encounters with Titan occurred on 2 August at 939,574 kilometres and 6 August at 837,216 kilometres. Revolution 13 began at apoapsis at 40.7 planetary radii on 11 August, and the rotation of Hyperion was documented on 16–17 August from a range of 600,000 kilometres.

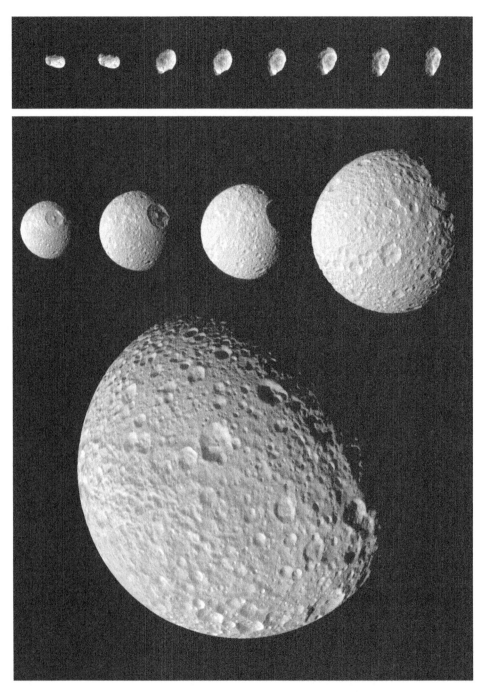

Imagery of Mimas on 2 August 2005 gave the best view yet of this battered moon. A sequence on 16–17 August 2005 showed the chaotic rotation of Hyperion (top).

Views of Rhea taken on 14 July 2005 (upper) and 1 August 2005 (lower).

Epimetheus on 14 July 2005 (upper left); Tethys on 27 June 2005 featuring the giant crater Odysseus (upper middle); the moonlet Pan in the Encke Division on 1 August (upper right); and the south polar region of Dione on 1 August 2005 (lower pair).

CHANNELS ON TITAN

The T_6 fly-by on 22 August 2005 at an altitude of 3,669 kilometres was the first close encounter since 16 April. This time the Composite Infrared Spectrometer was given priority from 11 hours prior to closest approach through to 19 hours afterwards. At closest approach, the first high-resolution imagery of the latitudes to the south of Xanadu was secured. On 30 June the Titan Atmospheric Model Working Group had discussed the results of the T_5 fly-by at the compromise altitude of 1,026 kilometres, and on 6 July the Navigation Team discussed options for the T_7 fly-by in September. In view of variations in the state of the atmosphere noted to date it was difficult to predict the density with any confidence, but when worst-case modelling established that the nominal altitude of 1,025 kilometres might not be safe the Working Group decided on 12 July to raise T_7 by 50 kilometres. Revolution 14 began with apoapsis at 37.8 planetary radii on 28 August. However, a software fault during the T_7 fly-by

A radar image of Titan taken on 7 September 2005 showing a 'contact' between the light and dark areas suggestive of a shoreline.

on 7 September denied the spacecraft one 'side' of its solid-state memory, and the other overflowed 12 minutes prior to closest approach and caused science activities to be curtailed 8 minutes into the radar imaging sequence. Nevertheless, the imagery that was returned was welcome because it showed the contact between a bright area and a very dark and evidently smooth terrain. The pattern of embayments, and the fact that channels on the bright area ended at the contact, suggested that this was a shoreline.[264] While there was still no firm evidence of liquid anywhere on the surface, the radar imagery from the T_a and T_3 encounters showed several types of channels. Although some were clearly drainage networks for rainfall runoff, others, including some in the T_7 imagery, were long and deep, had few tributaries, and ran for great distances.

A deeply incised channel in a radar image taken on 7 September 2005.

Gary Parker, a geologist at the University of Illinois at Urbana–Champaign, had made a study of terrestrial rivers. He argued that if river morphodynamics were truly understood at the physical level, then relations that give reasonable results on Earth should describe how flowing liquid methane on Titan transports disaggregated icy sediment.[265] Such variables are expressed as dimensionless relations developed from dimensioned parameters. "There are only three parameters that differ significantly on Earth and Titan," said Parker. "First is the acceleration due to gravity – on Titan this is about one-seventh the value on

Earth. Second is the viscosity of the flowing liquid – the viscosity of liquid methane on Titan is about one-fifth that of water on Earth. Third is the submerged specific gravity of the sediment – the value on Titan is about two-thirds that on Earth." Terrestrial rivers are characterised by dimensionless relations for the threshold of motion, the threshold of significant suspension, and the 'bank-full' hydraulic geometry. "This means that for the same discharge of liquid methane as to water, the channel characteristics on Titan should be remarkably similar to those on Earth. However, because of the smaller acceleration due to gravity, channel slopes on Titan should be wider, deeper and less steep than those on Earth." But there was a caveat, "The interaction of sunlight with a hydrocarbon-rich atmosphere could precipitate sticky compounds that would give streams on Titan a degree of cohesion that would make them behave differently." On the other hand, as noted earlier, it was likely that that haze particulates 'hardened' during their long, slow descent through the atmosphere.

Cruising

On 5 September 2005 Cassini flew within 52,000 kilometres of Pandora, the outer shepherd of the 'F' ring, showing it to be irregularly shaped, about 80 kilometres in size, with craters that were smoothed over by some loose material. Revolution 15 began with apoapsis at 41.5 planetary radii on 14 September. Approaching periapsis on 23 September, Cassini was able to image Calypso at a range of 91,000 kilometres with the moonlet, 24 kilometres in size, spanning about 35 pixels.

Pandora imaged on 5 September 2005 (left) and Calypso on 23 September 2005.

TETHYS FLY-BY

On 24 September 2005 Cassini flew by Tethys at an altitude of 1,500 kilometres on a trajectory that enabled it to peer into Ithaca Chasma at a high Sun elevation in order to study whether the process that formed this feature was associated with the impact that made the giant crater Odysseus. It was evident from impacts within the chasm that it was ancient. Landslides in large craters had modified their outlines and blanketed their floors with debris.

The south polar region of Tethys on 24 September 2005, with the southernmost part of
Ithaca Chasma showing scarps and ridges peppered with impacts indicating that it is
ancient (left), and landslides in large craters.

HYPERION UP CLOSE

Cassini passed by Hyperion at a range of 514 kilometres on 26 September 2005, revealing it to have a spongey appearance. The arcuate ridge called Bond–Lassell Dorsum proved to be circular, and the rim of an impact crater of a diameter only a little less than the longest dimension of the irregularly shaped moon, complete with a raised central mound. Landslides on the steep interior walls had submerged some of the small craters on the floor of the main cavity. A 200-metre-wide crater on a patch of the dark material that occupied a topographical hollow had created bright rays, demonstrating that the dark material was a thin mantle, perhaps only tens of metres thick, over bright material that was presumably representative of the icy moon's constitution.

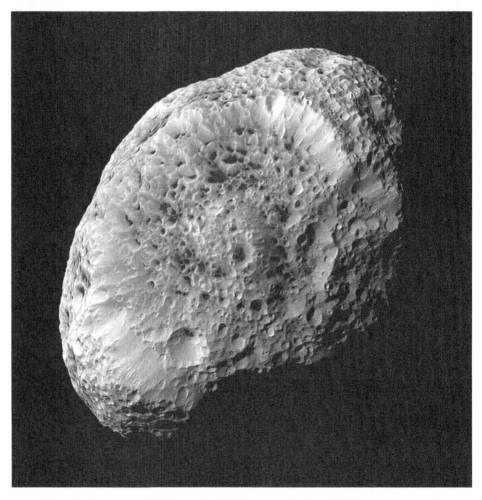

Hyperion imaged on 26 September 2005 at a resolution of 360 metres per pixel.

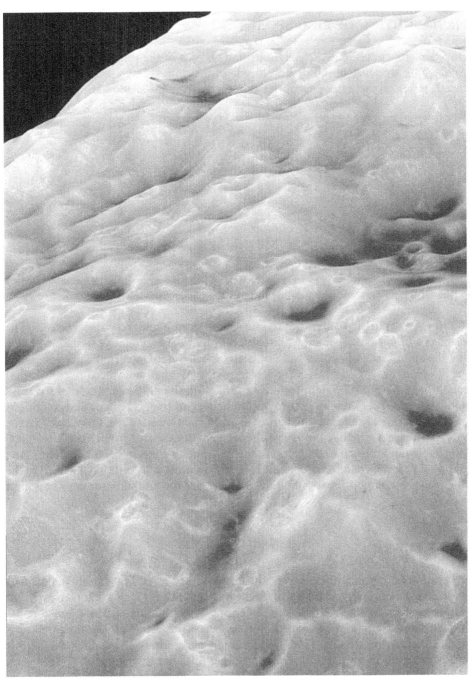

A small part of Hyperion imaged on 26 September 2005 at a resolution of 26 metres per pixel showing that dark material has collected in topographical lows.

FIRST SIGHTING OF RING SPOKES

On 5 September 2005 Cassini caught its first glimpse of the ring spoke phenomenon. An imaging sequence of the shadowed side of the rings showed several faint narrow radial spokes on the outer part of the 'B' ring, entering the shadow the planet cast on

Imagery taken over an interval of 27 minutes on 5 September 2005 provided Cassini's first glimpse of 'spokes'. They were on the outer part of the 'B' ring on the unilluminated side of the ring system, and heading towards the planet's shadow.

the rings.[266] "This is really a joy," said C.C. Porco, "and very unexpected." The visibility of ring spokes varied with the elevation of the Sun relative to the ring plane and the appearance of spokes had not been expected until later in the mission, when the illumination would be at a shallower angle. "We had convinced ourselves that conditions would not be right for seeing spokes on the illuminated side of the rings until about 2007," Porco said. "But this finding seems to be telling us that conditions on the dark side of the rings are almost as good right now for seeing spokes." In the Voyager imagery, the narrow wedge-shaped spokes extended from 10,000 to 20,000 kilometres out across the 'B' ring, but the new ones were just 3,500 kilometres long and 100 kilometres wide. By 2008, the spacecraft's orbit will be steeply inclined to the ring plane, enabling it to view the rings almost 'face on' and monitor the rotation of spokes.

DIONE FLY-BY

Revolution 16 began with apoapsis at 41.2 planetary radii on 2 October 2005, and there was a non-targeted fly-by of Titan at a range of 776,267 kilometres on 10 October. The next day, Cassini flew by Dione at an altitude of 500 kilometres. Its surface was heavily cratered, but in the bright wispy terrain it was fractured as a result of tectonism. In fact, there were multiple generations of fractures. Numerous fine parallel grooves were interrupted by larger irregular bright fractures. In several places, fractures interrupted deposits on the floors of craters. As B.J. Buratti of the Visual and Infrared Mapping Spectrometer team ventured, "Dione appears to be an older sibling of Enceladus. We think the cracks may be an older version of the 'tiger stripes' on Enceladus." The Ultraviolet Imaging Spectrograph detected water-ice.

Dione on 11 October 2005, with detail of fractures in the 'wispy' terrain.

The ice in the fractures was different to that on the adjacent terrain, possibly due to a variation in the grain size. Variations in albedo may have been due to downslope movement exposing 'clean' ice on steep flanks and darker material accumulating in topographic lows. The Voyagers had reported a concentration of magnetospheric ions near Dione's orbit, and the existence of the wispy terrain had given rise to speculation that this might indicate outgassing adding to the 'E' ring, but Cassini's magnetospheric and plasma science instruments found no sign of an envelope. Nevertheless, brief orbital resonances with Enceladus could have driven such activity in the past. As regards the magnetospheric ions near Dione's orbit, the Plasma Spectrometer found these to be neutrals that had been released by Enceladus and drifted outwards in the 'E' ring until, on reaching the most intense part of the magnetosphere, they had been ionised by electrons.[267]

TELESTO

Ninety minutes after Dione on 11 October 2005, Cassini passed Telesto at a range of 9,500 kilometres. At 24 kilometres in size, this Trojan satellite of Tethys was too small in the Voyager discovery imagery for surface detail to be resolved. This new imagery showed the surface to be remarkably smooth – apart from the large 'rocks' that traced the rims of craters, as if they had been left *in situ* as the finer material slid downslope into the cavities. This suggested an usually thick blanket of regolith for such a small predominantly icy body.

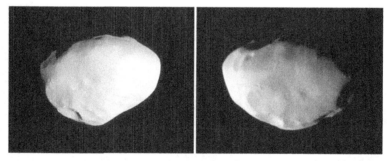

Views of Telesto taken on 11 October 2005 (left) and 25 December 2005.

DUNES ON TITAN

Revolution 17 began with apoapsis at 41 planetary radii on 20 October 2005. The T_8 fly-by on 28 October was at an altitude of 1,353 kilometres. This time, the radar imaged the dark Belet terrain to the west of Xanadu, which was revealed to contain a field of 100-metre-tall dunes. The individual dunes ran for hundreds of kilometres, and the field extended at least 1,000 kilometres along the equatorial zone.[268,269] The 'cat scratches' of T_3 were seen to be smaller dune fields. "It's bizarre," said R.D.

On 28 October 2005 Cassini's radar imaged the dark Belet area west of Xanadu, and found it to be a field of 100-metre-tall dunes.

Lorenz. "These images from a moon of Saturn look just like radar images of the deserts in Arabia and Namibia on Earth. Titan's atmosphere is thicker than Earth's, its gravity is lower, and its 'sand' is certainly different – everything's different except the physical process that forms the dunes and resulting landscapes."

Calculations showed that the variable tidal wind caused by Saturn's gravitation was combining with the west-to-east zonal circulation to create surface winds that would produce a sinuous pattern of longitudinal dunes that were aligned with the direction of the prevailing wind flow, except near elevated terrain that controlled the local wind direction.[270] This process could transfer fine material from the mid-latitudes into the equatorial zone. Furthermore, the presence of dunes implied the absence of persistent liquids on the surface in the equatorial zone that would serve as sand traps – a conclusion that seemed to conflict with the idea that Titan was a land of lakes.

RHEA FLY-BY

Revolution 18 began with apoapsis at 55.1 planetary radii on 13 November 2005, but there were no encounters with Titan this time around. On 26 November, Cassini made a fly-by of Rhea at a range of 500 kilometres; the only targeted study of this moon on the primary tour.

Rhea imaged on 26 November 2005.

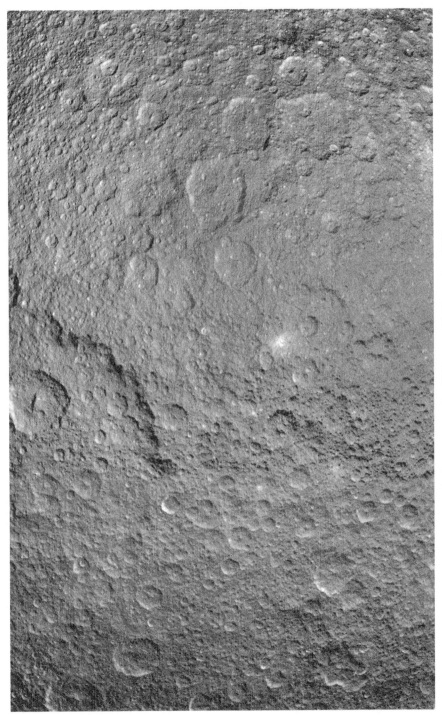

A close-up mosaic of Rhea on 26 November 2005.

GEYSERS ON ENCELADUS

A re-examination of imagery taken on 16 January 2005 from a range of 209,000 kilometres showed Enceladus as a crescent with a glow around its south polar region shining in forward-scattered sunlight. It was decided to make further observations, and the non-targeted fly-by on 27 November 2005 showed there to be jets from many sources that extended 500 kilometres into space. It was a discovery on a par with the erupting volcanoes on Jupiter's moon Io. As C.C. Porco put it, "For planetary explorers there is little that can compare to the sight of activity on another Solar System body." The great hope had been that Cassini would find a geyser in action, but this was beyond all expectation!

An analysis of an image of Enceladus taken on 16 January 2005, with a dark mask applied to occult the bright crescent, revealed evidence of plume activity in the south polar region (left and centre). Further imagery on 27 November 2005 found there to be multiple geysers venting material into space.

"I think what we're seeing", opined Andrew Ingersoll of JPL, "are ice particles in jets of water vapour that emanated from pressurised vents. To form the particles and carry them aloft the vapour must have a certain density, and this implies surprisingly warm temperatures for a cold body like Enceladus." As T.V. Johnson of JPL put it, "In some ways, Enceladus resembles a huge comet. But in this case the energy source for the geyser-like activity is believed to be internal heating, perhaps by radioactivity and gravitational tides, rather than the sunlight that causes cometary jets." It seemed that the jets were erupting from near-surface pockets of liquid water, in the same manner as the 'Old Faithful' geyser in the Yellowstone National Park on Earth. Furthermore, as Ingersoll pointed out, "Although other moons in the Solar System have liquid-water oceans covered by many kilometres of icy crust, in Enceladus the pockets of liquid might be no more than several tens of metres below the surface." The 'tiger stripes' that ran in parallel for 140 kilometres, spaced 40 kilometres apart, would appear to be fractures in the icy crust. Liquid water that made its way up through cracks to the surface would boil into space, with the stream of vapour carrying with it particles of ice and liquid. An analysis of the jets and plumes indicated that most of the particles fell back to give the south polar region its bright appearance. The Visual and Infrared Mapping Spectrometer found the surface water-ice to be in two forms. The fresh ice was crystalline, but this transformed to amorphous ice as it was 'aged' by the ambient radiation. Because this ageing occurs over an interval of decades or less, the southern polar region had evidently been geologically active very recently. Although this area was in darkness when the Voyagers visited, if the 'tiger stripes' had been on view at that time their appearance may well have been different.

As to the source of the heat that drove the geyser activity, Francis Nimmo of the University of California at Santa Cruz suggested that it was due to the moon's eccentric orbit: "Enceladus gets squeezed and stretched by tidal force as it orbits Saturn, and that mechanical energy is transformed into heat in the interior." Furthermore, resonances as the orbits of the moons evolved would produce periods of enhanced stress. In particular, Enceladus may well have had an intermittent resonance with Dione.[271] If the ice on Enceladus was laced with ammonia, this would act as an antifreeze, lowering the melting point of water-ice by 100°C and reducing its density sufficiently for neutral-buoyancy convection. The process of convection is the displacement of heat by the mechanical turnover of material. As a rising diapir broached the surface of the icy crust it would cause characteristic tectonism. It is clear that although Enceladus has had a long history of activity, this is currently restricted to the south polar region. A model developed by Nimmo and Robert Pappalardo of the University of Colorado indicated that when a deeply seated 'hot spot' developed, the moon would roll over to locate this at one of its axial poles.[272] The reason was simple: a rotating body is stable if more of its mass is close to the equator. Any redistribution of mass within the object would cause instability with respect to the axis of rotation. A reorientation would tend to locate excess mass at the equator and areas of low density at the poles. Nimmo and Pappalardo modelled two cases: one with the rising diapir located in the icy shell, and the other with the diapir in the underlying silicate core. If the diapir was in the icy shell, then the shell had to be relatively thick in order to maintain significant rigidity, and if it was in the core there could be no global subsurface ocean, since the core had to be coupled to the overlying ice in order for the moon to roll over.

If the presence of ammonia in the water-ice was facilitating diapiric convection, ammonia should be being vented. Although none was detected, the fact that nitrogen was present in the cloud over the south polar region, and indeed in the 'E' ring, indicated that the combination of solar ultraviolet and magnetospheric plasma was rapidly dissociating the ammonia.[273] The Visual and Infrared Mapping Spectrometer detected methane in the 'tiger stripes', indicating that, as team leader R.H. Brown of the University of Arizona at Tucson put it, "Methane has probably been locked up in Enceladus since the moon formed, and is now bubbling up through the vents." As the inner magnetosphere rotated rapidly with Saturn each slowly orbiting moon formed a 'hole' in the downwind plasma. In the case of the airless icy moons this hole had a sharp boundary, but the material vented by Enceladus made its hole less distinct. Data from the Magnetospheric Imaging Instrument showed significant variability on a time scale of days.[274] This explained an Ultraviolet Imaging Spectrograph observation. This instrument began to monitor Saturn in December 2003 to study its atmosphere, and in January 2004 it detected a massive cloud of atomic oxygen in the 'E' ring.[275] The first hypothesis was that the collision of two as-yet-unrecognised moonlets in this vicinity had released a puff of water-ice that had dissociated into its constituent hydrogen and oxygen.[276] It was belatedly realised that the source must have been an outburst from Enceladus.

There will be an opportunity to test the proposed model for the geyser activity on Enceladus when Cassini makes a 100-kilometre fly-by in March 2008, the final close encounter with this moon of the primary mission.

FEATURES NAMED ON TITAN

Revolution 19 began with apoapsis at 53.6 planetary radii on 11 December 2005. On 12 December the altitudes for the remaining Titan fly-bys of the primary mission were decided. Features on Titan had initially been referred to by nicknames, but in August the International Astronomical Union had announced several official names. The problem was *classifying* features that were poorly glimpsed through the haze, and whose character remained to be determined. The scheme was extended to include, for example, 'virga' for 'streak'.[277] The named features included: Xanadu, the bright 'continent'; Shangri-la, the dark area west of Xanadu; Shikoku Facula, an 'island' in central Shangri-la; Adiri, the bright area west of Shangri-la; and Belet, the dark area west of Adiri. Huygens had landed on the easternmost part of Adiri. On 24 December, Cassini imaged the leading hemisphere of Dione from a range of 150,000 kilometres. The T_9 fly-by of Titan on 26 December at an altitude of 10,400 kilometres gave a high-resolution mosaic that overlapped the eastern part of the region that had been visually imaged on T_a and by radar on T_3. Revolution 20 began with apoapsis at 48.3 planetary radii on 5 January 2006. The T_{10} fly-by on 15 January was at an altitude of 2,043 kilometres, and provided a view of the anti-Saturn hemisphere. In addition to imagery of albedo features in the Shangri-la area, at closest approach the Visual and Infrared Mapping Spectrometer took high-resolution data of Xanadu and Shangri-la, the latter including the infrared-bright, putative cryovolcanic feature seen on T_a and dubbed 'the snail', now named Tortola Facula by International Astronomical Union.

The leading hemisphere of Dione imaged on 24 December 2005.

Named features on the anti-Saturn hemisphere of Titan.

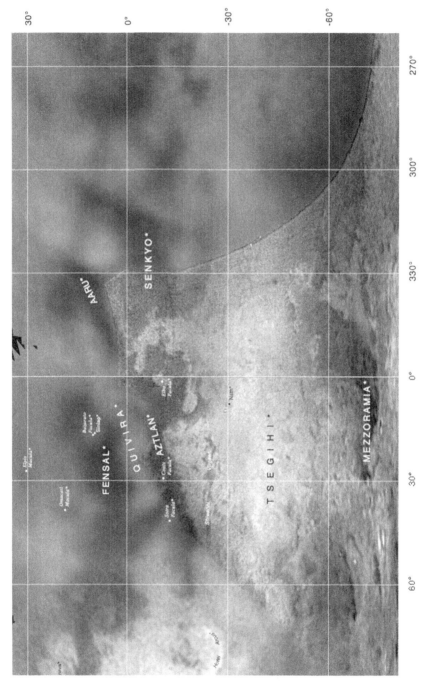

Named features on the Saturn-facing hemisphere of Titan.

SATURN BY 'RINGLIGHT'

On 23 January 2006, while Cassini was deep in Saturn's magnetotail, the Radio and Plasma Wave Spectrometer detected a strong burst of emission of the type known to be produced by lightning discharges. The spacecraft's position prevented it viewing the daylit hemisphere, but astronomers in France reported that a storm had appeared in 'storm alley' in the southern hemisphere. Cassini was told to watch for the storm on the night-time hemisphere by viewing in the illumination of sunlight reflected off the ring system, which was just sufficient to show detail on the planet. As Andrew Ingersoll put it, "This is the only large storm on the whole planet, it's in the right place, and it appeared at the right time to match the radio emissions, so it must be the right storm!"

JANUS AND EPIMETHEUS

On 21 January 2006 Janus and Epimetheus swapped orbits. Janus will occupy the low position until 2010. There is no danger of a collision during such exchanges, as the bodies can approach no closer than 15,000 kilometres. Revolution 21 began on 5 February with apoapsis at 68.2 planetary radii, which was the furthest that Cassini had been from Saturn since the conclusion of revolution 'a' in November 2004 at 78 planetary radii.

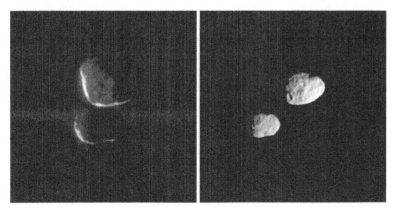

Janus and Epimetheus imaged on 25 December 2005 shortly before the exchange of orbits (left) and on 20 March 2006 after the exchange (right). Although the line of sight suggests the two moons were in very close proximity, both times Janus, which is rather larger, was further away.

DOES TITAN HAVE AN INTERNAL OCEAN?

The T_{11} fly-by on 27 February 2006 was at an altitude of 1,813 kilometres, and this time the highlight was the monitoring of the radio signal to measure the 'low degree coefficients' of Titan's gravity field, to enable a team led by Arvydas Kliore of JPL to investigate whether there is a subsurface liquid layer. In fact, this was the start of a campaign of four fly-bys – this one, T_{22} in December 2006, T_{33} in June 2007 and T_{38} in December 2007 – timed to examine the gravity field with the moon at specific positions in its orbit (and therefore under different tidal stresses) and with different encounter geometries. Two passes were to occur when the moon was near periapsis, and two (of which T_{11} was one) near apoapsis, and for each pair one was to be near-equatorial (as now) and the other as far from the equator as feasible. The objective was to measure the amount by which the shape of the moon flexed; if the moon has a layer of liquid it will flex more than if it is solid throughout. When the Galileo spacecraft made close fly-bys of Ganymede and Europa, it detected a weak magnetic field that Jupiter's magnetic field was inducing in the electrolyte-rich fluid of their subsurface oceans. Whatismore, as Jupiter's magnetic axis is significantly tilted with respect to its rotational axis, the axes of the induced fields rotate as the moons cross north and south of the magnetic equator. However, Saturn's magnetic axis is almost aligned with its rotational axis, making this type of observation more difficult. Nevertheless, the fact that Saturn's magnetosphere inflates and contracts with the solar wind and sometimes shrinks inside Titan's orbit means that it might be feasible to detect a subsurface ocean on Titan by observing how the moon reacts on transiting the magnetopause. However, this would simultaneously require: (1) the magnetosphere to shrink inside Titan's orbit, (2) Titan to be on the sunward side of its orbit, and (3) Cassini to be making a close fly-by. And, of course, to prove the point it would be necessary to make several such observations, which is not a very likely scenario.

Confirming the existence of a layer of liquid in Titan would resolve several issues: the orbital eccentricity, the nature of the cryovolcanism, and the source of methane in the atmosphere. Although some 50 per cent of the moon's mass is water, and there is evidence of cryovolcanism, there are three things about liquid water that make it a poor volcanic fluid: (1) it melts at 273K, which is rather high for Titan's interior; (2) it is denser than water-ice, making it difficult to move onto the surface; and (3) it has a low viscosity and therefore does not form relief features. An agent is required to lower the melting point, density and mobility of liquid water. As noted earlier, ammonia can lower the melting point of water-ice by 100°C, and make the resulting liquid neutrally buoyant with respect to water-ice and of a viscosity similar to that of basalt. As Titan accreted, the water-ice in the planetesimals would have melted, giving a layer of liquid beneath the icy crust. "One thing that Titan could not have done," noted J.I. Lunine, "is to have had a liquid layer which then froze solid, because the tidal dissipation rate goes way up as it solidifies, and that would have reduced the eccentricity of the orbit, making it more circular." The obvious explanation of the non-zero eccentricity is that the ammonia-water layer is still present. If at a shallow depth, it would tend to sustain cryovolcanism. Furthermore, modelling linked this to the evolution of the atmospheric methane.[278] This envisaged three primary outgassing episodes. The first was just after the interior differentiated to produce a dense rocky core, an ammonia-enriched liquid mantle and a crust of ice with methane in the form of clathrate hydrates. The residual accretional heat, the heat from radioactive elements in the core, and the ammonia acting as antifreeze aided the release of methane over the first several hundred million years, but it would rapidly have been photochemically destroyed. "The second episode around 2 billion years ago is even more interesting," said Lunine, a member of the team. "That was when convection began within the core. The core continued to heat up because it contained radioactive elements like uranium, potassium and thorium. On Earth these elements are concentrated in the crust, but on Titan they were in the core with the rock. So the core got hotter until finally it was soft enough for convection to start." This heat released methane, which outgassed to replenish the atmosphere for a time. The third episode began around 500 million years ago, when the icy crust itself became sufficiently soft to convect, causing diapirs to rise and induce surface activity. It has been proposed that cryovolcanism is related to the formation of crevasses in the underside of this crust.[279] As to the future, Lunine predicts, "Methane outgassing will cease within the next several hundred million years. The photochemistry will destroy the remaining surface methane, and the atmosphere will clear of haze."

If Titan is eventually found to be solid throughout, this would suit A.J.R. Prentice of Monash University in Victoria, Australia, who argues that Titan accreted directly from the solar nebula in the vicinity of Saturn's orbit and was soon captured, being hit in the process by several Rhea-sized native moons. He has predicted that probing Titan's gravity field will find mass anomalies attributable to these collisions.[280,281]

Revolution 22 began with apoapsis at 48.4 planetary radii on 9 March, and the T_{12} fly-by on 19 March at an altitude of 1,951 kilometres marked the first time that Cassini passed behind Titan as seen from Earth. The radio occultation was monitored in the X-Band, S-Band and Ka-Band. The consistency of the data by the various receiving stations indicated that the signal represented genuine structure in the ionosphere and neutral atmosphere. Interestingly, the profiles differed slightly during the ingress and egress events. The geometry of this fly-by also provided the first opportunity to use bistatic radar scattering to investigate the composition and texture of the surface by the strength and polarisation of the received signal. On its approach to Titan, the vehicle oriented itself to point its high-gain antenna to illuminate areas of the surface where the reflection would be able to be received by the Deep Space Network, but no specular reflections were observed that might have shown the presence of liquid on the surface.

MIMAS AND RHEA

At periapsis on 20 March 2006, Cassini snapped Mimas crossing Saturn's limb. The next day, it made a non-targeted fly-by of Rhea at a range of 82,000 kilometres. In addition to imaging the daylit hemisphere, it observed the hemisphere illuminated by 'Saturnshine' and gained the highest resolution view so far of this area. Since Rhea orbits just beyond the outer fringe of the 'E' ring, the Ultraviolet Imaging Spectrograph took measurements to investigate whether it was contributing material to the ring.

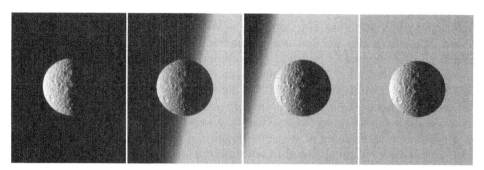

On 20 March 2006 Cassini caught Mimas starting a transit across Saturn's disk.

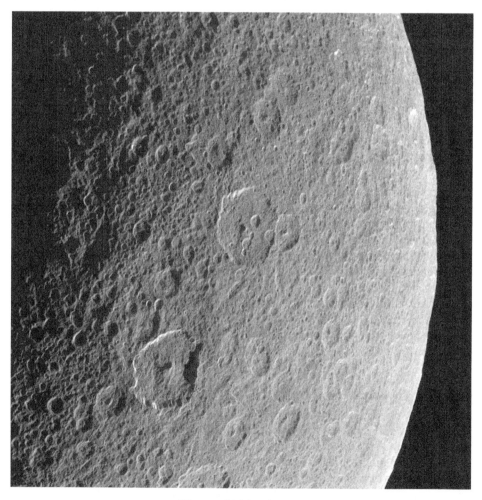

Rhea on 21 March 2006.

THE XANADU 'CONTINENT' ON TITAN

Revolution 23 began with apoapsis at 68.3 planetary radii on 9 April 2006. The T_{13} fly-by on 30 April was at an altitude of 1,855 kilometres. The radar undertook altimetry, radiometry, scatterometry and imaging. It showed Xanadu to stand above the surrounding dark terrain, making it, indeed, a continent. Although the dark area to the west was a sea of dunes, these were absent on the dark terrain to the east. The elevated ground was hilly, and deeply incised by river channels. As Steve Wall, the deputy team leader at JPL pointed out, "This land appears to be the only vast area *not* covered by organics. It has been washed clean, leaving very porous water-ice." Revolution 24 began at 48.4 planetary radii on 10 May. The T_{14} fly-by on 20 May at an altitude of 1,879 kilometres provided a second radio occultation.

POLYDEUCES INSPECTED

Earlier in the tour, Cassini had spotted the moonlet Polydeuces. On 22 May 2006 the spacecraft flew by it at a range of 64,057 kilometres, showing it to be somewhat elongated and about 3 kilometres on its longest dimension, but no surface detail was able to be resolved.

TITAN, LAND OF A THOUSAND LAKES

Revolution 25 began with apoapsis at 68.3 planetary radii on 10 June 2006. On 1 July Cassini reached the half-way point in its 4-year primary tour. The T_{15} fly-by on 2 July was at an altitude of 1,906 kilometres. Revolution 26 began at 48.4 planetary radii on 12 July. Cassini was to begin the process of rotating the apoapsis of its orbit from down-Sun to up-Sun in order to switch from investigating the magnetotail to imaging Saturn in full sunlight. Rather than perform this rotation within the plane of the Saturnian system, it had been decided to do it by simultaneously shortening the orbit and using a series of high-latitude passes of Titan to increase the inclination of the orbit. The process would then be reversed by extending the apoapsis while returning to the plane of the system. This would give unprecedented opportunities to view Saturn and its magnetosphere, but it would be a slow evolution whose duration would depend on how closely Cassini was able to approach Titan. Despite concern regarding the minimum safe altitude, it had been decided to try a pass at 950 kilometres for the T_{16} fly-by on 22 July. If all went well, then by February 2007 a series of such encounters should have increased the inclination to the planned maximum of 59.8 degrees, and the process would be put into reverse. The initial fly-bys provided an opportunity to take high-resolution radar imagery of Titan's far northern latitudes. Just as the south polar region had been in continuous sunlight since Cassini's arrival in the system, the north polar region had been in continuous darkness. The T_{16} imaging track contained dozens of well-defined dark patches up to 1 kilometre in size, some that were tens of

Radar imagery of Titan taken on 30 April 2006 showed narrow sinuous radar-bright channels on the western part of the elevated Xanadu 'continent'.

When Cassini's radar imaged Titan's dark northern latitudes on 21 July 2006 it found the first convincing evidence of hydrocarbon lakes: centred at 92°W, 80°N (upper) and at 18°W, 78°N (lower).

When Cassini's radar imaged Titan on 23 September and 9 October 2006 it found further evidence of hydrocarbon lakes: 65°W, 74°N (top left), 46°W, 73°N (top right), 343°W, 73°N (middle) and 357°W, 80°N (lower).

kilometres, and one that was approaching 100 kilometres. These were the darkest radar features yet seen, and dark usually signified either a smooth surface or a radar-absorbent material – and liquid hydrocarbon would be both. Furthermore, there were channels leading into or out from some patches with characteristics consistent with their having been created by flowing liquid.[282] Evidently, Titan was, as J.I. Lunine had put it, "a land of a thousand hydrocarbon lakes".

SATURN'S MAGNETIC FIELD

It is difficult to measure the length of Saturn's day since its visible surface rotates at different speeds at different latitudes, being fastest in the equatorial zone. During their fly-bys the Voyagers detected a periodicity in the rhythm of the radio emission, and this was taken to be the rate at which the planet's core was rotating. However, in 1997 the Paris Observatory announced a different radio periodicity. M.D. Desch of the Goddard Space Flight Center analysed the data collected by Cassini's Radio and Plasma Wave Spectrometer between 29 April 2003 and 10 June 2004 and found a periodicity of 10 hours 45 minutes 45 seconds (± 36 seconds); some 1 per cent longer than the Voyager time.[283] As team leader Donald Gurnett of the University of Iowa put it, "I don't think any of us could conceive of any process that would cause the

The trailing hemisphere of Dione on 24 July 2006 showing the mysterious elliptical feature in the Voyager imagery named Amata to be illusory.

rotation of the entire planet to actually slow down." Alex Dessler of the University of Arizona at Tucson had argued that the magnetic fields of the gas giants more resemble that of the Sun than Earth.[284],[285] The Sun's magnetic field does not rotate as a rigid body; its rotation period varies with latitude, being fastest at the equator. He suggested that the change in the periodicity meant that the part of Saturn's magnetic field that controlled the radio emissions had migrated to a higher latitude. If the magnetic axis had been significantly inclined to the rotational axis, it would have been simple to measure the rate at which the magnetic axis precessed. Nevertheless, by the half-way point in the tour the Dual-Technique Magnetometer had noted a periodicity in the magnetic field itself of 10 hours 47 minutes 6 seconds (± 40 seconds); even slower than the current radio periodicity.[286]

Cruising
On 24 July 2006, Cassini imaged the trailing hemisphere of Dione from a range of 263,000 kilometres. This recreated at higher resolution a Voyager-era perspective that showed the mysterious elliptical feature named Amata amid a knot of wispy streaks to be illusory. Revolution 27 began with apoapsis at 49.0 planetary radii on 4 August, and solar conjunction followed between 5 and 9 August. The plume from Enceladus's south pole was imaged by forward-scattered sunlight from a range of 2.2 million kilometres on 11 August. There was a non-targeted fly-by of Titan on 18 August at 340,595 kilometres. Revolution 28 began at 48.8 planetary radii on 28 August. The steepening of Cassini's orbital plane continued on 7 September with the T_{17} fly-by at 1,000 kilometres, during which the radar imagery revealed an impact crater of about 30 kilometres diameter with a central peak.

An image of Enceladus taken on 11 August 2006 showing the plume over its south polar region (left), and a radar image of Titan on 7 September 2006 centred at 70°W, 10°N, showing a 30-kilometre-diameter impact crater with a central peak.

THE RINGS IN SOLAR OCCULTATION

As revolution 29 began at 37.6 planetary radii on 17 September, Cassini transited Saturn's shadow, making the most prolonged solar occultation of the 4-year tour – occultations usually occurred when the spacecraft was near periapsis and lasted only an hour or so, but this time it was 2.2 million kilometres from Saturn and 15 degrees off the ring plane, with the result that it spent fully 12 hours in the shadow of the planet. This provided an opportunity to make a number of intriguing observations of the ring system in silhouette. The result was the discovery of a tenuous ring associated with Janus and Epimetheus, and another associated with Pallene. The 'G' ring was revealed to have a very distinct inner edge. The vast expanse of the 'E' ring was seen in its entirety. Wispy, finger-like projections were evident which extended into the ring from Enceladus, and probably comprised ice particles issued by the geysers at the moon's south pole. As a bonus, Earth, some 1.5 billion kilometres away, was viewed through the tenuous outer part of the ring system. Not since Voyager 1 took an image from beyond the orbit of Neptune showing Earth as (in the words of the late Carl Sagan) "a pale blue dot", have we had an opportunity to view ourselves from such a distance. As C.C. Porco opined, "The ever-widening view of our own little planet against the immensity of space is perhaps the greatest legacy of our interplanetary travels."

TITAN'S SEASONAL RAINS

The T_{18} fly-by on 23 September 2006 was at an altitude of 960 kilometres. On this occasion, the Visual and Infrared Mapping Spectrometer showed a vast tropospheric cloud occupying the observed range of longitudes (10° to 190°W) and latitudes (51° to 68°N) that appeared to be composed of ethane.[287] It is believed to be the result of the concentration of organic haze in the descending north polar vortex. Ethane raining into the lakes of methane would dissolve. If the temperature dipped low enough in the dead of winter, ethane would undoubtedly fall as snow, and might even build up an ice cap. The localisation of methane and ethane at the winter pole implied a seasonal cycle in which the volatile hydrocarbons migrated from one pole to the other over a Titanian year (which lasts 29.5 terrestrial years) and this, in turn, offered an explanation for the current paucity of clouds and absence of liquid on the surface at other latitudes.

All in all, with the evidence mounting for a subsurface ammonia-water ocean, the ongoing release by cryovolcanism of methane, and a seasonal cycle for atmospheric volatiles, the Cassini spacecraft and its Huygens probe had, between them, lifted the veil on Titan. By the end of the primary tour in June 2008 the high-resolution radar map will have covered some 20 per cent of the surface, and if the mission is extended it should be possible to monitor the onset of the northern summer, with storms forming in the tropics as the volatiles migrate south, possibly temporarily submerging the now inert Huygens probe on the floor of a lake of rainfall runoff.

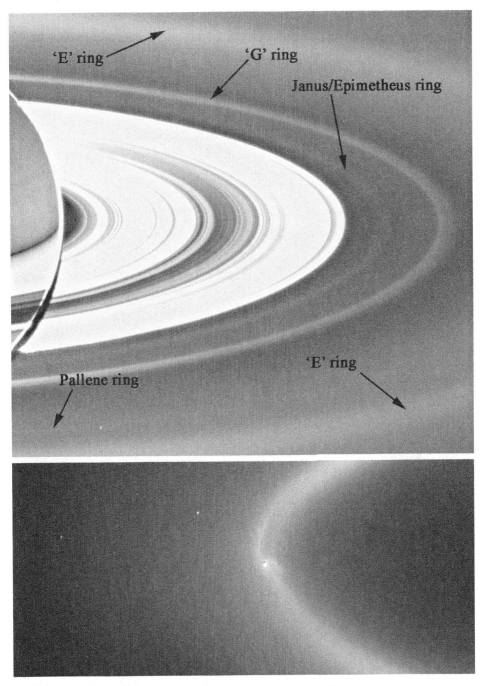

Detail of the ring system as imaged on 15 September 2006, together with a dramatic exposition of the jets on the south pole of Enceladus replenishing the densest section of the 'E' ring (lower).

Lakes on Titan near 35°W, 80°N.

Notes

Chapter 1: Saturn from afar

1 In fact, Laplace said that the rings could not be a *uniform* solid body unless most of their mass was near the inner rim, and there was no reason to believe that this might be the case.
2 'A spectroscopic proof of the meteoric constitution of Saturn's rings', J.E. Keeler. *Astron. J.*, vol. 1, p. 416, 1895.
3 'On the masses of Saturn's satellites', H. Jeffreys. *Mon. Not. Roy. Astron. Soc.*, vol. 113, p. 81, 1953.
4 'The new ring of Saturn', P. Guerin. *Sky and Telescope*, vol. 40, p. 88, 1970.

Chapter 2: First close look

1 'The Earth's magnetotail', N.F. Ness. *J. Geophys. Res.*, vol. 70, p. 2989, 1965.
2 'The Pioneer 11 high field fluxgate magnetometer', M.H. Acuna and N.F. Ness. *Space Sci. Inst.*, vol. 1, p. 177, 1975.
3 'Fast reconnaissance missions to the outer Solar System utilising energy derived from the gravitational field of Jupiter', G.A. Flandro. *Astronautica Acta*, vol. 12, p. 329, 1966.
4 'Images of Jupiter from the Pioneer 10 and 11 infrared radiometers: a comparison with visible and 5-micron images', G.S. Orton, A.P. Ingersoll, R.J. Terrile and S.J. Walton. *Icarus*, vol. 47, p. 145, 1981.
5 'Pathfinder to the rings – the Pioneer Saturn trajectory decision', M. Wolverton. *Quest*, vol. 7, no. 4, p. 5, 2000.
6 'The D and E-rings of Saturn', B.A. Smith. In *The Saturn system*, D.M. Hunten and D. Morrison (Eds.). NASA, Conf. Pub. CP-2068, p. 105, 1978.
7 'Preliminary results on the plasma environment at Saturn from the Pioneer 11 plasma analyser experiment', J.H. Wolfe *et al. Science*, vol. 207, p. 403, 1980.
8 'Saturn's magnetic field and magnetometer', E.J. Smith *et al. Science*, vol. 207, p. 407, 1980.
9 'Photometry and polarimetry of Saturn's rings from Pioneer 11', L.W. Esposito, J.P. Dilley and J.W. Fountain. *J. Geophys. Res.*, vol. 85, p. 5948, 1980.
10 'Impact of Saturn ring particles on Pioneer 11', D.H. Humes, R.L. O'Neil, W.H. Kinard and J.M. Alvarez. *Science*, vol. 207, p. 443, 1980.

11 'Radar observations of the rings of Saturn', R.M. Goldstein and G.A. Morris. *Icarus*, vol. 20, p. 249, 1973.

12 'Energetic particles in the inner magnetosphere of Saturn', J.A. Van Allen. In *Saturn*, T. Gehrels and M.S. Matthews (Eds.). University of Arizona Press, p. 281, 1984.

13 'Measurements of plasma, plasma waves and suprathermal charged particles in Saturn's inner magnetosphere', F.L. Scarf, L.A. Frank, D.A. Gurnett, L.J. Lanzerotti, A. Lazarus and E.C. Sittles. In *Saturn*, T. Gehrels and M.S. Matthews (Eds.). University of Arizona Press, p. 318, 1984.

14 'The magnetic field of Saturn: further studies of the Pioneer 11 observations', M.H. Acuna, N.F. Ness and J.E.P. Connerney. *J. Geophys. Res.*, vol. 85, p. 5675, 1980.

15 'Titan: a satellite with an atmosphere', G.P. Kuiper. *Ap. J.*, vol. 100, p. 378, 1944.

16 'On the possible detection of H_2 in Titan's atmosphere', L.M. Trafton. *Ap. J.*, vol. 175, p. 285, 1972.

17 'Titan: unidentified strong absorption in the photometric infrared', L.M. Trafton. *Icarus*, vol. 21, p. 175, 1974.

18 'Titan: polarimetric evidence for an optically thick atmosphere', J. Veverka. *Icarus*, vol. 18, p. 657, 1973.

19 'The polarisation of Titan', B. Zellner. *Icarus*, vol. 18, p. 661, 1973.

20 'Thermal radiation from Titan's atmosphere', J.J. Caldwell. In *Planetary satellites*, J.A. Burns (Ed.). University of Arizona, p. 348, 1977.

21 'Planetary radiation at infrared and millimetre wavelengths', F.J. Low. *Lowell Obs. Bull.*, vol. 6, p. 184, 1965.

22 'Infrared photometry of Titan', F.J. Low and G.H. Rieke. *Ap. J.*, vol. 190, L143, 1974.

23 'An inversion in the atmosphere of Titan', R.E. Danielson, J.J. Caldwell and D.R. Larach. *Icarus*, vol. 20, p. 437, 1973.

24 'The infrared spectra of Uranus, Neptune and Titan from 0.8–2.5 microns', U. Fink and H.P. Larson. *Ap. J.*, vol. 233, p. 1021, 1979.

25 'Titan's atmosphere and surface', D.M. Hunten. In *Planetary Satellites*, J.A. Burns (Ed.). University of Arizona, p. 430, 1977.

26 'A Titan atmosphere with a surface temperature of 200 K', D.M. Hunten. In *The Saturn system*, D.M. Hunten and D. Morrison (Eds.). NASA Conf. Pub. CP-2068, p. 127, 1978.

27 'Radius and brightness temperature observations of Titan at centimetre wavelengths by the Very Large Array', W.J. Jaffe, J.J. Caldwell and T.C. Owen. *Ap. J.*, vol. 242, p. 806, 1980.

28 'Greenhouse models for the atmosphere of Titan', J.B. Pollack. *Icarus*, vol. 19, p. 43, 1973.

29 'The greenhouse of Titan', C.E. Sagan. *Icarus*, vol. 18, p. 649, 1973.

30 'Polarimetry and photometry of Titan: Pioneer 11 observations and their implications for aerosol properties', M.G. Tomasko and P.H. Smith. *Icarus*, vol. 51, p. 6, 1982.

31 'A possible magnetic wake of Titan: Pioneer 11 observations', D.E. Jones, B.T. Tsututani, E.J. Smith, R.J. Walker and C.P. Sonett. *J. Geophys. Res.*, vol. 85, p. 5835, 1980.

32 'Scientific results from the Pioneer Saturn encounter: summary', A.G. Opp. *Science*, vol. 207, p. 401, 1980.

33 'The discovery of Janus, Saturn's tenth satellite', A. Dollfus. *Sky and Telescope*, vol. 34, p. 136, 1967.

34 'Saturn's syzgistic co-orbital satellites', B.A. Smith, H.J. Reitsema, J.W. Fountain and S.M. Larson. *Bull. Amer. Astron. Soc.*, vol. 12, p. 727, 1980.

35 'The 1966 observations of the co-orbiting satellites of Saturn S10 and S11', S.M. Larson, B.A. Smith, J.W. Fountain and H.J. Reitsema. *Icarus*, vol. 46, p. 175, 1981.

36 IAU Circ. No. 3475, 1980.

Chapter 3: Saturn revealed

1 'Voyager imaging experiment', B.A. Smith, G.A. Griggs, G.E. Danielson, A.F. Cook, M.E. Davies, G.E. Hunt, H. Masursky, L.A. Soderblom, T.C. Owen, C.E. Sagan and V.E. Suomi. *Space Sci. Rev.*, vol. 21, p. 103, 1977.

2 'The Voyager mission photopolarimeter experiment', C.F. Lillie, C.W. Hoard, K. Pang, D.L. Coffeen and J.L. Hansen. *Space Sci. Rev.*, vol. 21, p. 159, 1977.

3 'Ultraviolet spectrometer experiment for the Voyager mission', A.L. Broadfoot, B.R. Sandel, D.E. Shemansky, S.K. Atreya, T.M. Donahue, H.W. Moos, J.L. Bertaux, J.E. Blamont, J.M. Ajello, D.F. Strobel, J.C. McConnell, A. Dalgarno, R. Goody, M.B. McElroy and Y.L. Yung. *Space Sci. Rev.*, vol. 21, p. 183, 1977.

4 'The Voyager infrared spectroscopy and radiometry investigation', R. Hanel, B. Conrath, D. Gautier, P. Gierasch, S. Kumar, V. Kunde, P. Lowman, W. Maguire, J. Pearl, J. Pirraglia, C. Ponnamperuma and R. Samuelson. *Space Sci. Rev.*, vol. 21, p. 129, 1977.

5 'Infrared spectrometer for Voyager', R.A. Hanel *et al. Appl. Opt.*, vol. 19, p. 1391, 1980.

6 'A plasma wave investigation for the Voyager mission', F.L. Scarf and D.A. Gurnett. *Space Sci. Rev.*, vol. 21, p. 289, 1977.

7 'Planetary Radio Astronomy instrument for the Voyager missions', J.W. Warwick, J.B. Pearce, R.G. Peltzer and A.C. Riddle. *Space Sci. Rev.*, vol. 21, p. 309, 1977.

8 'The plasma experiment on the 1977 Voyager mission', H.S. Bridge, J.W. Belcher, R.J. Butler, A.J. Lazarus, A.M. Mavertic, J.D. Sullivam, G.L. Siscoe and V.M. Vasyluinas. *Space Sci. Rev.*, vol. 21, p. 259, 1977.

9 'The low-energy charged particle (LECP) experiment on the Voyager spacecraft', S.M. Krimigis, T.P. Armstrong, W.I. Axford, C.O. Bostrom, C.Y. Fan, G. Gloeckler and L.J. Lanzerotti. *Space Sci. Rev.*, vol. 21, p. 329, 1977.

10 'Cosmic ray investigation for the Voyager missions: energetic particle studies in the outer heliosphere and beyond', E.C. Stone, R.E. Vogt, F.B. McDonald, B.J. Teegarden, J.H. Trainor, J.R. Jokipii and W.R. Webber. *Space Sci. Rev.*, vol. 21, p. 355, 1977.

11 'Radio science investigations with Voyager', V.R. Eshleman, G.L. Tyler, J.D. Anderson, G. Fjeldbo, G.S. Levy, G.E. Wood and T.A. Croft. *Space Sci. Rev.*, vol. 21, p. 207, 1977.

12 'The Jupiter system through the eyes of Voyager 1', B.A. Smith *et al. Science*, vol. 204, p. 951, 1979.

13 'Jupiter's Great Red Spot: a free atmospheric vortex?', A.P. Ingersoll. *Science*, vol. 182, p. 1346, 1973.

14 'Pioneer 10 and 11 observations of the dynamics of Jupiter's atmosphere', A.P. Ingersoll. *Icarus*, vol. 29, p. 245, 1976.

15 'Rings of Saturn and Jupiter: so different, so similar', J.N. Cuzzi and J.A. Burns. International Symposium 'The Jovian System after Galileo. The Saturnian System before Cassini/Huygens', Nantes, France, May 1998.

16 'Io's interaction with the plasma torus', C.K. Goertz. *J. Geophys. Res.*, vol. 85, p. 2949, 1980.

17 'Discovery of currently active extraterrestrial volcanism', L.A. Morabito, S.P. Synnott, P.N. Kupferman and S.A. Collins. *Science*, vol. 204, p. 972, 1979.

18 'Melting of Io by tidal dissipation', S.J. Peale, P.M. Cassen and R.T. Reynolds. *Science*, vol. 203, p. 892, 1979.

19 'Heat flow from Io', D.L. Matson, G.A. Ransford and T.V. Johnson. *J. Geophys. Res.*, vol. 86, p. 1664, 1981.

20 'The magnetic field of Jupiter', M.H. Acuna and N.F. Ness. *J. Geophys. Res.*, vol. 81, p. 2917, 1976.

21 'The Galilean satellites and Jupiter: Voyager 2 imaging science results', B.A. Smith *et al. Science*, vol. 206, p. 927, 1979.

22 Voyager 1's Jovian preliminary results were published in *Science*, vol. 204, pp. 945–1008, 1979; *Nature*, vol. 280, pp. 725-806, 1979; and *Geophys. Res. Lett.*, vol. 7, pp. 1–68, 1980; Voyager 2's results appeared in *Science*, vol. 206, pp. 925–996, 1979; and a 'round up' of the Jovian encounters was published in *J. Geophys. Res.*, vol. 85, pp. 8123–8841, 1980.

23 'Planetary Radio Astronomy observations from Voyager 1 near Saturn', J.W. Warwick *et al. Science*, vol. 212, p. 239, 1981.

24 'Saturn's kilometric radiation: satellite modulation', M.D. Desch and M.L. Kaiser. *Nature*, vol. 292, p. 739, 1981.

25 'Saturnian kilometric radiation: source locations', M.L. Kaiser and M.D. Desch. *J. Geophys. Res.*, vol. 87, p. 4555, 1982.

26 'Source localisation of Saturn kilometric radio emission', A. Lecacheux and F. Genova. *J. Geophys. Res.*, vol. 88, p. 8993, 1983.

27 'Voyager measurement of the rotation period of Saturn's magnetic field', M.D. Desch and M.L. Kaiser. *Geophys. Res. Lett.*, vol. 8, p. 253, 1981.

28 'First Voyager view of the rings of Saturn', S.A. Collins *et al. Nature*, vol. 288, p. 439, 1980.

29 'Voyager observations of Saturn's rings', C.C. Porco. PhD Thesis, Caltech, Pasadena, 1983.

30 'The dynamical evolution of the Saturnian ring spokes', J.R. Hill and D.A. Mendis. *J. Geophys. Res.*, vol. 87, p. 7413, 1982.

31 'Impulsive radio discharges near Saturn', D.R. Evans, J.W. Warwick, J.B. Pearce, T.D. Carr and J.J. Schauble. *Nature*, vol. 292, p. 716, 1981.

32 'The source of Saturn's electrostatic discharges', D.R. Evans, J.H. Romig, C.W. Hord, K.E. Simmons, J.W. Warwick and A.L. Lane. *Nature*, vol. 299, p. 236, 1982.

33 'Saturn electrostatic discharges: properties and theoretical considerations', D.R. Evans, J.H. Romig and J.W. Warwick. *Icarus*, vol. 54, p. 267, 1983.

34 'A model for the formation of spokes in Saturn's ring', C.K. Goertz and G.E. Morfill. *Icarus*, vol. 53, p. 219, 1983.

35 'The evolution of spokes in Saturn's B-ring', E. Grün. G.E. Morfill, R.J. Terrile, T.V. Johnson and G.J. Schweln. *Icarus*, vol. 54, p. 227, 1983.

36 'Saturn as a radio source', M.L. Kaiser *et al.* In *Saturn*, T. Gehrels and M.S. Matthews (Eds.). University of Arizona Press, p. 378, 1984.

37 'Electrodynamic processes in the ring system of Saturn', D.A. Mendis, J.R. Hill, W.-H. Ip, C.K. Goertz and E. Grün. In *Saturn*, T. Gehrels and M.S. Matthews (Eds.). University of Arizona Press, p. 546, 1984.

38 'Stability of negatively charged dust grains in Saturn's ring plane', T.G. Northrop and J.R. Hill. *J. Geophys. Res.*, vol. 87, p. 6045, 1982.

39 'The dynamical evolution of Saturnian ring spokes', J.R. Hill and D.A. Mendis. *J. Geophys. Res.*, vol. 87, p. 7413, 1982.

40 'Saturn's electrostatic discharges: could lightning be the cause?', J.A. Burns, M.R. Showalter, J.N. Cuzzi and R.H. Durisen. *Icarus*, vol. 54, p. 280, 1983.

41 'Atmospheric storm explanation of Saturn's electrostatic disharges', M.L. Kaiser, J.E.P. Connerney and M.D. Desch. *Nature*, vol. 303, p. 50, 1983.

42 *The planet Saturn: a history of observation, theory and discovery*, A.F.O'D. Alexander. Faber and Faber, 1962.

43 'Theory of motion of Saturn's co-orbiting satellites', C.F. Yoder, G. Colombo, S.P. Synnott and K.A. Yoder. *Icarus*, vol. 53, p. 431, 1983.

44 'Discovering the rings of Uranus', J.L. Elliot, E. Dunham and R.L. Millis. *Sky and Telescope*, vol. 53, p. 412, 1977.

45 *Rings: discoveries from Galileo to Voyager*, J.L. Elliot and R. Kerr. MIT Press, 1984.

46 'Towards a theory for the Uranian rings', P. Goldreich and S. Tremaine. *Nature*, vol. 277, p. 97, 1979.

47 'Dynamical features in the northern hemisphere of Saturn from Voyager 1 images', G.E. Hunt, D. Godfrey, J.-P. Muller and R.F.T. Barrey. *Nature*, vol. 297, p. 132, 1982.

48 'Density waves in Saturn's rings', J.N. Cuzzi, J. Lissauer and F.H. Shu. *Nature*, vol. 292, p. 703, 1981.

49 See pp. 61–63 of this volume for a summary of these competing theories.

50 'The outer magnetosphere [of Saturn]', A.W. Schardt *et al.* In *Saturn*, T. Gehrels and M.S. Matthews (Eds.). University of Arizona Press, p. 416, 1984.

51 'Vertical distribution of scattering hazes in Titan's upper atmosphere', K. Rages and J.B. Pollack. *Icarus*, vol. 55, p. 50, 1983.

52 'Abundances of the elements in the Solar System', A.G.W. Cameron. *Space Science Reviews*, vol. 15, p. 121, 1973.

53 'Clathrate and ammonia hydrates at high pressure–application to the origin of methane on Titan', J.I. Lunine and D.J. Stevenson. *Icarus*, vol. 70, p. 61, 1987.

54 'The atmosphere of Titan: an analysis of the Voyager 1 radio occultation measurements', G.F. Lindal, G.E. Wood, H.B. Hotz and D.N. Sweetnam. *Icarus*, vol. 53, p. 348, 1983.

55 'Propane and methyl acetylene in Titan's atmosphere', W.C. Maguire, R.A. Hanel, D.E. Jennings, V.G. Kunde and R.E. Samuelson. *Nature*, vol. 292, p. 683, 1981.

56 'Mean molecular weight and hydrogen abundance of Titan's atmosphere', R.E. Samuelson, R.A. Hamel, V.G. Kunde and W.C. Maguire. *Nature*, vol. 292, p. 688, 1981.

57 'Laboratory investigation of the formation of unsaturated hydrocarbons in Titan's atmosphere', K.L. Kaiser, O. Asvany, C.C. Chiong, D. Rolland, Y.T. Lee, F. Stahl, P.v.R. Schleyer and H.F. Schaefer. European Geophysics Society XXV General Assembly, Nice, France, April 2000.

58 'Organic chemistry in the atmosphere [of Titan]', C.E. Sagan. In *The atmosphere of Titan*, D.M. Hunten (Ed.). NASA SP-340, p. 134, 1974.

59 'Tholins', C.E. Sagan and B.N. Khare. *Nature*, vol. 277, p. 102, 1979.

60 'The escape of H_2 from Titan', D.M. Hunten. *J. Atmos. Sci.*, vol. 30, p. 726, 1973.

61 'Titan's gas and plasma torus', A. Eviatar and M. Podolak. *J. Geophys. Res.*, vol. 88, p. 883, 1983.

62 'Whence comes the "Titan" hydrogen torus', D.E Shemansky and G.R. Smith. *EOS*, vol. 63, p. 1019, 1982.

63 'Ethane ocean on Titan', J.I. Lunine, D.J. Stevenson and Y.L. Yung. *Science*, vol. 222, p. 1229, 1983.

64 'Orbits of the Tethys Lagrangian bodies', H.J. Reitsema. *Icarus*, vol. 48, p. 140, 1981.

65 'The "braided" F-ring of Saturn', S.F. Demott. *Nature*, vol. 290, p. 454, 1981.

66 'A numerical study of Saturn's F-ring', M.R. Showalter and J.A. Burns. *Icarus*, vol. 52, p. 526, 1982.

67 'Orbits of Saturn's F-ring and its shepherding satellites', S.P. Synnott, R.J. Terrile and B.A. Smith. *Icarus*, vol. 53, p. 156, 1983.

68 'Clumps in Saturn's F-ring and their interaction with Prometheus', M.R. Showalter. International Symposium 'The Jovian System after Galileo. The Saturnian System before Cassini/Huygens', Nantes, France, May 1998.

69 'On the braids and spokes in Saturn's ring system', J.R. Hill and D.A. Mendis. *Moon and Planets*, vol. 24, p. 431, 1981.

70 'Drunken shepherds: random walk models for Pandora and Prometheus', L.W. Esposito. European Geophysical Society XXVII General Assembly Nice, France, A-05215, April, 2002.

71 'Ringed planets: still mysterious (Part 1)', J.N. Cuzzi. *Sky and Telescope*, vol. 68, no. 6, p. 511, December 1984.

72 'Particle size distribution in Saturn's rings from Voyager 1 radio occultation', E.A. Marouf, G.L. Tyler, H.A. Zebker, R.A. Simpson and V.R. Eshleman. *Icarus*, vol. 54, p. 189, 1983.

73 'Apparent thickness of Saturn's rings', A. Brahic and B. Sicardy. *Nature*, vol. 289, p. 447, 1981.

74 'Saturn's rings: structure, dynamics and particle properties', L.W. Esposito, J.N. Cuzzi, J.B. Holberg, E.A. Marouf, G.L. Tyler and C.C. Porco. In *Saturn*, T. Gehrels and M.S. Matthews (Eds.). University of Arizona Press, p. 463, 1984.

75 'Moonlets in Saturn's rings?', J. Lissauer, F.H. Shu and J.N. Cuzzi. *Nature*, vol. 292, p. 707, 1981.

76 'A simple model of Saturn's rings', M. Henon. *Nature*, vol. 293, p. 33, 1981.

77 During a 0.48-second exposure, there would otherwise have been 7 pixels of smear, which would have degraded the 'best' imagery to the point of being effectively useless.

78 'The origin of the E-ring of Saturn', J.R. Hill and D.A. Mendis. *EOS*, vol. 63, p. 1019, AGU Fall Meeting, San Francisco, California, December 1982.

79 'The E-ring of Saturn and its satellite Enceladus', K.D. Pang, C.C. Voge, J.W. Rhoads and J.M. Ajello. *Proc. Lunar Planet. Sci. Conf.*, p. 592, 1983.

80 'Saturn's rings: structure, dynamics and particle properties', L.W. Esposito, J.N. Cuzzi, J.B. Holberg, E.A. Marouf, G.L. Tyler and C.C. Porco. In *Saturn*, T. Gehrels and M.S. Matthews (Eds.). University of Arizona Press, p. 463, 1984.

81 'Theory, measurement and models of the upper atmosphere and ionosphere of Saturn', S.K. Atreya, J.H. Waitz, T.M. Donahue, A.F. Nagy and J.M. McConnell. In *Saturn*, T. Gehrels and M.S. Matthews (Eds.). University of Arizona Press, p. 239, 1984.

82 'Infrared observations of the Saturnian system from Voyager 1', R.A. Hanel *et al.* *Science*, vol. 212, p. 192, 1981.

83 'The helium abundance of Jupiter from Voyager 1', D. Gautier, B.J. Conrath, F.M. Flasar, R. Hanel, V. Kunde, A. Chedin and N. Scott. *J. Geophys. Res.*, vol. 86, p. 8713, 1981.

84 'Cosmogonical implications of elemental and isotopic abundances in atmospheres of the giant planets', D. Gautier and T.C. Owen. *Nature*, vol. 304, p. 691, 1983.

85 'On the convection and gravitational layering in Jupiter and in stars of small mass', E.E. Salpeter. *Ap. J.*, vol. 181, L.83, 1973.

86 'A calculation of Saturn's gravitational contraction history', J.B. Pollack *et al.* *Icarus*, vol. 30, p. 111, 1977.

87 'Comparative atmospheres of Jupiter and Saturn: deep atmospheric composition, clouds and vertical mixing', S.K. Atreya. International Symposium 'The Jovian System after Galileo. The Saturnian System before Cassini/Huygens', Nantes, France, May 1998.

88 'Helium and deuterium in Jupiter and Saturn', D.M. Hunten. International Symposium 'The Jovian System after Galileo. The Saturnian System before Cassini/Huygens', Nantes, France, May 1998.

89 'Hydrodynamic instability in the solar nebula in the presence of a planetary core', F. Perri and A.G.W. Cameron. *Icarus*, vol. 22, p. 416, 1974.

90 'Formation of giant planets', H. Mizuno. *Prog. Theor. Phys.*, vol. 64, p. 544, 1980.

91 'Dust to planetesimals: settling and coagulation in the solar nebula', S.J. Weidenschilling. *Icarus*, vol. 44, p. 172, 1980.

92 'On the origin and initial temperature of Jupiter and Saturn', V.S. Safronov and E.L. Ruskol. *Icarus*, vol. 49, p. 284, 1982.

93 'Evolution of giant gaseous protoplanets embedded in the primitive solar nebula', A.G.W. Cameron, W.M. DeCampli and P. Bodenheimer. *Icarus*, vol. 49, p. 298, 1982.

94 'From icy planetesimals to outer planets and comets', R. Greenberg, S.J. Weidenschilling, C.R. Chapman and D.R. Davis. *Icarus*, vol. 59, p. 87, 1984.

95 'Orbital resonances in the solar nebula: implications for planetary accretion', S.J. Weidenschilling and D.R. Davis. *Icarus*, vol. 62, p. 16, 1985.

96 'Comparisons of solar nebula models', S.J. Weidenschilling. In *Workshop on the origins of the solar systems*, J. Nuth and P. Sylvester (Eds.). Technical Report No. 88-04, Lunar and Planetary Institute, Houston, p. 31, 1988.

97 'The physics of planetesimal formation', S.J. Weidenschilling, B. Donn and P. Meakin. In *The formation and evolution of planetary systems*. H. Weaver and L. Danly (Eds.). Cambridge University Press, p. 131, 1989.

98 'Early stages of accumulation in the solar nebula', S.J. Weidenschilling. *Adv. Space Res.*, vol. 10, p. 101, 1990.

99 'Formation of planetesimals in the solar nebula', S.J. Weidenschilling and J.N. Cuzzi. In *Protostars and planets III*, E. Levy and J.I. Lunine (Eds.). Univ. of Arizona Press, p. 1031, 1993.

100 'Coagulation of grains in static and collapsing protostellar clouds', S.J. Weidenschilling and T.V. Ruzmaikina. *Ap. J.*, vol. 430, p. 713, 1994.

101 'Planetesimals from stardust', S.J. Weidenschilling. In *From stardust to planetesimals*, Y. Pendleton and A.G.G.M. Tielens (Eds.), ASP Conference Series, vol. 122, p. 281, 1997.

102 'Galileo Probe measurements of the deep zonal winds of Jupiter', D.H. Atkinson. In *The three Galileos: the man, the spacecraft, the telescope*, C. Barbieri, J.H. Rahe, T.V. Johnson and A.M. Sohus (Eds.). Kluwer Academic Press, p. 279, 1997.

103 'Atmospheric dynamics of Jupiter and Saturn', A.P. Ingersoll. International Symposium 'The Jovian System after Galileo. The Saturnian System before Cassini/Huygens', Nantes, France, May 1998.

104 'Compositional chemistry of Saturn's atmosphere', R.G. Prinn, H.P. Larson, J.J. Caldwell and D. Gautier. In *Saturn*, T. Gehrels and M.S. Matthews (Eds.). University of Arizona Press, p. 88, 1984.

105 'Clouds and aerosols in Saturn's atmosphere', M.G. Tomasko, R.A. West, G.S. Orton and V.G. Tejfel. In *Saturn*, T. Gehrels and M.S. Matthews (Eds.). University of Arizona Press, p. 150, 1984.

106 'Structure and dynamics of Saturn's atmosphere', A.P. Ingersoll, R.F. Beebe, B.J. Conrath and G.E. Hunt. In *Saturn*, T. Gehrels and M.S. Matthews (Eds.). University of Arizona Press, p. 195, 1984.

107 'Possible traversals of Jupiter's distant magnetic tail by Voyager and Saturn', F.L. Scarf. *J. Geophys. Res.*, vol. 84, p. 4422, 1979.

108 'Evidence for a distant ($>8,700$ Rj) Jovian magnetotail: Voyager 2 observations', R.P. Lepping, L.F. Burlaga, M.D. Desch and L.W. Klein. *Geophys. Res. Lett.*, vol. 9, p. 885, 1982.

109 'Observations of Jupiter's distant magnetotail and wake', W.S. Kurth, J.D. Sullivan, D.A. Gurnett, F.L. Scarf, H.S. Bridge and E.C. Sittler. *J. Geophys. Res.*, vol. 87, p. 10373, 1982.

110 'Radio emission signatures of Saturn's immersions in Jupiter's magnetic tail', M.D. Desch. *J. Geophys. Res.*, vol. 88, p. 6904, 1983.

111 'Physical properties of Saturn's rings', J.N. Cuzzi. *Planetary Rings Conf.*, Toulouse, France, August 1982.

112 'Ringed planets: still mysterious (Part 1)', J.N. Cuzzi. *Sky and Telescope*, vol. 69, no. 1, p. 19, January 1985.
113 'Voyager PPS stellar occultation of Saturn's rings', L.W. Esposito *et al. J. Geophys. Res.*, vol. 88, p. 8643, 1983.
114 'Photopolarimetry from Voyager 2: preliminary results on Saturn, Titan, and the rings', A.L. Lane *et al. Science*, vol. 215, p. 537, 1982.
115 'The structure of Saturn's rings: implications from the Voyager stellar occultation', L.W. Esposito, M. O'Callaghan and R.A. West. *Icarus*, vol. 56, p. 439, 1983.
116 'How tidal heating on Io drives the Galilean orbital resonance locks', C.F. Yoder. *Nature*, vol. 279, p. 767, 1979.

Chapter 4: The Titans

 1 'Surface compositions of the satellites of Saturn from infrared photometry', D. Morrison, D.P. Cruikshank, C.B. Pilcher and G.H. Rieke. *Ap. J.*, vol. 207, L.213, 1976.
 2 'Infrared spectra of the satellites of Saturn: identification of water ice on Iapetus, Rhea, Dione and Tethys', U. Fink, H.P. Larson, T.N. Gautier and R.R. Treffers. *Ap. J.*, vol. 207, L.63, 1976.
 3 'Evidence for frost on Rhea's surface', T.V. Johnson, G.J. Veeder and D.L. Matson. *Icarus*, vol. 24, p. 428, 1975.
 4 *Satellites of the Solar System*, W. Sandner. The Scientific Book Club, 1965.
 5 *Atlas of the Solar System*, P. Moore and G.E. Hunt. Mitchell Beazley in association with the Royal Astronomical Society, 1990.
 6 'Satellites of Saturn: geological perspective', D. Morrison, T.V. Johnson, E.M. Shoemaker, L.A. Soderblom, P. Thomas, J. Veverka and B.A. Smith. In *Saturn*, T. Gehrels and M.S. Matthews (Eds.). University of Arizona Press, p. 609, 1984.
 7 'Radii, albedos and 20-micron brightness temperatures of Iapetus and Rhea', R.E. Murphy, D.P. Cruikshank and D. Morrison. *Ap. J.*, vol. 177, L.93, 1972.
 8 'Albedos and densities of the inner satellites of Saturn', D. Morrison. *Icarus*, vol. 22, p. 51, 1974.
 9 'Lunar occultation of Saturn: the diameters of Tethys, Dione, Rhea, Titan and Iapetus', J.L. Elliot, J. Veverka and J. Goguen. *Icarus*, vol. 26, p. 389, 1975.
10 'The surfaces and interiors of Saturn's satellites', D.P. Cruikshank. *Rev. Geophys. Space Phys.*, vol. 17, p. 165, 1979.
11 'Sizes and densities of Saturn's satellites: a pre-Voyager analysis', D. Morrison. *Bull. Amer. Astron. Soc.*, vol. 12, p. 727, 1980.
12 'Gravity field of the Saturnian system from Pioneer and Voyager tracking data', J.K. Campbell and J.D. Anderson. *Bull. Amer. Astron. Soc.*, vol. 17, p. 697, 1985.
13 'Encounter with Saturn: Voyager 1 imaging science results', B.A. Smith, L.A. Soderblom, R. Beebe, J. Boyce, G. Briggs, A. Bunker, S.A. Collins, C.J. Hansen, T.V. Johnson, J.L. Mitchell, R.J. Terrile, M.H. Carr, A.F. Cook, J.N. Cuzzi, J.B. Pollack, G.E. Danielson, A.P. Ingersoll, M.E. Davies, G.E. Hunt, H. Masursky, E.M. Shoemaker and D. Morrison. *Science*, vol. 212, p. 163, 1981.
14 'A new look at the Saturn system: the Voyager 2 images', B.A. Smith, L.A. Soderblom, R. Batson, P. Bridges, J. Inge, H. Masursky, E.M. Shoemaker, R. Beebe, J. Boyce, G. Briggs, A. Bunker, S.A. Collins, C.J. Hansen, T.V. Johnson, J.L. Mitchell, R.J. Terrile, A.F. Cook, J.N. Cuzzi, J.B. Pollack, G.E. Danielson, A.P. Ingersoll, M.E. Davies and G.E. Hunt. *Science*, vol. 215, p. 504, 1982.

15 *Saturn*, T. Gehrels and M.S. Matthews (Eds.). University of Arizona Press, 1984.

16 *Satellites*, J.A. Burns and M.S. Matthews (Eds.). University of Arizona Press, 1986.

17 'Encounter with Saturn: Voyager 1 imaging science results', B.A. Smith *et al. Science*, vol. 212, p. 163, 1981.

18 'A new look at the Saturn system: the Voyager 2 images', B.A. Smith *et al. Science*, vol. 215, p. 504, 1982.

19 'Orbital resonances among Saturn's satellites', R. Greenberg. In *Saturn*, T. Gehrels and M.S. Matthews (Eds.). University of Arizona Press, p. 593, 1984.

20 'Evolution of satellite resonances by tidal dissipation', R. Greenberg. *Astron. J.*, vol. 78, p. 338, 1973.

21 'Satellites of the outer planets: their physical and chemical nature', J.S. Lewis. *Icarus*, vol. 15, p. 174, 1971.

22 'A new look at the Saturn system: the Voyager 2 images', B.A. Smith, L.A. Soderblom, R. Batson, P. Bridges, J. Inge, H. Masursky, E.M. Shoemaker, R. Beebe, J. Boyce, G. Briggs, A. Bunker, S.A. Collins, C.J. Hansen, T.V. Johnson, J.L. Mitchell, R.J. Terrile, A.F. Cook, J.N. Cuzzi, J.B. Pollack, G.E. Danielson, A.P. Ingersoll, M.E. Davies and G.E. Hunt. *Science*, vol. 215, p. 504, 1982.

23 A similar pattern of fractures on Mars's Phobos seems to be related to the impact that formed the large crater Stickney.

24 'Collisional history of the Saturn system', E.M. Shoemaker. *Saturn Conf.*, Tucson, Arizona, May 1982.

25 'Crater numbers and geological histories of Iapetus, Enceladus, Tethys and Hyperion', J.B. Plescia and J.M. Boyce. *Nature*, vol. 301, p. 666, 1983.

26 'Viscosity of the lithosphere of Enceladus', Q.R. Passey. *Icarus*, vol. 53, p. 105, 1983.

27 'Volcanic and igneous processes in small icy satellites', D.J. Stevenson. *Nature*, vol. 298, p. 142, 1982.

28 'The evolution of Enceladus', S.W. Squyres, R.T. Reynolds, P.M. Cassen and S.J. Peale. *Icarus*, vol. 53, p. 319, 1983.

29 'Tidal friction and Enceladus's anomalous surface', C.F. Yoder. AGU *EOS Trans.*, vol. 62, p. 939, 1981.

30 'Ring torques on Janus and the melting of Enceladus', J.J. Lissauer, S.J. Peale and J.N. Cuzzi. *Icarus*, vol. 58, p. 159, 1984.

31 'Saturn's small satellites: Voyager imaging results', P. Thomas, J. Veverka, D. Morrison, M. Davies and T.V. Johnson. *J. Geophys. Res.*, vol. 88, p. 8743, 1983.

32 'Orbital resonances among Saturn's satellites', R. Greenberg. In *Saturn*, T. Gehrels and M.S. Matthews (Eds.). University of Arizona Press, p. 593, 1984.

33 'Theory of Enceladus and Dione', W.H. Jefferys and L.M. Ries. *Astron. J.*, vol. 80, p. 876, 1975.

34 'Saturn's E-ring. I: CCD observations of March 1980', W.A. Baum *et al. Icarus*, vol. 47, p. 84, 1981.

35 'Saturn's E-ring and satellite Enceladus', K.D. Pang, C.C. Voge, J.W. Rhoads and J.M. Ajello. *Proc. Lunar Planet. Sci. Conf.*, p. 592, 1983.

36 'Evidence for an arc near Enceladus's orbit: a possible key to the origin of the 'E' ring', C. Roddier and F. Roddier. International Symposium 'The Jovian System after Galileo. The Saturnian System before Cassini/Huygens', Nantes, France, May 1998.

37 'Tectonics and geological history of Tethys', J.M. Moore and J.L. Ahern. *Lunar Planet. Sci. Conf.*, p. 538, 1982.

38 'Rheology of ices: a key to the tectonics of the ice moons of Jupiter and Saturn', J.P. Poirier. *Nature*, vol. 299, p. 683, 1982.

39 'The geology of Tethys', J.M. Moore and J.L. Ahern. *J. Geophys. Res.*, vol. 88, p. 577, 1983.

40 'The moons of Saturn', L.A. Soderblom and T.V. Johnson. In *The planets*, B.C. Murray. (Ed.) W.H. Freeman, p. 95, 1983.

41 'Crater numbers and geological histories of Iapetus, Enceladus, Tethys and Hyperion', J.B. Plescia and J.M. Boyce. *Nature*, vol. 301, p. 666, 1983.

42 'A new look at the Saturn system: the Voyager 2 images', B.A. Smith *et al. Science*, vol. 215, p. 504, 1982.

43 'Six-colour photometry of Iapetus, Titan, Rhea, Dione and Tethys', M. Noland, J. Veverka, D. Morrison, D.P. Cruikshank, A.R. Lazarewicz, N.D. Morrison, J.E. Elliot, J. Goguen and J.A. Burns. *Icarus*, vol. 23, p. 334, 1974.

44 'Voyager photometry of Saturn's satellites', B.J. Buratti, J. Veverka and P. Thomas. NASA TM-85127, p. 41, 1982.

45 'Voyager photometry of Rhea, Dione, Tethys, Enceladus and Mimas', B.J. Buratti and J. Veverka. *Icarus*, vol. 58, p. 254, 1983.

46 'The geology of Dione', J.B. Plescia. *Icarus*, vol. 56, p. 255, 1983.

47 'The tectonic and volcanic history of Dione', J.M. Moore. *Icarus*, vol. 59, p. 205, 1984.

48 'The plains and lineaments of Dione', J.M. Moore. *Proc. Lunar Planet. Sci. Conf.*, p. 511, 1983.

49 'Volcanic and igneous processes in small icy satellites', D.J. Stevenson. *Nature*, vol. 298, p. 142, 1982.

50 'Interactions of planetary magnetospheres with icy satellite surfaces', A.F. Cheng, P.K. Haff, R.E. Johnson and L.J. Lanzerotti. In *Satellites*, J.A. Burns and M.S. Matthews (Eds.). University of Arizona Press, p. 403, 1986.

51 'Six-colour photometry of Iapetus, Titan, Rhea, Dione and Tethys', M. Noland, J. Veverka, D. Morrison, D.P. Cruikshank, A.R. Lazarewicz, N.D. Morrison, J.L. Elliot, J. Goguen and J.A. Burns. *Icarus*, vol. 23, p. 334, 1974.

52 'Implications of Voyager data for energetic ion erosion of icy satellites of Saturn', L.J. Lanzerotti, C.G. Maclennan, W.L. Brown, R.E. Johnson, L.A. Barton, C.T. Reimann, J.W. Garrett and J.W. Boring. *J. Geophys. Res.*, vol. 88, p. 8765, 1983.

53 'Geomorphology of Rhea: implications for geologic history and surface processes', J.M. Moore, V.M. Horner and R. Greeley. *J. Geophys. Res.*, vol. 90, p. 785, 1985.

54 *Planetary landscapes*, R. Greeley. Allen & Unwin p. 235, 1987; Greeley cites a personal communication by R. Pike and P.D. Spudis, but (according to Spudis) this observation was never formally published.

55 'The geomorphologic features on Rhea', J.M. Moore and V.M. Horner. *Lunar Planet. Sci. Conf.*, p. 560, 1984.

56 'Crater densities and geological histories of Rhea, Dione, Mimas and Tethys', J.B. Plescia and J.M. Boyce. *Nature*, vol. 295, p. 285, 1982.

57 'Titan', T.C. Owen. In *The planets*, B.C. Murray (Ed.). W.H. Freeman, p. 84, 1983.

58 'On the origin of Titan's atmosphere', T.C. Owen. *Planet. Space Sci.*, vol. 48, p. 747, 2000.

59 'The tides in the sea of Titan', C.E. Sagan and S.F. Dermott. *Nature*, vol. 300, p. 731, 1982.

60 'Oceans on Titan', F.M. Flasar. *Science*, vol. 221, p. 55, 1983.

61 'Post-accretional evolution of Titan's surface and atmosphere', J.I. Lunine and D.J. Stevenson. *Ap. J.*, vol. 238, p. 357, 1982.

62 'Titan's atmosphere from ISO observations: temperature, composition and detection of water vapour', A. Coustenis, A. Salama, E. Lellouch, Th. Encrenaz, Th. de Graauw,

G.L. Bjoraker, R.E. Samuelson, D. Gautier, H. Feuchtgruber, M.F. Kessler and G.S. Orton. American Astronomical Society, Division of Planetary Sciences Meeting, Madison, October 1998.

63 'Evidence for a strong $^{15}N/^{14}N$ enrichment in Titan's atmosphere from millimetre observations', T. Hidayat and A. Marten. *Ann. Geophys.*, vol. 16, p. C988, 1998.

64 'On the volatile inventory of Titan from isotopic abundances in nitrogen and methane', J.I. Lunine, Y.L. Yung and R.D. Lorenz. *Planet. Space Sci.*, vol. 47, p. 1291, 1999.

65 'Nitrogen isotope fractionation and its consequences for Titan's atmospheric evolution', H. Lammer, W. Stumptner, G.J. Molina-Cuberos, S.J. Bauer and T.C. Owen. European Geophysics Society XXV General Assembly, Nice, France, April 2000.

66 'Impact erosion history of Titan', A.-L. Tsai and W.-H. Ip. European Geophysics Society XXV General Assembly, Nice, France, April 2000.

67 At the present time, Titan's surface temperature is 92 K, due to a 21 K greenhouse effect, which is offset in part by a 9 K anti-greenhouse effect due to the haze absorbing about 60 per cent of insolation.

68 'Stable methane hydrate above 2 GPa and the source of Titan's atmospheric methane', J.S. Loveday, R.J. Nelmes, M. Guthrie, S.A. Belmonte, D.R. Allan, D.D. Klug, J.S. Tse and Y.P. Handa. Letters to *Nature*, 5 April 2001.

69 'Analytic stability of Titan's climate: sensitivity to volatile inventory', R.D. Lorenz, C.P. McKay and J.I. Lunine. *Planet. Space Sci.*, vol. 47, p. 1503, 1999.

70 P.H. Smith, M.T. Lemmon, R.D. Lorenz, J.J. Caldwell, L.A. Sromovsky and M.D. Allison. See STScI-PR94-55, December 1994.

71 'Titan's surface, revealed by HST imaging', P.H. Smith, M.T. Lemmon, R.D. Lorenz, L.A. Sromovsky, J.J. Caldwell and M.D. Allison. *Icarus*, vol. 119, p. 336, 1996.

72 'Titan: a world seen but darkly', J.I. Lunine. In *Our world: the magnetism and thrill of planetary exploration*, S.A. Stern (Ed.). Cambridge University Press, p. 135, 1999.

73 'Spatially resolved images of Titan by means of adaptive optics', M. Combes, L. Vapillon, E. Gendron, A. Coustenis, O. Lai, R. Wittemberg and R. Sirdey. *Icarus*, vol. 129, p. 482, 1997.

74 'Titan's surface from spectra and images', A. Coustenis, E. Gendron, O. Lai, B. Schmitt, M. Combes, J.-P. Veran, E. Lellouch, L. Vapillon, P. Rannou, M. Cabane, C. McKay, J.-P. Maillard, Th. Fusco. European Geophysics Society XXV General Assembly, Nice, France, p. 1888, April 2000.

75 'Adaptive optics images of Titan at 1.3 and 1.6 microns at the CFHT', A. Coustenis, E. Gendron, O. Lai, J.-P. Veran, M. Combes, J. Woillez, Th. Fusca and L. Mugnier. *Icarus*, vol. 154, p. 501, 2002.

76 'Expected surface of Titan compared with observed surfaces of the Galileans', R.D Lorenz. International Symposium 'The Jovian System after Galileo. The Saturnian System before Cassini/Huygens', Nantes, France, May 1998.

77 'Admissible heights of the local roughness on Titan's landscape', V.I. Dimitrov and A. Bar-Nun. International Symposium 'The Jovian System after Galileo. The Saturnian System before Cassini/Huygens', Nantes, France, May 1998.

78 'The life, death and afterlife of a raindrop on Titan', R.D. Lorenz. *Planet. Space Sci.*, vol. 41, p. 647, 1993.

79 'Transient clouds in Titan's lower atmosphere', C.A. Griffith, T.C. Owen, G.A. Miller and T.R. Geballe. *Nature*, vol. 395, p. 575, 1998.

80 'Titan's clouds', C.A. Griffiths, T.R. Geballe, J.L. Hall and T.C. Owen. European Geophysics Society XXV General Assembly, Nice, France, April 2000.

81 'Titan: High-Resolution Speckle Images from the Keck Telescope', S.G. Gibbard, B.

Macintosh, D. Gavel, C.E. Max, I. de Pater, A.M. Ghez, E.F. Young and C.P. McKay. *Icarus*, vol. 139, p. 189, 1999.

82 'Optical constants of solid ethane from 0.4 to 2.5 microns', B.N. Khare, W.R. Thompson, C.E. Sagan, E.T. Arakawa and J.J. Lawn. *Bull. Am. Astron. Soc.*, vol. 22, p. 1033, 1990.

83 'Titan's surface model: water ice covered with tholin deposits?', P. Rannou, A. Coustenis, B. Schmitt, E. Lellouch, C.P. McKay and M. Cabane. International Symposium 'The Jovian System after Galileo. The Saturnian System before Cassini/Huygens', Nantes, France, May 1998.

84 'Impacts and cratering on Titan: a pre-Cassini view', R.D. Lorenz. *Planet. Space Sci.*, vol. 45, p. 1009, 1997.

85 'The dark side of Iapetus and the atmosphere and surface of Titan', T.C. Owen, D.P. Cruikshank, C. Dalle Ore, T.R. Roush, R. Meier, T. Geballe, C.A. Griffiths, C. de Bergh, N. Biver, A. Marten, B.A. Smith, R.J. Terrile, H.E. Matthews and Y. Yung. European Geophysics Society XXV General Assembly, Nice, France, April 2000.

86 'Tidal effects of disconnected hydrocarbon seas on Titan', S.F. Dermott and C.E. Sagan. *Nature*, vol. 374, p. 238, 1995.

87 'The interior of Titan', D.J. Steavenson. *Symposium on Titan*, Toulouse, France, September 1991; ESA SP-338, p. 29, 1992.

88 'Hiding Titan's ocean: densification and hydrocarbon storage in an icy regolith', K.J. Kossacki and R.D. Lorenz. *Planet. Space Sci.*, vol. 44, p. 1029, 1996.

89 'Evidence for water vapor in Titan's atmosphere from ISO/SWS data', A. Coustenis, E. Lellouch, Th. Encrenaz and A. Salama. *Astron. & Astrophys.*, vol. 336, L.85, 1998.

90 'ISO observations and tentative detection of water on Titan', A. Coustenis, Th. Encrenaz, A. Salama, E. Lellouch, D. Gautier, M.F. Kessler, Th. de Graauw, R.E. Samuelson, G.L. Bjoraker, G. Orton and R. Wittemberg. International Symposium 'The Jovian System after Galileo The Saturnian System before Cassini/Huygens', Nantes, France, May 1998.

91 'Molecular synthesis in simulated reducing planetary atmospheres', C.E. Sagan and S.L. Miller. *Astron. J.*, vol. 65, p. 499, 1960.

92 'Prebiotic ribose synthesis: a critical analysis', R. Shapiro. *Orig. Life Evol. Biosph.*, vol. 18, p. 71, 1988.

93 'Cyanoacetylene in prebiotic synthesis', R.A. Sanchez, J.P. Ferris and L.E. Orgel. *Science*, vol. 154, p. 784, 1966.

94 'Molecular analysis of tholins produced under simulated Titan conditions', B.N. Khare, C.E. Sagan, S. Shrader and E.T. Arakawa. *Bull. Amer. Astron. Soc.*, vol. 14, p. 714, 1982.

95 'Destination Titan', S. Mirsky. *Astronomy*, p. 42, November 1997.

96 'Organic chemistry on Titan surface interactions', W.R. Thompson and C.E. Sagan. In *Proceedings of the Symposium on Titan*, ESA Special Publication 338, p. 167, 1992.

97 'Titan under a red giant Sun: a new kind of "habitable" world', R.D. Lorenz, J.I. Lunine and C.P. McKay. *Geophys. Res. Lett.*, vol. 24, p. 2905, 1997.

98 'Asteroid collisions: effective body strength and efficiency of catastrophic disruption', D.R. Davis, C.R. Chapman, R. Greenberg and S.J. Weidenschilling. *Proc. Lunar Planet. Sci. Conf.*, p. 146, 1983.

99 'Hyperion: collisional disruption of a resonant satellite', P. Farinella, A. Miloni, A.M. Nobili, P. Paolicchi and V. Zappala. *Icarus*, vol. 54, p. 353, 1983.

100 'The radius and albedo of Hyperion', D.P. Cruikshank. *Icarus*, vol. 37, p. 307, 1979.

101 'The surface composition and radius of Hyperion', D.P. Cruikshank and R.H. Brown. *Icarus*, vol. 50, p. 82, 1982.

102 'Voyager photometry of Hyperion: rotation rate', P. Thomas, J. Veverka, D. Wenkert, G.E. Danielson and M.E. Davies. *Planetary Sciences Conference*, Ithaca, New York, July 1983.

103 'Hyperion: 13-day rotation from Voyager data', P. Thomas, J. Veverka, D. Wenkert, G.E. Danielson and M.E. Davies. *Nature*, vol. 307, p. 717, 1984.

104 'Hyperion: analysis of Voyager observations', P. Thomas and J. Veverka. *Icarus*, vol. 64, p. 414, 1985.

105 'The chaotic rotation of Hyperion', J. Wisdom, S.J. Peale and F. Mignard. *Icarus*, vol. 58, p. 137, 1984.

106 'The rotational light curve of Hyperion during 1983', J. Goguen, H. Hammel, D.P. Cruikshank and W.K. Hartmann. *Bull. Amer. Astron. Soc.*, vol. 15, p. 854, 1983.

107 The discovery of Iapetus's chaotic rotation was confirmed by James Klavetter, who published his results through the American Astronomical Society.

108 'UVB photometry of Iapetus', R.L. Millis. *Icarus*, vol. 18, p. 247, 1973.

109 'UVB photometry of Iapetus: results from five apparitions', R.L. Millis. *Icarus*, vol. 31, p. 81, 1977.

110 'The two faces of Iapetus', D. Morrison, T.J. Jones, D.P. Cruikshank and R.E. Murphy. *Icarus*, vol. 24, p. 157, 1975.

111 The Iapetus aspect of the novel was elided when Stanley Kubrick made the movie version of this story.

112 'The albedo asymmetry of Iapetus', S.W. Squyres and C.E Sagan. *Bull. Amer. Astron. Soc.*, vol. 14, p. 739, 1982.

113 'Albedo asymmetry of Iapetus', S.W. Squyres and C.E Sagan. *Nature*, vol. 303, p. 782, 1983.

114 'Variation of the photometric function over the surface of Iapetus', J. Goguen, D. Morrison and M. Tripicco. *Proc. Planet. Satellites Conf.*, Ithaca, New York, July 1983.

115 'The dark side of Iapetus', D.P. Cruikshank, J.F. Bell, M.J. Gaffrey, R.H. Brown, R Howell, C. Beerman and M. Rognstad. *Icarus*, vol. 53, p. 90, 1983.

116 'Eight-colour photometry of Hyperion, Iapetus and Phoebe', D.J. Tholen and B. Zellner. *Icarus*, vol. 53, p. 341, 1982.

117 'Near-infrared colorimetry of J6 Himalia and S9 Phoebe: a summary of 0.3 to 2.2 micron reflectances', J. Degewij, D.P. Cruikshank and W.K. Hartmann. *Icarus*, vol. 44, p. 541, 1980.

118 See *Planetary landscapes*, R. Greeley. Allen & Unwin, p. 243, 1987; Greeley attributes the Phoebe hypothesis to Steve Soter of Cornell University who presented it at a conference in 1974.

119 'Small bodies and their origins', W.K. Hartmann. In *The new Solar System*, J.K. Beatty and A. Chaikin (Eds.). Cambridge University Press (Third Edition), p. 251, 1990.

120 'The composition and origin of the Iapetus dark material', J.F. Bell, D.P. Cruikshank and M.J. Gaffey. *Icarus*, vol. 61, p. 192, 1985.

121 'Physics and chemistry of comets', D. Boice and W. Huebner. *Encyclopedia of the Solar System*, P.R. Weissman, L.-A. McFadden and T.V. Johnson (Eds.). Academic Press, 1999.

122 'The dark side of Iapetus and the atmosphere and surface of Titan', T.C. Owen, D.P. Cruikshank, C. Dalle Ore, T.R. Roush, R. Meier, T.R. Geballe, C.A. Griffiths, C. de Bergh, N. Biver, A. Marten, B.A. Smith, R.J. Terrile, H.E. Matthews and Y. Yung. European Geophysics Society XXV General Assembly, Nice, France, April 2000.

123 'Decoding the Domino: the dark side of Iapetus', T.C. Owen. *Icarus*, vol. 149, p. 160, 2001.

124 'A new look at the Saturn system: the Voyager 2 images', B.A. Smith, L.A. Soderblom, R. Batson, P. Bridges, J. Inge, H. Masursky, E.M. Shoemaker, R. Beebe, J. Boyce, G. Briggs, A. Bunker, S.A. Collins, C.J. Hansen, T.V. Johnson, J.L. Mitchell, R.J. Terrile,

A.F. Cook, J.N. Cuzzi, J.B. Pollack, G.E. Danielson, A.P. Ingersoll, M.E. Davies and G.E. Hunt. *Science*, vol. 215, p. 504, 1982.

125 'How big is Iapetus', J. Veverka, J. Burt, J.E. Elliot and J. Goguen. *Icarus*, vol. 33, p. 301, 1978.

126 'Saturn's icy satellites: thermal and structural models', K. Ellsworth and G. Schubert. *Icarus*, vol. 54, p. 490, 1983.

127 'Iapetus size, topography and dark-side surface structures', T. Denk, K.-D. Matz, T. Roatsch, G. Neukum and R. Jaumann. European Geophysics Society XXV General Assembly, Nice, France, April 2000.

128 'Crater numbers and geological histories of Iapetus, Enceladus, Tethys and Hyperion', J.B. Plescia and J.M. Boyce. *Nature*, vol. 301, p. 666, 1983.

129 'Phoebe: Voyager 2 observations', P. Thomas. J. Veverka, D. Morrison, M. Davies and T.V. Johnson. *J. Geophys. Res.*, vol. 88, p. 8736, 1983.

130 'Near-infrared colorimetry of J6 Himalia and S9 Phoebe: a summary of 0.3 to 2.2 micron reflectances', J. Degewij, D.P. Cruikshank and W.K. Hartmann. *Icarus*, vol. 44, p. 541, 1980.

131 *Satellites of the Solar System*, W. Sandner. The Scientific Book Club, p. 95, 1965.

132 'Discovery of 12 satellites of Saturn exhibiting orbital clustering', B. Gladman, J.J. Kavelaars, M. Holman, P.D. Nicholson, J.A. Burns, C.A. Hergenrother, J.-M. Petit, B.G. Marsden, R. Jacobson, W. Gray and T. Grav. *Nature*, vol. 412, p. 6843, 2001.

Chapter 5: Cassini–Huygens

1 'Cassini: Saturn Orbiter and Titan Probe', ESA/NASA Assessment Study, ESA REF SCI(85)1, 1985.

2 *The Space Shuttle – roles, missions and accomplishments*, D.M. Harland. Wiley–Praxis, 1998.

3 'Cassini: Report on the Phase 'A' study', ESA REF SCI(88)5, 1988.

4 'The Visible and Infrared Mapping Spectrometer for the Cassini mission', E. Miller *et al. Society of Photo-optical Instrumentation Engineers* (SPIE), vol. 2803, p. 206, 1996.

5 'Surface temperature retrieval on Titan: from Voyager IRIS to Cassini CIRS', R. Courtin, and S.J. Kim. International Symposium 'The Jovian System after Galileo. The Saturnian System before Cassini/Huygens', Nantes, France, May 1998.

6 'Optical design of the Ultraviolet Imaging Spectrograph for the Cassini mission to Saturn', W.E. McClintock, G.M. Lawrence, R.A. Kohnert and L.W. Esposito. *Optical Engineering*, vol. 32, p. 3038, 1993.

7 'The HDAC – A Lyman-alpha Photometer for Atomic D/H Measurements', H. Lauche, M. Ludwig, G. Lawrence, J. Maki, U. Keller and L.W. Esposito. International Union of Geodesy and Geophysics, 1995.

8 'Cassini UVIS observations of Saturn's rings', L.W. Esposito, J.E. Colwell and W.E. McClintock. *Planetary and Space Sciences*, vol. 46, p. 1221, 1998.

9 On 1 May 2001, Charles Elachi took over as JPL's director, following E.C. Stone's retirement.

10 'Predicted responses of the Cassini Plasma Spectrometer (CAPS) and Ion Neutral Mass Spectrometer (INMS) during fly-bys of Titan', D.T. Young, J.H. Waite, M. Blanc, T.E. Cravens, R. Goldstein and S. Maurice. International Symposium 'The Jovian System after Galileo. The Saturnian System before Cassini/Huygens', Nantes, France, May 1998.

11 'Cassini Plasma Spectrometer investigation', D.T. Young, B.L. Barraclough, J.-J. Berthelier, M. Blanc, J.L. Burch, A.J. Coates, R. Goldstein, M. Grande, T.W. Hill, J.M. Illiano, M.A. Johnson, R.E. Johnson, R.A. Baragiola, V. Kelha, D. Linder, D.J. McComas, B.T. Narheim, J.E. Nordholt, A. Preece, E.C. Sittler, K.R. Svenes, S. Szalai, K. Szego, P. Tanskanen and K. Viherkanto. Proceedings from 'Cassini/Huygens: A Mission to the Saturnian Systems', *SPIE*, vol. 2803, p. 118, 1996.

12 'Cassini Plasma Spectrometer Investigation', D.T. Young, B.L. Barraclough, J.-J. Berthelier, M. Blanc, J.L. Burch, A.J. Coates, R. Goldstein, M. Grande, T.W. Hill, J.M. Illiano, M.A. Johnson, R.E. Johnson, R.A. Baragiola, V. Kelha, D. Linder, D.J. McComas, B.T. Narheim, J.E. Nordholt, A. Preece, E.C. Sittler, K.R. Svenes, S. Szalai, K. Szego, P. Tanskanen and K. Viherkanto. In *Measurement techniques for space plasmas: particles*, R.E. Pfaff, J.E. Borovsky and D.T. Young (Eds.). AGU Geophysical Monograph no. 102, p. 237, 1998.

13 'The Cassini Ion Mass Spectrometer', D.J. McComas, J.E. Nordholt, J.-J. Berthelier, J.M. Illiano and D.T. Young. In *Measurement techniques for space plasmas: particles*, R.E. Pfaff, J.E. Borovsky and D.T. Young (Eds.). AGU Geophysical Monograph no. 102, 1998.

14 'The Cassini Ion Mass Spectrometer: performance metrics and techniques', J.E. Nordholt, D.M. Burr, H.O. Funsten, D.J. McComas, D.M. Potter, K.P. McCabe, J.-J. Berthelier, J.M. Illiano, D.T. Young and R. Goldstein. In *Measurement techniques for space plasmas: particles*, R.E. Pfaff, J.E. Borovsky and D.T. Young (Eds.). AGU Geophysical Monograph no. 102, 1998.

15 'Cassini mission and the CAPS/IBS instrument', J.H. Vilppola. Licentiate Thesis, University of Oulu, Finland, November 1998.

16 'Simulations of the response function of a Plasma Ion Beam Spectrometer for the Cassini mission to Saturn', J.H. Vilppola, P.J. Tanskanen, H. Huomo and B.L. Barraclough. *Rev. Sci. Instrum.*, vol. 67, 1996.

17 'The Electron Spectrometer for the Cassini spacecraft', A.J. Coates, C. Alsop, A.J. Coker, D.R. Linder, A.D. Johnstone, R.D. Woodliffe, M. Grande, A. Preece, S. Burge, D.S. Hall, B. Narheim, K. Svenes, D.T. Young, J.R. Sharber and J.R. Scherrer, *J. Brit. Interplanetary. Soc.*, vol. 45, p. 387, 1992.

18 'The Cassini CAPS Electron Spectrometer', D.R. Linder, A.J. Coates, R.D. Woodliffe, C. Alsop, A.D. Johnstone, M. Grande, A. Preece, B. Narheim, K. Svenes, and D.T. Young. In *Measurement techniques for space plasmas: particles*, R.E. Pfaff, J.E. Borovsky and D.T. Young (Eds.). AGU Geophysical Monograph no. 102, p.257, 1998.

19 'Predicted responses of the Cassini Plasma Spectrometer (CAPS) and Ion Neutral Mass Spectrometer (INMS) during fly-bys of Titan', D.T. Young, J.H. Waite, M. Blanc, T.E. Cravens, R. Goldstein, and S. Maurice. International Symposium 'The Jovian system after Galileo. The Saturnian system before Cassini/Huygens', Nantes, France, May 1998.

20 'Experimental and theoretical investigations on the Cassini RPWS antennas', H.O. Rucker, W. Macher and S. Albrecht. In *Planetary radio emissions IV*, H.O. Rucker, S.J. Bauer and A. Lecacheux (Eds.). Osterreichischen Akademie der Wissenschaften, Wien, p. 327, 1997.

21 'The Langmuir probe part of the RPWS investigation on Cassini, magnetospheres of the outer planets', J.W. Wahlund, R. Bostrom, G. Gustafsson, D.A. Gurnett and W.S. Kurth. Paris, France, 9–14 August, 1999.

22 'Lightning activity on Titan: can Cassini detect it?', H. Lammer, T. Tokano, G. Fischer, W. Stumptner, G.J. Molina-Cuberos, K. Schwingenschuh and H.O. Rucker. *Planet. Space Sci.*, vol. 49, p. 561, 2001.

23 'Detection capability of Cassini for thundercloud generated lightning discharges on Titan', H. Lammer, T. Tokano, G. Fischer, G.J. Molina-Cuberos, W. Stumptner, K. Schwingenschuh and H.O. Rucker. In *Planetary radio emissions V*, H.O. Rucker, M.L. Kaiser and Y. Leblanc (Eds.). Austrian Academy of Sciences Press, Vienna, p. 261, 2001.

24 'Detectability of possible Titan lightning by Cassini/RPWS', H. Lammer, T. Tokano, G. Fischer, W. Stumptner, G.J. Molina-Cuberos, K. Schwingenschuh and H.O. Rucker. European Geophysical Society Meeting, Nice, France, March 2001.

25 'Detection capability of Cassini for thundercloud-generated lightning discharges on Titan', H. Lammer, T. Tokano, G. Fischer, G.J. Molina-Cuberos, W. Stumptner, K. Schwingenschuh and H.O. Rucker. Planetary and Solar Radio Emissions V Workshop, Graz, Austria, April 2001.

26 'INCA: the ion neutral camera for energetic neutral atom imaging of the Saturnian magnetosphere', D.G. Mitchell, A.F. Cheng, S.M. Krimigis, E.P. Keath, S.E. Jaskulek, B.H. Mauk, R.W. McEntire, E.C. Roelof, D.J. Williams, K.C. Hsieh and V.A. Drake. *Optical Engineering*, vol. 32, p. 3096, 1993.

27 'The Ion Neutral Camera for the Cassini mission to Saturn and Titan', D.G. Mitchell *et al.* In *Measurement techniques in space plasmas: fields*, R.E. Pfaff, J.E. Borovsky and D.T. Young (Eds.). AGU Geophysical Monographs, 103, p. 281, 1998.

28 'Operation of the dual magnetometer on Cassini: science performance', M.W. Dunlop, M.K. Dougherty, S. Kellock and D.J. Southwood, *Planet. Space Sci.*, vol. 47, p. 1389, 1999.

29 'The Cassini magnetic field science investigation', M.K. Dougherty *et al.* Intl. Symposium 'The Jovian System after Galileo. The Saturnian System before Cassini/Huygens', Nantes, France, May 1998.

30 'The Cassini magnetic field investigation', M.K. Dougherty, S. Kellock, D.J. Southwood, A. Balogh, M. Barlow, T. Beek, M.W. Dunlop, R. White, E.J. Smith, L. Wigglesworth, M. Fong, R. Marquedant, B.T. Tsurutani, B. Gerlach, G. Musmann, K.-H. Glassmeier, H. Hartje, M. Rahm, I. Richter, C.T. Russell, D. Huddelston, R.C. Snare, G. Erdos, S. Szalai, F.M. Neubauer, S.W.H. Cowley and G.L. Siscoe. *Space Sci. Rev.*, in review, 2001.

31 'A simulation of chemical trails produced by Huygens' entry into Titan's atmosphere', D. Luz and F. Hourdin. European Geophysics Society XXV General Assembly, Nice, France, April 2000.

32 'The Huygens Probe: science, payload and mission overview', J.-P. Lebreton and D.L. Matson. ESA SP-1177, August 1997.

33 'Titan exploration by the Huygens Probe', J.-P. Lebreton. International Symposium 'The Jovian System after Galileo. The Saturnian System before Cassini–Huygens', Nantes, France, May 1998.

34 'ASI to Jupiter: results. HASI to Titan: expected results', A. Seiff and the ASI team, M. Fulchignoni and the HASI team. International Symposium 'The Jovian System after Galileo. The Saturnian System before Cassini/Huygens', Nantes, France, May 1998.

35 'Detection of daily clouds on Titan', C.A. Griffith, J.L. Hall and T.R. Geballe. *Science*, vol. 290, p. 509, 2000.

36 'Search for lightning on Titan', H. Lammer, G.J. Molina-Cuberos, T. Tokano, W. Stumptner, W. Macher, H.O. Ruckner and K. Schwingenschuh. European Geophysics Society XXV General Assembly, Nice, France, April 2000.

37 'Detection of daily clouds on Titan', C.A. Griffith, J.L. Hall and T.R. Geballe. *Science*, vol. 290, p. 509, 2000.

38 'The life, death and afterlife of a raindrop on Titan', R.D. Lorenz. *Planetary & Space Science*, vol. 41, p. 647, 1993.

39 'Cassini maneuver experience: launch and early cruise', T.D. Goodson, D.L. Gray, Y. Hahn and F. Peralta. AIAA Guidance, Navigation and Control Conference, Boston, Massachusetts, 10-12 August 1998.

40 'Cassini orbit determination from launch to the first Venus fly-by', D.C. Roth, M.D. Guman, R. Ionasescu and A.H. Taylor. AIAA/AAS Astrodynamics Specialist Conference, Boston, Massachusetts, 10–12 August 1998.

41 'Constraints on Venus lightning from Cassini RPWS observations and consequences for Saturn and Titan observations', P. Zarka, D.A. Gurnett, G.B. Hospodarsky, M.L. Kaiser, W.S. Kurth and R. Manning. European Geophysics Society XXV General Assembly, Nice, France, April 2000.

42 'Non-Detection at Venus of High-Frequency Radio Signals Characteristic of Terrestrial Lightning', D.A. Gurnett, P. Zarka, R. Manning, W.S. Kurth, G.B. Hospodarsky, T.F. Averkamp, M.L. Kaiser and W.M. Farrell. *Nature*, vol. 409, p. 313, 2001.

43 'An upper limit on radio intensities of Venus lightning from the Cassini Radio and Plasma Wave Instrument', D.A. Gurnett, P. Zarka, W.S. Kurth, R. Manning, G.B. Hospodarsky, M.L. Kaiser and W.M. Farrell. American Geophysical Union Meeting, San Francisco, December, 1998.

44 The possibility of lightning on Venus was first prompted by low-frequency signals detected by a Venera lander in 1978, and a few years later by the Pioneer Venus Orbiter.

45 'Cassini orbit determination from first Venus fly-by to Earth fly-by', M.D. Guman, D.C. Roth, R. Ionasescu, T.D. Goodson, A.H. Taylor and J.B. Jones. AIAA Paper 00-168, presented at the AAS Spaceflight Mechanics Meeting, Clearwater, FL, 23-26 January, 2000.

46 'The first year of dust measurements with the Cassini dust detector', A. Graps *et al.* European Geophysics Society XXV General Assembly, Nice, France, April 2000.

47 'Cassini encounter with Earth: planetary magnetic field measurements', E.J. Smith *et al.* European Geophysics Society XXV General Assembly, Nice, France, April 2000.

48 'Magnetic field measurements from the Cassini Earth swing-by', D.J. Southwood *et al.* European Geophysics Society XXV General Assembly, Nice, France, April 2000.

49 'An overview of the Cassini radio and plasma wave Science results from the Earth swingby', W.S. Kurth, D A. Gurnett, G.B. Hospodarsky, M.L. Kaiser, J.W. Wahlund, P. Canu and P. Zarka. American Geophysical Union Meeting, San Francisco, December 1999.

50 'An overview of the Cassini radio and plasma wave science results from the Earth swingby', W.S. Kurth, D.A. Gurnett, G.B. Hospodarsky, M.L. Kaiser, J.-E. Wahlund, P. Cani and P. Zarka. European Geophysics Society XXV General Assembly, Nice, France, April 2000.

51 'An overview of observations by the Cassini Radio and Plasma Wave investigation at Earth', W.S. Kurth, G.B. Hospodarsky, D.A. Gurnett, M.L. Kaiser, J.-E. Wahlund, A. Roux, P. Canu, P. Zarka and Y. Tokarev. *J. Geophys. Res.*, vol. 106, p. 30,239, 2001.

52 'Observations of two complete substorm cycles during the Cassini Earth swing-by: Cassini magnetometer data in a global context', H. Khan, S.W.H. Cowley, E. Kolesnikova, M. Lester, M.J. Brittnacher, T.J. Hughes, W.J. Hughes, W.S. Kurth, D.J. McComas, L. Newitt, C.J. Owen, G.D. Reeves, H.J. Singer, C.W. Smith, D.J. Southwood and J.F. Watermann. *J. Geophys. Res.*, vol. 106, p. 30,141, 2001.

53 'Cassini electron spectrometer observations during Earth swingby', A.J. Coates *et al.* European Geophysics Society XXV General Assembly, Nice, France, April 2000.

54 'CAPS electron spectrometer measurements in the plasmasphere and ionosphere', K.R. Svenes *et al.* European Geophysics Society XXV General Assembly, Nice, France, April 2000.

55 'Observations during the Earth fly-by', D. Young *et al.* European Geophysics Society XXV General Assembly, Nice, France, April 2000.

56 CAPS, MAG, MIMI, RPWS and UVIS remained on after the fly-by to investigate the Earth's magnetotail.

57 'In-flight performance and first observations at Venus and Earth of Magnetospheric Imaging Instrument (MIMI) on Cassini/Huygens', S.M. Krimigis. European Geophysics Society XXV General Assembly, Nice, France, April 2000.

58 'First results of the energetic particle detector MIMI/LEMMS aboard Cassini during Earth fly-by', N. Krupp, S. Livi, A. Lagg, S.M. Krimigis, D.G Mitchell, T.P. Armstrong, I. Dandouras and D.C. Hamilton. European Geophysics Society XXV General Assembly, Nice, France, April 2000.

59 'ENA images from Cassini MIMI/INCA during the Earth fly-by', E.C. Roelof, D.G. Mitchell and S.M. Krimigis. European Geophysics Society XXV General Assembly, Nice, France, April 2000.

60 'First composition measurements of energetic neutral atoms', A.T.Y. Lui, D.J. Williams, E.C. Roelof, R.W. McEntire and D.C. Mitchell. *Geophys. Res. Lett.*, vol. 23, p. 2641, 1996.

61 'Cassini and Wind stereoscopic observations of Jovian non-thermal radio emissions', M.L. Kaiser, P. Zarka, W.S. Kurth, G.B. Hospodarsky and D.A. Gurnett. European Geophysics Society XXV General Assembly, Nice, France, April 2000.

62 'Cassini and Wind stereoscopic observations of Jovian nonthermal radio emissions: measurement of beam widths', M.L. Kaiser, P. Zarka, W.S. Kurth, G.B. Hospodarsky and D.A. Gurnett. *J. Geophys. Res.*, vol. 105, p. 16,053, 2000.

63 'Simultaneous observations of Jovian radio emissions by Cassini and Wind', M.L. Kaiser, W.S. Kurth, G.B. Hospodarsky and D.A. Gurnett. American Geophysical Union Meeting, Boston, May–June, 1999.

64 'The VIMS/Cassini observations of the Moon', R.H. Brown *et al. Proc. Lunar Planet. Sci. Conf.*, p. 1794, 2000.

65 'Cassini radio detection and ranging (RADAR): Earth and Venus observations', R.D. Lorenz *et al. J. Geophys. Res.*, vol. 106, p. 30271, 2001.

66 'Cassini maneuver experience: finishing Inner Cruise', T.D. Goodson, D.L. Gray, Y. Hahn and F. Peralta. AAS/AIAA Space Flight Mechanics Meeting, Clearwater, Florida, 23–26 January 2000.

67 'Cassini UVIS cruise science results', L. Esposito, C. Barth. J. Colwell, G. Lawrence, W. McClintock, I. Stewart and C. Hansen. European Geophysics Society XXV General Assembly, Nice, France, April 2000.

68 *Jupiter odyssey: the story of NASA's Galileo mission*, D.M. Harland. Springer–Praxis, 2001.

69 'Satellite tour design at Saturn using an Excel spreadsheet', J.C. Smith. AAS/AIAA Astrodynamics Specialist Conference, Paper AAS 95-324, Halifax, Nova Scotia, Canada, 14-17 August 1995.

70 'Incorporating icy satellite fly-bys in the Cassini orbital tour', A. Wolf. AIAA/AAS Space Flight Mechanics Meeting, Paper AAS 98-106, Monterey, CA, 9–11 February 1998.

71 In terms of planning Cassini's orbital tour, a 'targeted' fly-by was one contrived to approach within 1,000 kilometres of a satellite. Encounters at a ranges of around 100,000 kilometres were regarded as 'non-targeted'.

72 'Description of three candidate Cassini satellite tours', J.C. Smith. AAS-98-106 AAS/AIAA Space Flight Mechanics Meeting, Monterey, California, 9-11 February 1998.

73 Some asteroids appear to be clumps of 'pristine' matter left over from the solar nebula,

but others are 'thermally evolved' suggesting that they were once deep in the hot interiors of larger body that were shattered by collisions.

74 'Huygens communications link enquiry board report: findings, recommendations and conclusions', D.C.R. Link *et al.* ESA, December 2000.

75 'Cassini and Galileo imaging of the Jovian atmosphere,' T.V. Johnson. AGU Spring Meeting, 2001.

76 'Motions in the interiors and atmospheres of Jupiter and Saturn', A.P. Ingersoll and D. Pollard. *Icarus*, vol. 52, p. 62, 1982.

77 'Cassini CIRS observations in the Jovian environment', F.M. Flasar *et al.* AGU Spring Meeting, 2001.

78 'Cassini CIRS observations of thermal waves on Jupiter', R.K. Achterberg, B.J. Conrath, F.M. Flasar, A.A. Simon-Miller, C.A. Nixo and B. Bezard. Division of Planetary Sciences Meeting, November 2001.

79 'Cassini, Galileo and ground-based observations of Jupiter's thermal emissions during the joint spacecraft encounter', G. Orton *et al.* AGU Spring Meeting, 2001.

80 'An overview of the results from the Cassini VIMS instrument during the Jupiter fly-by', R.H. Brown *et al.* European Geophysical Society XXVI General Assembly, March 2001.

81 'On the presumed capture origin of Jupiter's outer satellites', T.A. Heppenheimer. *Icarus*, vol. 24, p. 172, 1975.

82 'Near-infrared colorimetry of J6 Himalia and S9 Phoebe: a summary of 0.3 to 2.2 micron reflectances', J. Degewij, D.P. Cruikshank and W.K. Hartmann. *Icarus*, vol. 44, p. 541, 1980.

83 'Himalia – first disk-resolved observations of an outer Jovian satellite', T. Denk, T. Roatsch, J. Oberst, G. Neukum, C.C. Porco, P.C Thomas, P. Helfenstein and S.W. Squyres. 26th General Assembly, Nice, France, AAI 3361, 25–30 March 2001.

84 'Near-infrared spectroscopy of Himalia – an irregular Jovian satellite', R.H. Brown *et al.* *Proc. Lunar Planet. Sci. Conf.*, p, 2001, 2002.

85 'Discovery of 12 satellites of Saturn exhibiting orbital clustering', B. Gladman, J.J. Kavelaars, M. Holman, P.D. Nicholson, J.A. Burns, C.A. Hergenrother, J.-M. Petit, B.G. Marsden, R. Jacobson, W. Gray and T. Grav. *Nature*, vol. 412, p. 6843, 2001.

86 'Jovian dust streams revisited: Cassini dust detector at Jupiter', S. Kempf *et al.* European Geophysical Society XXVII General Assembly Nice, France, A-02407, 21–26 April 2002.

87 'Radio and plasma wave observations from Cassini and Galileo spacecraft during the Cassini fly-by of Jupiter', D. Gurnett *et al.* AGU Spring Meeting, 2001.

88 'High-resolution observations of low-frequency Jovian radio emissions by Cassini', W.S. Kurth, G.B. Hospodarsky, D.A. Gurnett, A. Lecacheux, P. Zarka, M.D. Desch, M.L. Kaiser and W.M. Farrell. In *Planetary Radio Emissions V*, H.O. Rucker, M.L. Kaiser and Y. Leblanc (Eds.). Austrian Academy of Sciences Press, Vienna, p. 15, 2001.

89 'Control of Jupiter's radio emission and aurorae by the solar wind', D.A. Gurnett, W.S. Kurth, G.B. Hospodarsky, A.M. Persoon, P. Zarka, A. Lecacheux, S.J. Bolton, M.D. Desch, W.M. Farrell, M.L. Kaiser, H.-P. Ladreiter, H.O. Rucker, P. Galopeau, P. Louarn, D.T. Young, W.R. Pryorand and M.K. Dougherty. *Nature*, vol. 415, p. 985, 2002.

90 'The dusk flank of Jupiter's magnetosphere', W.S. Kurth, D.A. Gurnett, G.B. Hospodarsky, W.M. Farrell, A. Roux, M.K. Dougherty, S.P. Joy, M.G. Kivelson, R.J. Walker, F.J. Crary and C.J. Alexander. *Nature*, vol. 415, p. 991, 2002.

91 'Plasma observations during the Cassini fly-by of Jupiter', D. Young *et al.* AGU Spring Meeting, 2001.

92 'Cassini magnetometer measurements in the Jovian environment', M.K. Dougherty *et al.* AGU Spring Meeting, 2001.

93 'Cassini UVIS observations of Jupiter's auroral variability', W.R. Pryor, A.I.F. Stewart, L.W. Esposito, D.E. Shemansky, J.M. Ajello, R.A. West, A.J. Jouchoux, C.J. Hansen, W.E. McClintock, J.E. Colwell, B.T. Tsurutani, N. Krupp, F.J. Crary, D.T. Young, J.T. Clarke, J.H. Waite, D. Grodent, W.S. Kurth, D.A. Gurnett and M.K. Dougherty. *Nature*, submitted, 2002.

94 'Cassini UVIS observations of Jupiter's auroral variability', R. Pryor, A.I.F. Stewart, L.W. Esposito, A.J. Jouchoux, W.E. McClintock, J.E. Colwell, D.E. Shemansky, J.M. Ajello, R.A. West, C.J. Hansen, B.T. Tsurutani, N. Krupp, F. Crary, D. Young, J.H. Waite, D. Grodent, J.T. Clarke, W.S. Kurth, D.A. Gurnett and M.K. Dougherty. Presented at the Division of Planetary Sciences Meeting, November 2001.

95 'HST/Cassini campaign of Jupiter auroral observations', J.T. Clarke. AGU Spring Meeting, 2001.

96 'Cassini VIMS observations of the Galilean satellites', T.B. McCord *et al.* European Geophysical Society XXVII General Assembly Nice, France, A-00859, 21–26 April 2002.

97 'Cassini and Galileo satellite imaging results; giant north polar plume on Io', A.S. McEwen, P. Geissler, M.J.S. Belton, C.C. Porco, R. Pappalardo, T.V. Johnson, S. Squyres, D. Williams and J.W. Head. AGU Spring Meeting, 2001.

98 R.H. Howell. International Jupiter Watch Satellite Newsletter No. 3, 2001.

99 'Mapping the thermal output of Io's hot spots with the Galileo near-infrared mapping spectrometer', W. Smythe, R. Lopes-Gauthier and L. Kamp. European Geophysical Society XXVI General Assembly, March 2001.

100 'Cassini imaging of auroral emissions on the Galilean satellites', P. Geissler, A.S. McEwen and C.C. Porco. AGU Spring Meeting, 2001.

101 'Cassini imaging of visible aurorae on Io and Europa', P. Geissler, A.S McEwen and C.C. Porco. Presented at the Division of Planetary Sciences Meeting, November 2001.

102 'Cassini and Galileo observations of Io and satellite eclipses', A.S. McEwen, P. Geissler, R. Pappalardo, M. Milazzo, C.C. Porco, M.J.S. Belton, T.V. Johnson, S. Squyres, P. Thomas, T. Roatsch, G. Neukum, J. Veverka, E. Turtle, D. Simonelli and P. Schuster. European Geophysical Society XXVI General Assembly, March 2001.

103 'Imaging the Jovian magnetosphere in energetic neutral atoms with the Cassini/Huygens Magnetospheric Imaging Instrument Ion and Neutral Camera', D.G. Mitchell, S.M. Krimigis and B.H. Mauk. AGU Spring Meeting, 2001.

104 'A nebula of gases from Io surrounding Jupiter', S.M. Krimigis *et al. Nature*, vol. 415, p. 994, 2002.

105 'MIMI/CHEMS observations of Jovian pickup and magnetospheric ions during the Cassini fly-by of Jupiter', D.C. Hamilton *et al.* AGU Spring Meeting, 2001.

106 'Cassini RADAR/Radiometer and VLA observations of Jupiter's synchrotron emission', M.A. Jansen *et al.* 5th International Workshop on Planetary and Solar Radio Emissions in Graz, Austria, April 2001.

107 'Ultra-relativistic electrons in Jupiter's radiation belts', S.J. Bolton *et al. Nature*, vol. 415, p. 987, 2001.

108 'Observations of Jovian synchrotron emission', S.J. Bolton *et al.* AGU Spring Meeting, 2001.

109 Several missions in Jovian space are in various stages of planning. In addition to spacecraft which will either orbit or land on Europa, the INSIDE Jupiter mission would fly within 4,000 kilometres of the planet. Although Galileo's probe's reported that the radiation belts are much less intense very close in to Jupiter, the INSIDE Jupiter would be exposed to more radiation than expected in its planned polar orbit, which hooks 'through' the intense radiation belts above the planet's equator.

110 'Observations in Jupiter's vicinity with the Magnetospheric Imaging Instrument (MIMI) during Cassini/Huygens fly-by (October 2000–March 2001)', S.M. Krimigis *et al.* American Geophysical Union meeting in Boston, Massachusetts, May 2001.

111 'Huygens Communications Link Enquiry Board Report: Findings, recommendations and conclusions', D.C.R. Link *et al.* ESA, December 2000.

112 'Direct measurement of winds on Titan', T. Kostiuk. Geophys. Res. Lett. vol. 28, p. 2361, 2001.

113 'Lowering V-infinity by 500 m/s and recovering T18-5', N.J. Strange. JPL IOM 312/00.E-01, 3 January 2001.

114 'Orbit determination results for Huygens probe delivery', D.C. Roth. JPL IOM 312.A-04-01, 30 March 2001.

115 'Manoeuvre simulation results for Huygens probe delivery (orig. plan)', T.D. Goodson. JPL IOM 312.H-01-07, 25 June 2001.

116 'Modifying the Cassini T18-5 tour for a post T10 Huygens probe delivery', Y. Hahn. JPL IOM 312.H-01-01, 12 January 2001.

117 'Feasibility of a post-Iapetus Huygens probe delivery', F. Peralta. JPL IOM 312/01/.D-01, 24 January 2001.

118 'Scenario 2: high-altitude delivery', J.B. Jones. Report to the Huygens Recovery Task Force's 5th Meeting, 17 May 2001.

119 'High-altitude delivery description', J.B. Jones, N.J. Strange, T. Goodson, D. Roth, Y. Hahn and F. Peralta. Technical Note to supplement the Huygens Recovery Task Force's Final Report issued on 27 July 2001.

120 The three improvised Titan fly-bys were designated 'T$_a$', 'T$_b$' and 'T$_c$' to distinguish them from the original numerical sequence. For convenience, it was decided to retain this sequence for the rest of the T18-5 tour, so when Cassini rejoined the planned tour it would do so at 'T$_3$'.

121 'Huygens Recovery Task Force Final Report', K. Clausen *et al.* ESA HUY-RP-12241, issued on 27 July 2001.

122 'Transient clouds in Titan's lower atmosphere', C.A. Griffith, T.C. Owen, G.A. Miller and T.R. Geballe. *Nature*, vol. 395, pp. 575–578, 1998.

123 'Evidence for clouds on Titan from HST WFPC-2', R.D. Lorenz, M.T. Lemmon and P.H. Smith. American Astronomical Society, Division of Planetary Sciences meeting in Padua, Italy, 10–15 October 1999.

124 'Titan's clouds', C.A. Griffith, T.R. Geballe, J.L. Hall and T.C. Owen. European Geophysics Society XXV General Assembly, Nice, France, April 2000.

125 'Detection of daily clouds on Titan', C.A. Griffith, J.L. Hall and T.R. Geballe. *Science*, vol. 290, pp. 509–512, 2000.

126 'Direct detection of variable tropospheric clouds near Titan's south pole', M.E. Brown, A.H. Bouchez and C.A. Griffith. *Nature*, vol. 420, pp. 795–797, December 2002.

127 'Titan's clouds from Gemini and Keck adaptive optics imaging', H.G. Roe, I. de Pater, B.A. Macintosh and C.P. McKay. *Ap. J.*, vol. 581, pp. 1399–1406, 20 December 2002.

128 'Seasonal changes in Titan's haze 1992-2002 from HST observations', R.D. Lorenz, P.H. Smith and M.T. Lemmon. *Geophys. Res. Lett.*, vol. 31, #10 L10702, 2004.

129 'Seasonal evolution of Titan's dark polar hood: midsummer disappearance observed by the Hubble Space Telescope', R.D. Lorenz, M.T. Lemmon and P.H. Smith. *Mon. Not. R. Astron. Soc.*, vol. 369, pp. 1683–1687, 2006.

130 'A wind origin for Titan's haze structure', P. Rannou, F. Hourdin and C.P. McKay. *Nature*, vol. 418, pp. 853–856, August 2002.

131 'Discovery of temperate latitude clouds on Titan,' H.G. Roe, A.H. Bouchez, C.A. Trujillo, E.L. Schaller and M.E. Brown. *Ap. J.*, vol. 618, L49-52, 1 January 2005.

132 'The tides in the sea of Titan', C.E. Sagan and S.F. Dermott. *Nature*, vol. 300, pp. 731–733, 1982.

133 'Radar reflectivity of Titan', D.O. Muhleman, A.W. Grossman, B.J. Butler and M.A. Slade. *Science*, vol. 248, pp. 975–980, May 1990.

134 'Titan: 1–5 m photometry and spectrophotometry, and a search for variability', K.S. Noll and R.F. Knacke. *Icarus*, vol. 101, pp. 272–281, 1993.

135 M.T. Lemmon, E. Karkoschka and M. Tomasko. IAU Circ. No. 5624, 28 September 1992.

136 'Titan's rotation: Surface feature observed', M.T. Lemmon, E. Karkoschka and M. Tomasko. *Icarus*, vol. 103, pp. 329–332, 1993.

137 'Evidence for surface heterogeneity on Titan', C.A. Griffith. *Nature*, vol. 364, pp. 511–514, 1993.

138 'Titan's surface revealed by HST imaging', P.H. Smith, M.T. Lemmon, R.D, Lorenz, L.A. Sromovsky, J.J. Caldwell and M.D. Allison. *Icarus*, vol. 119, pp. 336–349, 1996.

139 'Evidence for the exposure of water-ice on Titan's surface', C.A. Griffith, T.C. Owen, T.R. Geballe, J. Rayner and P. Rannou. *Science*, vol. 300, pp. 628–630, April 2003.

140 'Radar evidence for liquid surfaces on Titan', D.B. Campbell, G.J. Black, L.M. Carter and S.J. Ostro. *Science*, vol. 302, pp. 431–434, October 2003.

141 'The glitter of distant seas', R.D. Lorenz. *Science*, vol. 302, pp. 403-404, October 2003.

142 'No oceans on Titan from the absence of near-infrared specular reflection', R.A. West, M.E. Brown, S.V. Salinas, A.H. Bouchez and H.G. Roe. *Nature*, vol. 436, pp. 670–672, 2005.

143 MPEC 203-G39 (8 April 2003)

144 'Solar wind dynamic pressure and electric field as the main factors controlling Saturn's aurorae', F.J. Crary *et al. Nature*, vol. 433, pp. 720–722, February 2005.

145 'Morphological differences between Saturn's ultraviolet aurorae and those of Earth and Jupiter', J.T. Clarke *et al. Nature*, vol. 433, pp. 717–719, February 2005.

146 'An Earth-like correspondence between Saturn's auroral features and radio emissions', W.S. Kurth *et al. Nature*, vol. 433, pp. 722–725, February 2005.

147 'Ulysses discovers dust emissions from Jupiter and probable interstellar grains', E. Grun *et al. Nature*, vol. 362, p. 428, 1993.

148 'Io as a source of the Jovian dust streams', A. Graps *et al. Nature*, vol. 405, p. 48, 2000.

149 'Interactions of Saturnian dust streams with the solar wind', S. Kempf *et al. Workshop on Dust in Planetary Systems*, Kauai, Hawaii, p. 4052, 26–30 September 2005.

150 'High-velocity streams of dust originating from Saturn', S. Kempf *et al. Nature*, vol. 433, pp. 289–291, 2005.

151 'Composition of Saturnian stream particles', S. Kempf *et al. Science*, vol. 307, pp. 1274–1276, 2005.

152 'Saturn's temperature field from high-resolution middle-infrared imaging', G.S. Orton and P. Yanamandra-Fisher. *Science*, vol. 307, pp. 696-698, February 2005.

153 'Saturn's equatorial winds decreasing', A. Sanchez-Lavega, S. Perez-Hoyos, J.F. Rojas, R. Hueso and R. French. *Nature*, vol. 423, pp. 623-625, June 2003.

154 'Temperatures, winds, and composition in the Saturnian system', F.M. Flasar *et al. Science*, vol. 307, pp. 1247–1251, 2005.

155 C.C. Porco *et al.* IAU Circ. No. 8389, 16 August 2004.

156 J.M. Bauer, D.P. Simonelli and B.J. Buratti. IAU Circ. No. 8279, 30 January 2004.

157 'The Saturn system as observed by Cassini's Ultraviolet Imaging Spectrograph', L.W. Esposito *et al. Proc. Lunar Planet. Sci. Conf.*, 1586, 14–18 March 2005.

158 In October 2004 the International Astronomical Union announced its revised nomenclature for Phoebe based on the legend of *Jason and the Argonauts*. The names of the 24 largest craters were announced on 24 February 2005.

159 'Compositional maps of Saturn's moon Phoebe from imaging spectroscopy', R.N. Clark *et al. Nature*, vol. 435, pp. 66-69, 2005.

160 'Temperatures, winds, and composition in the Saturnian system', F.M. Flasar *et al. Science*, vol. 307, pp. 1247-1251, 2005.

161 'Is Phoebe a Kuiper Belt refugee?' editorial commentary in the 5 May 2005 issue of *Nature*.

162 'Saturn's moon Phoebe as a captured body from the outer Solar System', T.V. Johnson and J.I. Lunine. *Nature*, vol. 435, pp. 69-71, 2005.

163 'Titan's atmosphere: an early Cassini ISS perspective', R. West *et al*. American Astronomical Society, Division of Planetary Sciences meeting in Louisville, Kentucky, 8–12 November 2004.

164 C.C. Porco *et al*. IAU Circ. No. 8401, 9 September 2004.

165 C.C. Porco *et al*. IAU Circ. No. 8432, 8 November 2004.

166 Allowing for Pacific Daylight Time being 7 hours behind Greenwich and a one-way light-time of 1 hour 24 minutes, a spacecraft event occurring at 23:47 GMT became known at 18:11 at JPL.

167 'Cassini magnetometer observations during Saturn Orbit Insertion', M.K. Dougherty *et al. Science*, vol. 307, pp. 1266-1270, 2005.

168 'Cassini's discoveries in the D and G rings', M.M. Hedman *et al*. American Astronomical Society, Division of Planetary Sciences meeting in Cambridge, England, 4–9 September 2005.

169 'Dynamics of Saturn's magnetosphere from MIMI during Saturn Orbit Insertion', S.M. Krimigis *et al. Science*, vol. 307, pp. 1270-1273, 2005.

170 'Ultraviolet imaging spectroscopy shows an active Saturnian system', L.W. Esposito *et al. Science*, vol. 307, pp. 1251-1255, 2005.

171 'Cassini VIMS at Saturn: The first 6 months', R.H. Brown *et al. Proc. Lunar Planet. Sci. Conf.*, 1166, 14–18 March 2005.

172 'Temperatures, winds, and composition in the Saturnian system', F.M. Flasar *et al. Science*, vol. 307, pp. 1247-1251, 2005.

173 C.C. Porco *et al*. IAU Circ. No. 8401, 9 September 2004.

174 'Saturn's rings: pre-Cassini status and mission goals', J.N. Cuzzi *et al. Space Sci. Rev.*, vol. 104, pp. 209-251, 2002.

175 'HST observations of spokes in Saturn's B ring', C.A. McGhee *et al. Icarus*, vol. 173, pp. 508-521, 2005.

176 'Where are the spokes in Saturn's B ring?' C.J. Mitchell and M. Mihaly Horanyi. American Astronomical Society, Division of Planetary Sciences meeting in Cambridge, England, 4–9 September 2005.

177 'Occultation of Xi Ceti by Saturn's rings', J.E. Colwell and L.W. Esposito. American Astronomical Society, Division of Planetary Sciences meeting in Louisville, Kentucky, 8–12 November 2004.

178 'Iapetus: Dark-side origin', T. Denk and G. Neukum. *Proc. Lunar Planet. Sci. Conf.*, 1660, 13–17 March 2000.

179 'First Cassini ISS observations of Iapetus', T. Denk *et al*. American Astronomical Society, Division of Planetary Sciences meeting in Louisville, Kentucky, 8–12 November 2004.

180 C.C. Porco *et al*. IAU Circ. No. 8432, 8 November 2004.

181 'Cassini imaging science: initial results on Saturn's rings and small satellites', C.C. Porco *et al. Science*, vol. 307, pp. 1226–1236, 2005.

182 'S/2004 S 5: A new co-orbital companion for Dione', C.D. Murray *et al. Icarus*, vol. 179, p. 222, 2005. Bull. American Astron. Soc., vol. 37, p.702, 2005.

183 'Astrometry and dynamics of Polydeuces, a new co-orbital satellite of Dione', N.J. Cooper *et al.* American Astronomical Society, Division of Planetary Sciences meeting in Cambridge, England, 4–9 September 2005.

184 'The orbits of Saturn's small satellites derived from combined historic and Cassini imaging observations', J.N. Spitale *et al. Astron. J.*, vol. 132, p. 692, 2006.

185 'Ion Neutral Mass Spectrometer results from the first fly-by of Titan', J.H. Waite *et al. Science*, vol. 308, p. 982, May 2005.

186 'Titan's magnetic field signature during the first Cassini encounter', H. Backes *et al. Science*, vol. 308, pp. 992–995, 2005.

187 'CAPS, no evidence of a nitrogen torus at Titan's orbit', D.B. Reisenfeld *et al. Proc. Lunar Planet. Sci. Conf.*, 1887, 14–18 March 2005.

188 'Aging of Titan's aerosols', V.I. Dimitrov and A. Bar-Nun. *Icarus*, vol. 156, pp. 530–538, 2002.

189 'Hardening of Titan's aerosols by their charging' V.I. Dimitrov and A. Bar-Nun. *Icarus*, vol. 166, pp. 440–443, 2003.

190 'Tidal winds on Titan caused by Saturn', T. Tokano and F. Neubauer. *Icarus*, vol. 158, pp. 499–515, 2002.

191 'Titan's winds blow north at noon', R.D. Lorenz, *Astronomy.com*, August 2002.

192 'Release of volatiles from a possible cryovolcano from near-infrared imaging of Titan', C. Sotin *et al. Nature*, vol. 435, pp. 786–789, June 2005.

193 *Lifting Titan's veil – Exploring the giant moon of Saturn*, Ralph Lorenz and Jacqueline Mitton. Cambridge University Press p. 150, 2002.

194 'Shape and thermal modeling of the possible cryovolcanic dome Ganesa Macula on Titan', C.D. Neish, R.D. Lorenz and D.P. O'Brien. American Astronomical Society, Division of Planetary Sciences meeting in Cambridge, England, 4–9 September 2005.

195 'Resurfacing of Titan by ammonia-water cryomagma', G. Mitri *et al.* Eos Trans. AGU, 86 (52), Fall Meeting, San Francisco, 5–9 December 2005.

196 'Radar surprises from Titan', an interview with R.D. Lorenz in the 17 November 2004 issue of *Astrobiology Magazine*.

197 'Impact craters on Titan? Cassini Radar view', C.A. Wood *et al. Proc. Lunar Planet. Sci. Conf.*, 1117, 14–18 March 2005.

198 'Flows on the surface of Titan as revealed by the Cassini Radar', R.M.C. Lopes *et al.* American Astronomical Society, Division of Planetary Sciences meeting in Cambridge, England, 4–9 September 2005.

199 'Saturn's F-ring and its retinue', C.D. Murray *et al.* American Astronomical Society, Division of Planetary Sciences meeting in Cambridge, England, 4–9 September 2005.

200 'How Prometheus creates structures in Saturn's F ring', C.C. Porco *et al. Nature*, vol. 437, pp. 1326–1329, October 2005.

201 'Imaging of Titan from the Cassini spacecraft', C.C. Porco *et al. Nature*, vol. 434, pp. 159–168, 2005.

202 'Imaging of Titan from the Cassini spacecraft', C.C. Porco *et al. Nature*, vol. 434, pp. 159–168, March 2005.

203 The radar was scheduled for only 25 of the 44 fly-bys of the primary mission, during which time it was hoped to map 25 per cent of the surface at high resolution. Of course, extending this coverage may well become the objective of an extended mission.

204 'The Cassini UVIS stellar probe of the Titan atmosphere', D.E. Shemansky *et al. Science*, vol. 308, pp. 978–982, May 2005.

205 'Geology, ages and topography of Saturn's satellite Dione observed by the Cassini ISS camera', R. Wagner *et al. Proc. Lunar Planet. Sci. Conf.*, 1805, 13–17 March 2006.

206 'Cassini imaging science: Initial results on Phoebe and Iapetus', C.C. Porco *et al. Science*, vol. 307, pp. 1237–1242, 2005.

207 'First imaging results from the Iapetus B/C fly-by of the Cassini spacecraft', T. Denk *et al. Proc. Lunar Planet. Sci. Conf.*, 2268, 14–18 March 2005

208 'Iapetus: Size, topography, surface structures, craters', T. Denk, K.-D. Matz, T. Roatsch, U. Wolf, R.J. Wagner, G. Neukum and R. Jaumann. *Proc. Lunar Planet. Sci. Conf.*, 1596, 13–17 March 2000.

209 'Iapetus: shape, craters, dark side', T. Denk, G. Neukum, T. Roatsch, K.-D. Matz, U. Wolf, R.J. Wagner and R. Jaumann. American Astronomical Society, Division of Planetary Sciences meeting in Pasadena, California, 23–27 October 2000.

210 'Shapes of the Saturnian icy satellites', P.C. Thomas *et al. Proc. Lunar Planet. Sci. Conf.*, 1639, 13–17 March 2006.

211 'Radar detection of Iapetus', G.J. Black, D.B. Campbell, L.M. Carter and S.J. Ostro. *Science*, vol. 304, p. 553, 2004.

212 'Cassini Visual and Infrared Mapping Spectrometer observations of Iapetus: Detection of CO2', B.J. Buratti *et al. Ap. J.*, vol. 622, L149-L152, 2005.

213 'Spectrophotometry and organic matter on Iapetus. 1: Composition model', P.D. Willson and C. Sagan. *J. Geophys. Res.*, vol. 100, pp. 7531–7537, April 1995.

214 'CIRS/Cassini first results from Saturn', F.M. Flasar. Scientific colloquium, Goddard Space Flight Center, 11 February 2005.

215 'Cassini CIRS observations of Iapetus's thermal emission', J.R. Spencer *et al. Proc. Lunar Planet. Sci. Conf.*, 2305, 14–18 March 2005.

216 'Iapetus and Phoebe as measured by the Cassini UVIS', A.R. Hendrix and C.J. Hansen. *Proc. Lunar Planet. Sci. Conf.*, 2272, 14–18 March 2005.

217 'Surface colours of Iapetus and Hyperion as derived from Cassini ISS data, and implications for the global albedo dichotomy origin', T. Denk *et al.* American Astronomical Society, Division of Planetary Sciences meeting in Cambridge, England, 4–9 September 2005.

218 For simplicity the timing of the events in the Huygens descent will be specified in terms of GMT, Earth Received Time.

219 'An overview of the descent and landing of the Huygens probe on Titan', J.-P. Lebreton *et al. Nature*, vol. 438, pp. 758–764, December 2005.

220 On the original design, a specifically aligned medium-gain antenna would have picked up the probe's transmission, and the high-gain would have been maintained pointing at Earth in order to provide real-time forwarding of the probe's transmission, with the recording serving as a backup.

221 'Imaging of Titan from the Cassini spacecraft', C.C. Porco *et al. Nature*, vol. 434, pp. 159–168, March 2005.

222 'The vertical wind profile of winds on Titan', M.K. Bird *et al. Nature*, vol. 438, pp. 800–802, December 2005.

223 'In situ measurements of the physical characteristics of Titan's environment', M. Fulchignoni *et al. Nature*, vol. 438, pp. 785–791, December 2005.

224 The microphone is known to have been working because it picked up airflow acoustics and the sounds of the probe's equipment, such as pumps.

225 'The Cassini-Huygens Probe Gas Chromatograph Mass Spectrometer (GCMS) experiment: First results', H. Niemann *et al. Proc. Lunar Planet. Sci. Conf.*, 1663, 14–18 March 2005.

226 'The abundances of constituents of Titan's atmosphere from the GCMS instrument on the Huygens probe', H.B. Niemann *et al. Nature*, vol. 438, pp. 779–784, December 2005.

227 'Complex organic matter in Titan's atmospheric aerosols from in situ pyrolysis and analysis,' G. Israel *et al. Nature*, vol. 438, pp. 796–799, December 2005.

228 'Rain, winds and haze during the Huygens probe's descent to Titan's surface', M.G. Tomasko *et al. Nature*, vol. 438, pp. 765–778, December 2005.

229 'A soft solid surface on Titan as revealed by the Huygens Surface Science Package', J.C. Zarnecki *et al. Nature*, vol. 438, pp. 792–795, December 2005.

230 'Thermal interactions of the Huygens probe with the Titan environment: Constraint on near-surface wind', R.D. Lorenz. *Icarus*, vol. 182, pp. 559–566, June 2006.

231 'Bistatic observations of Titan's surface with Huygens probe radio signal', M. Perez-Ayucar *et al. J. Geophys. Res.*, vol. 111, No. E7, E07001, 25 July 2006.

232 In 'The characteristics of Titan's atmospheric physical properties by the Huygens Atmospheric Structure Instrument (HASI)', M. Fulchignoni *et al, Space Sci. Rev.*, vol. 104, pp. 395–431, March 2002, the intended landing site was given as 167.36°E (i.e. 192.64°W), 10.12°N, but this was before the revision designed to overcome the relay problem.

233 'Overview of the coordinated ground-based observations of Titan during the Huygens mission', O. Witasse *et al. J. Geophys. Res.*, 27 July 2006, reported that the VLBI data put the landing site at 192.20°W, 10.20°S.

234 'Crateriform structures on Titan', C.A. Wood *et al. Proc. Lunar Planet. Sci. Conf.*, 1659, 13–17 March 2006.

235 'VIMS evidence for palimpsests on Titan suggests limits on widespread precipitation', R.M. Nelson *et al.* American Astronomical Society, Division of Planetary Sciences meeting in Louisville, Kentucky, 8–12 November 2004.

236 'Mapping of Titan: first results from the Cassini radar', E.R. Stofan *et al. Proc. Lunar Planet. Sci. Conf.*, 1714, 14–18 March 2005.

237 'Cryovolcanic features on Titan's surface as revealed by the Cassini radar,' R.M. Lopez *et al. Proc. Lunar Planet. Sci. Conf.*, 1885, 14–18 March 2005.

238 'Channels and fan-like features on Titan surface imaged by the Cassini radar', F. Paganelli. *Proc. Lunar Planet. Sci. Conf.*, 2150, 14–18 March 2005.

239 'Impact craters on Titan? Cassini radar view', C.A. Wood. *Proc. Lunar Planet. Sci. Conf.*, 1117, 14–18 March 2005.

240 'Cassini radar views the surface of Titan', C. Elachi *et al. Science*, vol. 308, pp. 970–974, 2005.

241 'Titan's atmospheric temperatures, winds and composition', F.M. Flasar *et al. Science*, vol. 308, pp. 975–978, 2005.

242 'The evolution of Titan's mid-latitude clouds', C.A. Griffith *et al. Science*, vol. 310, p. 474–477, 2005.

243 P. Rannou *et al.* COSPAR, p .1272, 18–25 July 2004.

244 'Geographical control of Titan's mid-latitude clouds', H.G. Roe *et al. Science*, vol. 310, p. 477–479, 2005.

245 'The deep winds of Saturn: First measurements of the zonal wind field near the two-bar level', K.H. Baines *et al. Bull. Amer. Astron. Soc.*, vol. 37, p. 658, 2005.

246 'Physical characteristics and possible accretionary origins for Saturn's small satellites', C.C Porco *et al. Proc. Lunar Planet. Sci. Conf.*, 2289, 13–17 March 2006.

247 J.W. Barnes *et al* in the 7 October 2005 issue of *Science*.

248 'Titan's enigmatic 5-micron-bright terrain', J.W. Barnes *et al*. *Proc. Lunar Planet. Sci. Conf.*, 2319, 13–17 March 2006.

249 C.C. Porco *et al*. IAU Circ. No. 8524, 6 May 2005.

250 'Cassini radio occultations of Saturn's rings: Scattered signal and particle sizes', F. Thomson *et al*. American Astronomical Society, Division of Planetary Sciences meeting in Cambridge, England, 4–9 September 2005.

251 'Cassini RSS occultation observations of density waves in Saturn's rings', C.A. McGhee *et al*. American Astronomical Society, Division of Planetary Sciences meeting in Cambridge, England, 4–9 September 2005.

252 'Cassini-VIMS observations of stellar occultations by Saturn's rings', P.D. Nicholson *et al*. American Astronomical Society, Division of Planetary Sciences meeting in Cambridge, England, 4–9 September 2005.

253 'Size of particles and clumps in Saturnian rings inferred from Cassini UVIS occultations', M. Sremcevic *et al*. American Astronomical Society, Division of Planetary Sciences meeting in Cambridge, England, 4–9 September 2005.

254 '100-meter-diameter moonlets in Saturn's A ring from observations of 'propeller' structures', M.S. Tiscareno *et al*. *Nature*, vol. 440, pp. 648–650, March 2006.

255 It was fortunate that at this point in the mission the combination of Saturn's position in its orbit around the Sun and the inclination of its system with respect to its orbit, put the south polar regions of its moons in continuous sunlight. If the southern pole of Enceladus had been in darkness (as it was in the Voyager era) we would not have discovered the 'tiger stripes' and realised their significance!

256 'Geological features and terrains on Enceladus as seen by Cassini ISS', P. Helfenstein *et al*. American Astronomical Society, Division of Planetary Sciences meeting in Cambridge, England, 4–9 September 2005.

257 'Patterns of fractures and tectonic convergence near the south pole of Enceladus', P. Helfenstein *et al*. *Proc. Lunar Planet. Sci. Conf.*, 2182, 13–17 March 2006.

258 'Discovery of a dynamic atmosphere at Enceladus from Cassini magnetometer observations', M.K. Dougherty *et al*. *Proc. Lunar Planet. Sci. Conf.*, 1585, 13–17 March 2006.

259 'CDA measurements of Saturn's E-ring', R. Srama *et al*. American Astronomical Society, Division of Planetary Sciences meeting in Cambridge, England, 4–9 September 2005.

260 'Cassini dust measurements at Enceladus and implications for the origin of the E-ring', F. Spahn *et al*. *Science*, March 2006.

261 'The Saturnian satellites in the near-infrared: Absolute photometry at ring plane crossing', T.W. Momary *et al*. *Icarus*, vol. 148, pp. 397–406, 2000.

262 'Keck near-IR observations of Saturn's main rings, E and G rings, bracketing Earth's August 1995 ring plane crossing', G. Verbanac, I. de Pater, M. Showalter and J. Lissauer. Geophysical Research Abstracts, vol. 7, A01365, European Geophysical Union in Vienna, Austria, April 2005.

263 'The south polar hot spots on Enceladus', J.R. Spencer *et al*. *Proc. Lunar Planet. Sci. Conf.*, 2252, 13–17 March 2006.

264 'Cassini radar's third and fourth looks at Titan', C. Elachi *et al*. *Proc. Lunar Planet. Sci. Conf.*, 1252, 13–17 March 2006.

265 'Comparative application of dimensionless bankfull hydraulic relations for Earth and Titan', G. Parker. Eos Trans. AGU, 86 (52), Fall Meeting, San Francisco, 5–9 December 2005.

266 'Saturn's spokes: lost and found', C. J. Mitchell *et al*. *Science*, vol. 311, pp. 1587–1589, March 2006.

267 'Ion source rate within Saturn's inner magnetosphere: Cassini results', E.C. Sittler *et al.* American Astronomical Society, Division of Planetary Sciences meeting in Cambridge, England, 4–9 September 2005.

268 'Cassini radar's third and fourth looks at Titan', C. Elachi *et al. Proc. Lunar Planet. Sci. Conf.*, 1252, 13–17 March 2006.

269 'Radar imaging of giant longitudinal dunes: Namib desert (Earth) and the Belet sand sea (Titan)', R.D. Lorenz *et al. Proc. Lunar Planet. Sci. Conf.*, 1249, 13–17 March 2006.

270 'The sand seas of Titan: Cassini observations of longitudinal dunes', R.D. Lorenz *et al. Science*, vol. 312, pp. 724–727, May 2006.

271 'Viscoelectric models of tidal heating in Enceladus', M.N. Ross and G. Schubert. *Icarus*, vol. 78, pp. 90–101, March 1989.

272 'Diapir-induced reorientation of Saturn's moon Enceladus', F. Nimmo and R.T. Pappalardo. *Nature*, vol. 441, pp. 614–616, June 2006.

273 'Plasma near Titan and Saturn's icy satellites and the E-ring: Recent Cassini CAPS results', F. Crary and D. Young. American Astronomical Society, Division of Planetary Sciences meeting in Cambridge, England, 4–9 September 2005.

274 'Enceladus's varying imprint on the magnetosphere of Saturn', G.H. Jones *et al. Science*, March 2006.

275 'The Saturn system as observed by Cassini's Ultraviolet Imaging Spectrograph', L.W. Esposito *et al. Proc. Lunar Planet. Sci. Conf.*, 1586, 14–18 March 2005.

276 'Ultraviolet imaging spectroscopy shows an active Saturnian system', L.W. Esposito *et al. Science*, vol. 307, pp. 1251-1255, 2005.

277 'The earliest names on Titan: Nomenclature system for one more world', T.C. Owen, K. Aksnes, R. Beebe, J. Blue, A. Brahic, G. A. Burba, B. A. Smith, and V. G. Tejfel. *Proc. Vernadsky Brown Microsymposium no. 42*, Moscow, Russia, abs. 52, 10–12 October 2005.

278 'Episodic outgassing as the source of atmospheric methane on Titan', G. Tobie, J.I. Lunine and C. Sotin. *Nature*, vol. 440, pp. 61–64, March 2006.

279 'Resurfacing of Titan by ammonia-water cryomagma', G. Mitri *et al.* Eos Trans. AGU, 86 (52), Fall Meeting, San Francisco, 5–9 December 2005.

280 'Titan at the time of the Cassini spacecraft first flyby: A prediction for its origin, bulk chemical composition and internal physical structure', A.J.R. Prentice. Unpublished paper written in 2004 (arxiv.org/pdf/astro-ph/0602512).

281 'Saturn's icy moon Rhea: A prediction for its bulk chemical composition and physical structure at the time of the Cassini spacecraft first flyby', A.J.R. Prentice. Pub. Astron. Soc. of Australia, vol. 23, pp. 1–11, 2006.

282 'The lakes of Titan', E.R. Stofan *et al. Science*, vol. 445, pp. 61–64, 4 January 2007.

283 'An overview of recent results from the Cassini radio and plasma wave science investigation', D. Gurnett *et al. Bull. Amer. Astron. Soc.*, vol. 37, p. 625, 2005.

284 'Differential rotation of the magnetic fields of gaseous planets', A.J. Dessler. *Geophys. Rev. Lett.*, vol. 12, pp. 299–302, May 1985.

285 'Magnetic reversals of Jupiter and Saturn', D.H. Hathaway and A.J. Dessler. *Icarus*, vol. 67, pp. 88–95, July 1986.

286 'A regular period for Saturn's magnetic field that may track its internal rotation', G. Giampieri, M.K. Dougherty, E.J. Smith and C.T. Russell. *Nature*, vol. 441, pp. 62–64, May 2006.

287 'Evidence for a polar ethane cloud on Titan', C.A. Griffith *et al. Science*, vol. 313, pp. 1620–1622, 15 September 2006.

Acronyms

AACS	Attitude and Articulation Control Subsystem
ACC	Accelerometer
ACC-E	Accelerometer, external
ACC-I	Accelerometer, internal
ACP	Aerosol Collector and Pyrolyser
AM	Articulation Mechanism
API	Acoustic Properties sensor
API-S	Acoustic Properties (sounder) sensor
API-V	Acoustic Properties (velocity) sensor
ASI	Italian Space Agency
BPS	Bipropellant Propulsion System
CAPS	Cassini Plasma Spectrometer
CCD	charge-coupled device
CDA	Cosmic Dust Analyser
CDS	Command and Data Subsystem
CEM	Channel Electron Multiplier
CHEMS	Charge-Energy Mass Spectrometer
CIRS	Composite Infrared Spectrometer
CON	Conductivity sensor
CNES	Centre National de la Recherche Scientifique
CRAF	Comet Rendezvous and Asteroid Fly-by
CRS	Cosmic Ray Science
CSA	Charge-Sensitive Amplifier
CSIRO	Commonwealth Scientific and Industrial Research Organisation
DA	Dust Analyser
DEN	Fluid Density sensor
DISR	Descent Imager and Spectral Radiometer
DPU	Data Processing Unit
DSM	Deep Space Manoeuvre
DSN	Deep Space Network
DVD	Digital Versatile Disk

DWE	Doppler Wind Experiment
ELS	Electron Spectrometer
ESA	European Space Agency
EUV	Extreme-Ultraviolet Spectrometer
FDS	Flight Data Subsystem
FGM	Fluxgate Magnetometer
FUV	Far-Ultraviolet Spectrometer
GCMS	Gas Chromatograph Mass Spectrometer
HASI	Huygens Atmospheric Structure Instrument
HDAC	Hydrogen Deuterium Absorption Cell
HGA	High-Gain Antenna
HRD	High-Rate Detector
HSP	High-Speed Photometer
IAU	International Astronomical Union
IBS	Ion Beam Spectrometer
IFC	In-Flight Calibrator
IFOV	instantaneous field of view
IMS	Ion Mass Spectrometer
INCA	Ion and Neutral Camera
INMS	Ion and Neutral Mass Spectrometer
IRIS	Infrared Radiometer, Interferometer and Spectrometer
ISS	Imaging Science System
JIVE	Joint Institute for VLBI in Europe
JPL	Jet Propulsion Laboratory
KBO	Kuiper Belt Object
LECP	Low-Energy Charged Particle Experiment
LEMMS	Low-Energy Magnetospheric Measurement System
LGA	Low-Gain Antenna
MAG	Dual-Technique Magnetometer
MAG	Magnetic Fields Experiment
MIMI	Magnetospheric Imaging Instrument
NASA	National Aeronautics and Space Administration
NIMS	Near-Infrared Mapping Spectrometer
NRAO	National Radio Astronomy Observatory
ODM	Orbiter Deflection Manoeuvre
OTM	Orbit Trim Manoeuvre
PER	Fluid Permittivity sensor
PLS	Plasma Science Experiment
PPS	Photopolarimeter System
PPS	Power and Pyro Subsystem
PRA	Planetary Radio Astronomy
PSE	Probe Support Equipment
PTM	Probe Targeting Manoeuvre
PWS	Plasma Wave System
R/	ring

RADAR	Titan Radar Mapper
REF	Refractometer sensor
RPWS	Radio and Plasma Wave Spectrometer
RSS	Radio Science Subsystem
RTG	Radioisotope Thermo-electric Generator
S/	satellite
SOI	Saturn Orbit Insertion
SSP	Surface Science Package
TCM	Trajectory Correction Manoeuvre
THP	Thermal Properties sensor
TIL	Tilt sensor
TNO	Trans-Neptunian Object
TRW	Thompson Ramo and Wooldridge
UVIS	Ultraviolet Imaging Spectrograph
UVS	Ultraviolet Spectrometer
VIMS	Visual and Infrared Mapping Spectrometer
VLBI	Very Long Baseline Interferometry
V/SHM	Vector/Scalar Helium Magnetometer

Facts and figures

SATURN

Diameter (km)	120,540 (equatorial)
($\times$ Earth)	9.45
Mass (kg)	5.685×10^{26}
($\times$ Earth)	95.16
Volume ($\times$ Earth)	745
Mean density (g/cm^3)	0.69
Rotational period (hours)	10.6562
(days)	0.44
Escape velocity (km/s)	35.49
Mean surface temperature (K)	88 (1-bar level)
Semi-major axis (AU)	9.53707
Orbital eccentricity	0.054151
Orbital inclination (deg)	2.4845
Obliquity of axis (deg)	26.73
Gravity ($\times$ Earth)	1.12
Orbital period (years)	29.425
Synodic period (days)	378
Albedo	0.47
Atmospheric components	97% hydrogen, 3% helium, 0.05% methane

Rings

D
Planetocentric distance (km)	67,000–74,500
Radial width (km)	7,500

C
Planetocentric distance (km)	74,500–92,000
Radial width (km)	17,500

Maxwell gap
Planetocentric distance (km)	87,500
Radial width (km)	270

B

Planetocentric distance (km)	92,000–117,500
Radial width (km)	25,500

Cassini Division

Planetocentric distance (km)	117,500–122,200
Radial width (km)	4,700

Huygens gap

Planetocentric distance (km)	117,680
Radial width (km)	285–440

A

Planetocentric distance (km)	122,200–136,800
Radial width (km)	14,600

Encke Division

Planetocentric distance (km)	133,570
Radial width (km)	325

Keeler gap

Planetocentric distance (km)	136,530
Radial width (km)	35

F

Planetocentric distance (km)	140,210
Radial width (km)	30–500

G

Planetocentric distance (km)	165,800–173,800
Radial width (km)	8,000

E

Planetocentric distance (km)	180,000–480,000
Radial width (km)	300,000

The rings are only a few hundred metres thick. The particles are centimetres to decametres in size and are ice (some may be covered with ice), but with traces of silicate and carbon minerals. There are four main ring groups and three more faint, narrow ring groups separated by 'gaps' and 'divisions'.

Daphnis

Discovered	2005, Cassini; C.C. Porco *et al.*
Orbital period (days)	0.594
Semi-major axis (km)	136,550
Diameter (km)	7

Note: Daphnis is in the Keeler gap.

Pan

Discovered	1980, Voyager 2; 1990, M. Showalter
Orbital period (hours)	13.8
(days)	0.575

Semi-major axis (km)	133,600
Diameter (km)	20 km
Orbital eccentricity	0.0
Orbital inclination (deg)	0.0
Density (g/cm^3)	0.5

Note: Pan is in Encke's Division.

Atlas

Discovered	1980, Voyager 1, R. Terrile
Orbital period (hours)	14.4
(days)	0.602
Semi-major axis (km)	137,700
Diameter (km)	38 × 34 × 28
Orbital eccentricity	0.002
Orbital inclination (deg)	0.3
Density (g/cm^3)	0.5
Albedo	0.4

Note: Atlas shepherds the periphery of the 'A' ring.

Prometheus

Discovered	1980, Voyager 1, S. Collins, D. Carlson
Orbital period (hours)	14.7
(days)	0.613
Rotational period	Synchronous
Semi-major axis (km)	139,400
Diameter (km)	145 × 95 × 68
Orbital eccentricity	0.003
Orbital inclination (deg)	0.0
Albedo	0.6

Note: Prometheus is the 'F' ring's inner shepherd.

Pandora

Discovered	1980, Voyager 1, S. Collins, D. Carlson
Orbital period (hours)	15
(days)	0.629
Rotational period	Synchronous
Semi-major axis (km)	141,700
Diameter (km)	114 × 82 × 62
Orbital eccentricity	0.004
Orbital inclination (deg)	0.06
Albedo	0.5

Note: Pandora is the 'F' ring's outer shepherd.

Epimetheus

Discovered	1979, Pioneer 11, R. Walker, J. Fountain, S. Larson
Orbital period (hours)	16.7
(days)	0.695
Rotational period	Synchronous
Semi-major axis (km)	151,422
Diameter (km)	$196 \times 192 \times 150$
Orbital eccentricity	0.007
Orbital inclination (deg)	0.34
Albedo	0.3

Note: Epimetheus is co-orbital with Janus.

Janus

Discovered	1966, A. Dollfus
Orbital period (hours)	16.7
(days)	0.695
Rotational period	Synchronous
Semi-major axis (km)	151,472
Diameter (km)	$144 \times 108 \times 98$
Orbital eccentricity	0.009
Orbital inclination (deg)	0.14
Albedo	0.5

Note: Janus is co-orbital with Epimetheus.

Mimas

Discovered	1789, W. Herschel
Orbital period (hours)	22.6
(days)	0.942
Semi-major axis (km)	185,600
Diameter (km)	396
Orbital eccentricity	0.020
Orbital inclination (deg)	1.53
Mass (kg)	3.7×10^{19}
($\times$ Moon)	5.2×10^{-4}
Density (g/cm^3)	1.15
Albedo	0.8

Methone

Discovered	2004, Cassini; C.C. Porco *et al.*
Orbital period (days)	1.01
Semi-major axis (km)	194,000
Diameter (km)	3

Pallene

Discovered	2004, Cassini; C.C. Porco *et al.*
Orbital period (days)	1.14

Semi-major axis (km)   211,000
Diameter (km)   4

Enceladus

Discovered	1789, W. Herschel
Orbital period (days)	1.370
Semi-major axis (km)	238,000
Diameter (km)	498
Orbital eccentricity	0.004
Orbital inclination (deg)	0.01
Mass (kg)	7.3×10^{19}
($\times$ Moon)	1.0×10^{-3}
Density (g/cm^3)	1.6
Albedo	0.95

Tethys

Discovered	1684, J.D. Cassini
Orbital period (days)	1.888
Semi-major axis (km)	294,670
Diameter (km)	1,050
Orbital eccentricity	0.0
Orbital inclination (deg)	1.1
Mass (kg)	6.3×10^{20}
($\times$ Moon)	1.0×10^{-2}
Density (g/cm^3)	1.0
Albedo	0.8

Telesto

Discovered	1980, Voyager 1; H. Reitsema *et al.*
Orbital period (days)	1.888
Semi-major axis (km)	294,670
Diameter (km)	$34 \times 28 \times 26$
Orbital eccentricity	0.001
Orbital inclination (deg)	1.158
Albedo	0.7

Note: Telesto librates about Tethys's leading (L4) Lagrangian point.

Calypso

Discovered	1980, D. Pascu *et al.*
Orbital period (days)	1.888
Rotational period	Synchronous
Semi-major axis (km)	294,670
Diameter (km)	$34 \times 22 \times 22$
Orbital eccentricity	0.001
Orbital inclination (deg)	1.473
Albedo	0.9

Note: Calypso librates about Tethys's trailing (L5) Lagrangian point.

Dione

Discovered	1684, J.D. Cassini
Orbital period (days)	2.737
Semi-major axis (km)	377,400
Diameter (km)	1,120
Orbital eccentricity	0.002
Orbital inclination (deg)	0.02
Mass (kg)	1.1×10^{21}
(× Moon)	1.4×10^{-2}
Density (g/cm^3)	1.45
Albedo	0.6

Helene

Discovered	1980, J. Lecacheux, P. Laques
Orbital period (days)	2.737
Rotational period	Synchronous
Semi-major axis (km)	377,400
Diameter (km)	$36 \times 32 \times 30$
Orbital eccentricity	0.005
Orbital inclination (deg)	0.2
Albedo	0.6

Note: Helene librates about Dione's leading (L4) Lagrangian point.

Polydeuces

Discovered	2004, Cassini; C.C. Porco *et al.*
Orbital period (days)	2.74
Semi-major axis (km)	377,400
Diameter (km)	5

Note: Polydeuces librates about Dione's trailing (L5) Lagrangian point.

Rhea

Discovered	1672, J.D. Cassini
Orbital period (days)	4.518
Semi-major axis (km)	527,000
Diameter (km)	1,530
Orbital eccentricity	0.001
Orbital inclination (deg)	0.35
Mass (kg)	2.31×10^{21}
(× Moon)	3.4×10^{-2}
Density (g/cm^3)	1.25
Albedo	0.6

Titan

Discovered	1655, C. Huygens
Orbital period (days)	15.945
Semi-major axis (km)	1,222,000

Diameter (km)	5,140
Orbital eccentricity	0.029
Orbital inclination (deg)	0.33
Mass (kg)	1.346×10^{23}
($\times$ Moon)	1.83
Density (g/cm^3)	1.88
Albedo	0.2

Hyperion

Discovered	1848, W.C. Bond, G.P. Bond, W. Lassell
Orbital period (days)	21.277
Rotational period	Chaotic
Semi-major axis (km)	1,484,000
Diameter (km)	$360 \times 280 \times 236$
Orbital eccentricity	0.104
Orbital inclination (deg)	0.4
Mass (kg)	1.59×10^{21}
Density (g/cm^3)	0.6
Albedo	0.3

Iapetus

Discovered	1671, J.D. Cassini
Orbital period (days)	79.330
Semi-major axis (km)	3,561,000
Diameter (km)	1,436
Orbital eccentricity	0.028
Orbital inclination (deg)	14.72
Mass (kg)	1.88×10^{21}
($\times$ Moon)	2.6×10^{-2}
Density (g/cm^3)	1.2
Albedo	0.08

Phoebe

Discovered	1898, W.H. Pickering
Orbital period (days)	550.45, retrograde
Semi-major axis (km)	12,960,000
Diameter (km)	220
Orbital eccentricity	0.163
Orbital inclination (deg)	150
Mass (kg)	1.0×10^{19}
Density (g/cm^3)	1.63
Albedo	0.05

Outer irregular satellites (including Phoebe), in order of distance from Saturn

Name	Designation	Mean semi-major axis (km)	Mean inclination (deg)	Size (km)
Kiviuq	S/2000S5	11,110,000	46.16	16
Ijiraq	S/2000S6	11,125,000	46.74	12
Phoebe	–	12,944,300	174.8	240
Paaliaq	S/2000S2	15,200,000	45.13	22
Skathi	S/2000S8	15,539,000	152.7	8
Albiorix	S/2000S11	16,182,000	33.98	32
–	S/2004S11	16,950,000	41.0	6
Erriapo	S/2000S10	17,342,000	34.45	10
Siarnaq	S/2000S3	17,531,000	45.56	40
–	S/2006S8	17,610,000	155.6	6
Tarvos	S/2000S4	17,982,000	33.51	15
–	S/2006S4	18,105,000	172.7	6
–	S/2004S19	18,217,125	153.3	8
Mundilfari	S/2000S9	18,418,000	167.5	7
–	S/2004S13	18,450,000	167.4	6
–	S/2004S17	18,600,000	166.6	4
–	S/2006S6	18,600,000	162.9	6
–	S/2004S15	18,750,000	156.9	6
–	S/2006S1	18,981,135	154.2	6
Narvi	S/2003S1	19,007,000	145.8	7
–	S/2004S10	19,350,000	167.0	6
Suttungr	S/2000S12	19,459,000	175.8	7
–	S/2004S12	19,650,000	164.0	5
–	S/2004S18	19,650,000	147.4	7
–	S/2004S7	19,800,000	165.1	6
–	S/2004S9	19,800,000	157.6	5
Thrymr	S/2000S7	19,941,000	176.0	7
–	S/2004S14	19,950,000	162.7	6
–	S/2006S3	21,132,000	150.8	6
–	S/2004S16	22,200,000	163.0	4
–	S/2004S8	22,200,000	168.0	6
–	S/2006S7	22,290,000	166.9	6
–	S/2006S2	22,350,000	148.4	7
Ymir	S/2000S1	23,041,000	173.1	18
–	S/2006S5	23,190,000	166.5	6

Notes: (1) Having run out of Titans from classical mythology, the International Astronomical Union drew upon Norse, Inuit and Gallic giants. (2) Spelling is as amended in IAU Circ. No. 8471 (21 January 2005).

Further reading

Guide to the planets
Patrick Moore
The Scientific Book Club, 1955

The planet Jupiter
Bertrand M. Peek
Faber and Faber, 1958

The planet Saturn: a history of observation, theory and discovery
A.F.O'D. Alexander
Faber and Faber, 1962 (reissued by Dover Publications in 1980)

Satellites of the Solar System
Werner Sandner
The Scientific Book Club, 1965

The atmosphere of Titan
Donald M. Hunten
NASA SP-340, 1973

Jupiter
Tom Gehrels (Ed)
University of Arizona Press, 1976

Pioneer Odyssey
Richard O. Fimmel, William Swindell and Eric Burgess
NASA SP-349, 1977

Planetary satellites
Joseph A. Burns (Ed)
University of Arizona Press, 1977

The Saturn system
Donald M. Hunten and David Morrison (Eds)
NASA CP-2068, 1978

Voyage to Jupiter
David Morrison and Jane Samz
NASA SP-439, 1980

Pioneer: first to Jupiter, Saturn and beyond
Richard O. Fimmel, James Van Allen and Eric Burgess
NASA SP-446, 1980

Voyager: story of a space mission
M. Poynter and A.L. Lane
Atheneum, 1981

Imaging Saturn: the Voyager flights to Saturn
Henry S. F. Cooper
Holt, Rinehart and Winston, 1982

Satellites of Jupiter
David Morrison (Ed)
University of Arizona Press, 1982

Voyages to Saturn
David Morrison
NASA SP-451, 1982

Distant encounters: the exploration of Jupiter and Saturn
Mark Washburn
Harcourt Brace, 1983

Rings: discoveries from Galileo to Voyager
James Elliot and Richard Kerr
MIT Press, 1984

Saturn
Tom Gehrels and Mildred Shapley Matthews (Eds)
University of Arizona Press, 1984

Planetary rings
Richard Greenberg and Andre Brahic (Eds)
University of Arizona Press, 1984

Voyager 1 and 2 atlas of six saturnian satellites
Raymond M. Batson
NASA SP-474, 1984

Satellites
Joseph A. Burns and Mildred Shapley Matthews (Eds)
University of Arizona Press, 1986

Planetary landscapes
Ron Greeley
Allen and Unwin, 1985 (updated 1987)

Planets beyond: discovering the outer Solar System
Mark Littmann
John Wiley, 1988 (updated 1990)

Journey into space: the first thirty years of space exploration
Bruce Murray
W.W. Norton, 1989

The new Solar System
J. Kelly Beatty and Andrew Chaikin (Eds)
Sky Publishing and Cambridge University Press, Third Edition, 1990

Pale blue dot: a vision of the human future in space
Carl Sagan
Headline, 1995

Asteroids: their nature and utilisation
Charles T. Kowal
Wiley-Praxis, 1996

The three Galileos: the man, the spacecraft, the telescope
Cesare Barbieri, Jürgen H. Rahe and Torrence V. Johnson and Anita M. Sohus
(Eds). Kluwer Academic Press, 1997

Uranus: the planet, rings and satellites
Ellis D. Miner
Wiley-Praxis, 1998

Planetary astronomy from ancient times to the third millennium
Ronald A. Schorn
Texas A&M University Press, 1998

Exploring the Moon: the Apollo Expeditions
David M. Harland
Springer-Praxis, 1999

Our world: the magnetism and thrill of planetary exploration
S. Alan Stern (Ed)
Cambridge University Press, 1999

Encyclopedia of the Solar System
Paul R. Weissman, Lucy-Ann McFadden and Torrence V. Johnson (Eds)
Academic Press, 1999

Titan: the Earth-like moon
Athéna Coustenis and Fred Taylor
Series on Atmospheric, Oceanic and Planetary Physics, World Scientific, 1999

Jupiter Odyssey: the story of NASA's Galileo mission
David M. Harland
Springer-Praxis, 2000

The worlds of Galileo: the inside story of NASA's mission to Jupiter
Michael Hanlon
St. Martin's Press, 2001

Mission Jupiter: the spectacular journey of the Galileo spacecraft
Daniel Fischer
Copernicus Books, 2001

The Earth in context: a guide to the Solar System
David M. Harland
Springer-Praxis, 2001

Neptune: the planet, rings and satellites
Ellis D. Miner and Randii R. Wessen
Springer-Praxis, 2002

Lifting Titan's veil: exploring the giant moon of Saturn
Ralph Lorenz and Jacqueline Mitton
Cambridge University Press, 2002

The Cassini–Huygens Mission
C.T. Russell (Ed.)
Springer, 2003

Titan Unveiled
Ralph Lorenz and Jacqueline Mitton
Princeton University Press, 2007

Index

Printing: Mercedes-Druck, Berlin
Binding: Stein+Lehmann, Berlin

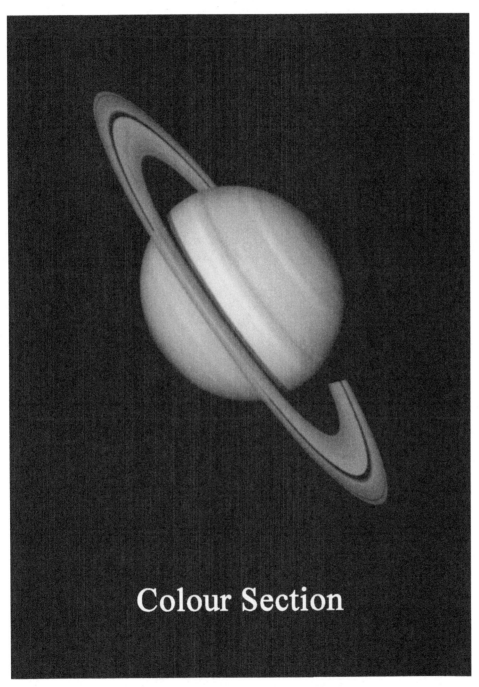

Colour Section

A false-colour view of Saturn taken by Voyager 2.

Ready: the Cassini spacecraft with the Huygens probe attached.

The Cassini mission sets off.

An artist's depiction of Cassini firing its bipropellant propulsion system while north of the ring plane during the Saturn Orbit Insertion manoeuvre.

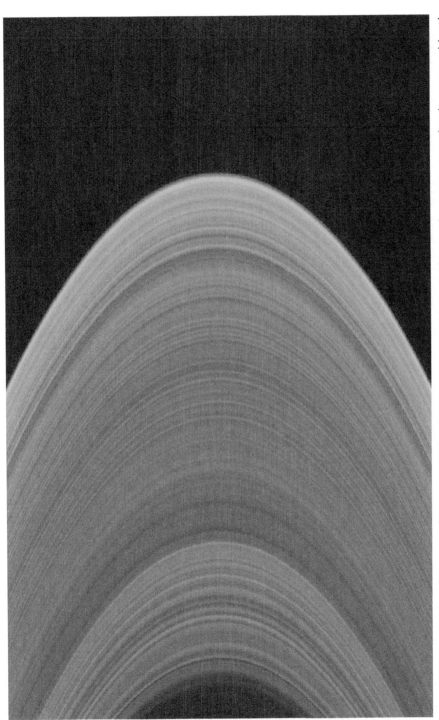

A view by the Ultraviolet Imaging Spectrograph immediately after the Saturn Orbit Insertion manoeuvre showing compositional variations (left to right) from Cassini's Division out across the 'A' ring.

A view by the Ultraviolet Imaging Spectrograph immediately after the Saturn Orbit Insertion manoeuvre showing compositional variations across (left to right) the outer part of the 'C' ring and inner part of the 'B' ring.

The span of Cassini's ring system (top), gravitational clumps in the 'A' ring inferred from stellar occultations (upper), temperature variations across the rings (lower left), infrared data showing variations in the sizes of the particles (lower right and bottom left) and a Voyager 2 study suggesting variation in the composition of the rings (bottom right).

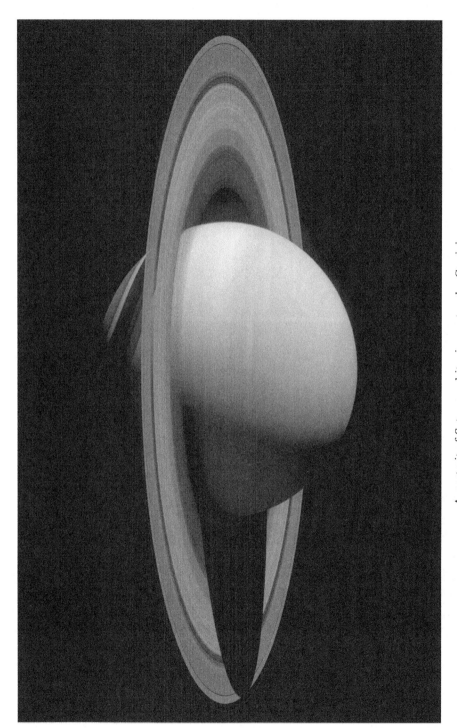

A portrait of Saturn and its ring system by Cassini.

A preliminary panoramic mosaic of imagery taken by Huygens during its parachute descent, plus analyses of specific topographic features.

Full-horizon views by Huygens at different altitudes during its descent.

Quadrant views by Huygens at different altitudes during its descent.

Downward-looking views by Huygens at different altitudes during its descent.

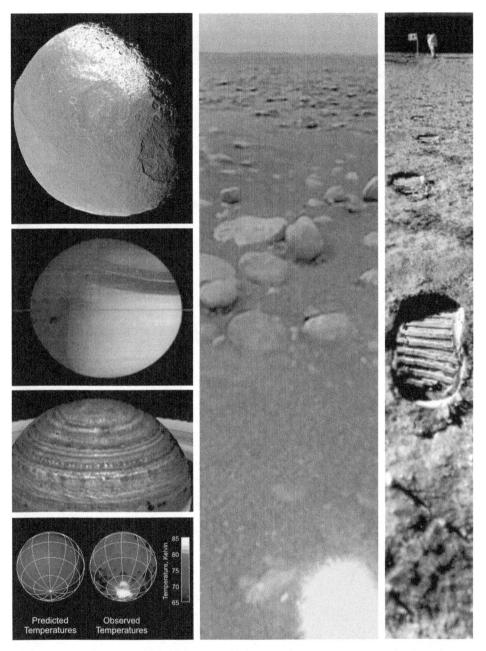

Cassini Regio on Iapetus (upper left), infrared views of Saturn's deep clouds (centre left), the thermal anomaly at the south polar region of Enceladus (lower left), and a mosaic of the oblique and side-looking views from Huygens on Titan's surface, with a matching perspective from human lunar exploration.

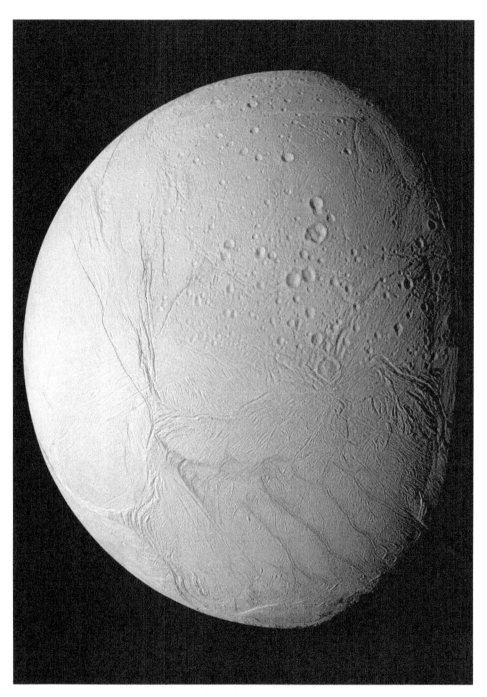

The 'tiger stripes' at the south polar region of Enceladus.

False-colour views of the plume produced by the geysers at the south polar region of Enceladus (upper) and the 'dragon storm' on Saturn.

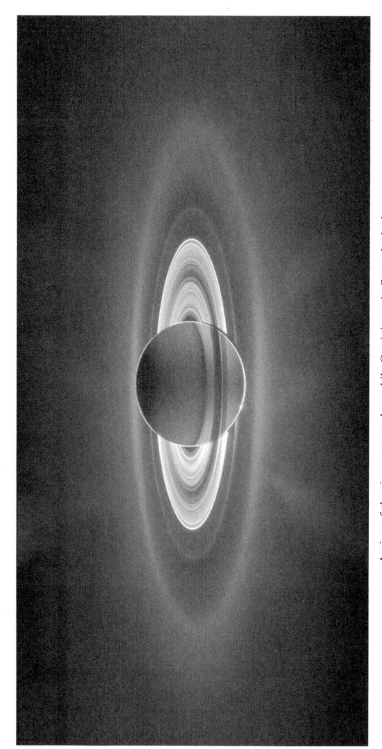

A view of the ring system taken while Cassini was in Saturn's shadow.

CPSIA information can be obtained
at www.ICGtesting.com
Printed in the USA
BVOW07s1256271117
501354BV00003B/45/P